Precision Programming of Roving Robots

Project-Based Fundamentals of Wheeled, Legged and Hybrid Mobile Robots

Francis Nickols
Yueh-Jaw Lin

ASME PRESS

Library of Congress Cataloging-in-Publication Data

Names: Francis Nickols &Yueh-Jaw Lin author.
Title: Precision programming of roving robots / Francis Nickols &Yueh-Jaw Lin.
Description: New York : ASME Press, 2019- | Includes bibliographical references and index. Contents: Project-Based Fundamentals of Wheeled, Legged and Hybrid Mobile Robots.
Identifiers: LCCN:2017049008 | ISBN:9780791861912
Subjects: LCSH: Machinery--Design and construction. | Robots--Design and construction.
Classification: LCC TJ230 .P625 2019 | DDC 620/.0042--dc23
LC record available at https://lccn.loc.gov/2017049008

Series Editor's Preface

Dear Reader,

The ASME Press Robotics Engineering Book Series continues to present research and engineering accomplishments in the key areas of robotics. I am delighted to introduce a book by Dr. Francis Nickols and Dr. Yueh-Jaw Lin from the University of Nottingham Ningbo China. Entitled *Precision Programming of Roving Robots*, this book has launched a new area of our robotics publications, namely textbook supplements for undergraduate students. As specified in the subtitle, this book concerns with project-based fundamentals of wheeled, legged and hybrid mobile robots.

The authors has a notable experience of undergraduate and graduate teaching that is multiplied by engineering work in the automotive industry. A distinguished feature of the book is a mechatronics-based approach to mobile robots that is presented in a form of a lab operation manual to assist students with their projects and, thus, to comprehend a lecture course on roving robots. Remarkably, the book focuses on precision sensorless actuation of robot movements. Hence, the learning outcome of the book is the precision programming of robots to perform accurate behavior pattern movements.

The book can serve as a supplement to undergraduate courses on robotics engineering that allows students to learn course material "in action" by doing actual projects in the lab environment.

The Robotics Engineering Book Series is committed to both assisting eminent scholars in publishing their findings and introducing outstanding new researchers and educators to the world of books in Robotics. We look forward to close collaboration with researchers, engineers and academia around the world.

Vladimir V. Vantsevich, ScD, PhD, ASME Fellow
Founding Editor, ASME Press Robotics Engineering Book Series
May 11, 2019

ASME Press Robotics Engineering Book Series

Contents

Preface

Project-Based fundamentals of wheeled, legged and hybrid robots

This book is designed primarily as a mechatronics and robotics textbook supplement for undergraduate students. More precisely it serves as a project-based laboratory operations manual for an academic course related to mobile roving robots. The book explains how to programme precision behaviour pattern motions in a ground plane using these three roving robots shown below.

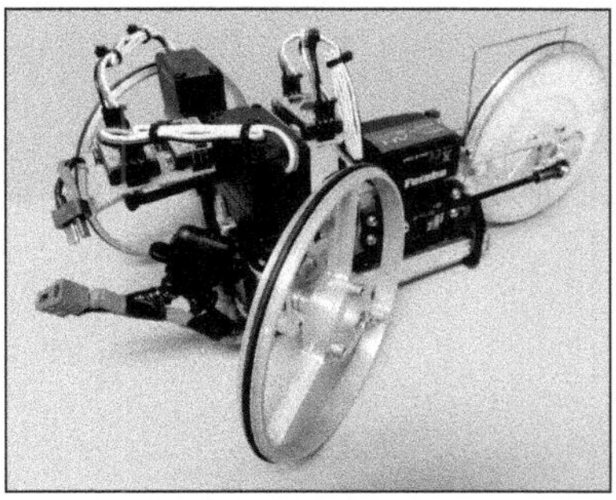

The 'Tricycle' Robot
A 3-wheeled Robot

The 'Rickshaw' Robot
A 2-wheeled, 2-legged Robot

The 'Hexapod' Robot
A six-legged omnidirectional Robot

The three roving robots are graded in complexity from the all-wheeled Tricycle Robot through to the hybrid wheeled and legged Rickshaw Robot up to the most complex of the three, the all-legged Hexapod Robot. The precision programming of behaviour patterns for the three robots is demonstrated, evaluated and measured with a 'figure-of-eight' pattern that is marked out on to a 4.8metre × 2.4 metre table as shown below. Robots are programmed to travel consecutively from node 1 to node 2, then nodes, 5, 4, 3, 2, 5, 6 and back to node 1. Target 'bulls-eyes' at each node are given accuracy marks from 1 to 10 such that 10 marks are given for the robot stopping at the centre of the bulls-eye. *Sensor-less* dead-reckoning precision programming is used to programme the robot to travel through the figure-of-eight path. The book does not cover the design and programming of robots with sensors that measure quantities in the outside observed world. This is an important aspect since the authors believe that sensorless behaviour patterns are fundamental to actuated reactionary patterns that are built into biological systems. Sensors such as vision and touch sensors serve mainly to learn, to correct, or to fine-tune, built-in behaviour patterns.

Based on such a premise, this book solely concentrates on sensorless actuation of robot motion. For example, we as human beings are exercising sensorless behaviour patterns all the time; for example, the act of getting off our chair to stand up is to execute a particular actuated routine. Yes, your senses adjust the routine based on height of the chair and other factors, but largely it is a set muscular routine that is executed. So this is the flavour of the book, i.e. precision sensorless behaviour pattern actuation. You can imagine that this means your robot is blind-folded. As soon as your blind-folded robot can perform behaviour patterns accurately then sensors can be added further to improve its performance. Such programmed actuation is fundamental to robotics and forms the basis of the learning outcomes from this book.

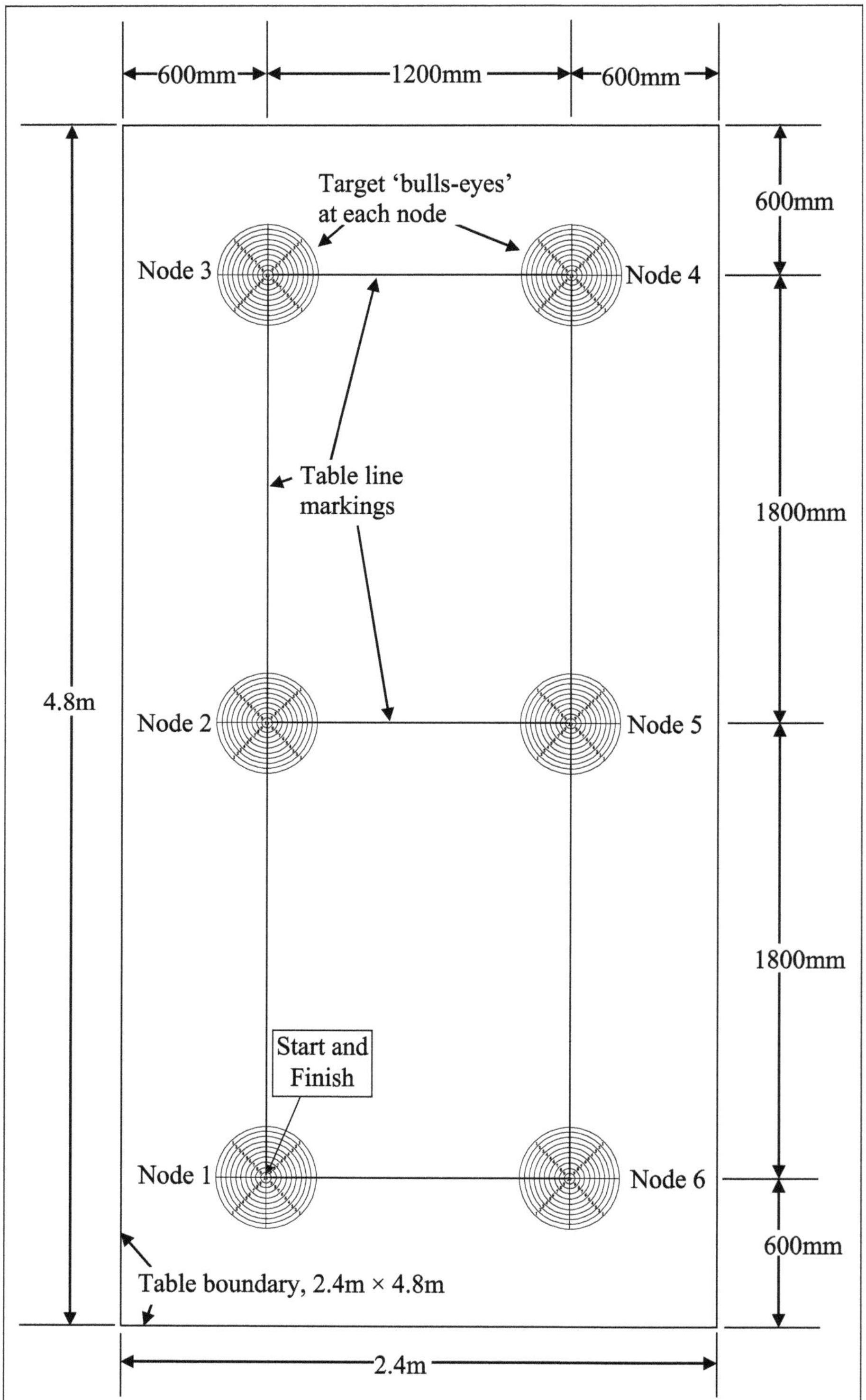

Diagram of the 'Vehicle Motion in a Ground Plane' table' showing nodes and target 'bulls-eyes'. This table is used to evaluate robot motion computer programmes.

The book is designed primarily as a mechatronics and robotics experiential and project-based teaching and learning platform for tertiary students. It is also graded in that starting with the Tricycle Robot, students build up their knowledge and experience of programming to be able to tackle the Rickshaw Robot and finally the most complex robot, i.e., the Hexapod Robot. As regards the academic level, the book is aimed at university and college students. However, nowadays the authors are seeing robotics curricula extending down into schools so this book can also be very useful for giving ideas to teachers at school level. Furthermore, the book provides useful ideas for driverless vehicles and robots as well as for educationalists who are developing practical project-based teaching and learning modules.

The book covers the concept of a robot turning about an Instantaneous Centre of Rotation, ICofR, and how a wheeled and/or legged robot is programmed to achieve such motion in a ground plane. Programming concepts move on to creating behaviour and motion patterns in legged robots eventually leading to the design of a hierarchical central pattern generator, CPG, for the hexapod robot. Early chapters are concerned with wheeled vehicles and how Taylor series equations are used to simplify complex inverse kinematic control equations. Wheeled vehicles serve as a smooth introduction to legged motion which suddenly becomes complex. Understanding wheeled motion lessens the sudden jump in understanding the complexity of programming legged motion.

The real time programming code is executed on Parallax.com Basic Stamp2 series microcomputers using the PBasic language which means 'Parallax Basic'. The language is very easy to learn and you get your robot up and running and operational very fast. Even though the programming language is straightforward, almost like writing short English instructions, you can develop complex real-time, parallel processing, interconnected, multi-cored microcomputer systems tailored to your own application. The book gives the code at all stages of the robot motion and behaviour pattern programming. Since the code is given in PBasic it is easy for students and readers to alter the code to use an alternative microcomputer to their liking. The choice of microcomputer and programming language is like politics and religion meaning that every robot constructor and programmer has his or her own views together with preferences as to the best hardware and software to use.

The robots featured in this book have all been created by the author and used in experiential project-based teaching and learning classes. Hence the book is tried and tested. The authors see themselves as teachers, mentors and coaches. This means that students are taught with both theoretical and experiential learning concepts and then as they gain proficiency, mentoring is gradually lessened. Students then begin to exhibit problem solving abilities, sometimes in novel ways, thus becoming promising inventors and autonomous learners and only needing teachers as sounding boards. This has certainly been the case when using the Tricycle Robot, the Rickshaw Robot and the Hexapod Robot in our classroom.

Acknowledgements

The authors would like to thank Dr. Mani Le Vasan, an expert in pedagogy, for her help, advice, persistence and encouragement in getting this book written. In fact, the book would always have been on 'Planet Procrastination' until she directed 'nose to the grindstone'. Not only has Mani, a prolific author herself, been a tower of strength in the book's creation but she has also been an inspirational coach who has constantly provided guidance and gems of advice every step of the way.

1. Basic motion of wheeled vehicles in a ground plane (A forerunner to legged vehicles)

1.1 Introduction

This chapter explores basic theories related to creating vehicle motion in a ground plane, e.g. a road, using wheels whose axles are always parallel to the ground plane. Wheels that lean, such as with bicycles, are not considered. The idea is to generate vehicle motion concepts based on wheels because wheeled motion, acting as a forerunner to legged motion, is more easily grasped than starting with more complex vehicle motion using legs. Chapter 3 follows up with the real-time programming of an actual wheeled vehicle so the reader can see theory in practice. Later, in chapter 10, these concepts are translated into equivalent motion of a robot using legs such that wheeled vehicle motion is entirely replaced with legs. However, we first start with understanding the simplified motion of a single wheel.

1.2 Simplified motion of a single wheel

Consider a rolling wheel in a vertical plane with a low friction axle and high friction between tyre and road which is in a horizontal ground plane as shown in the figure 1.1 below. The

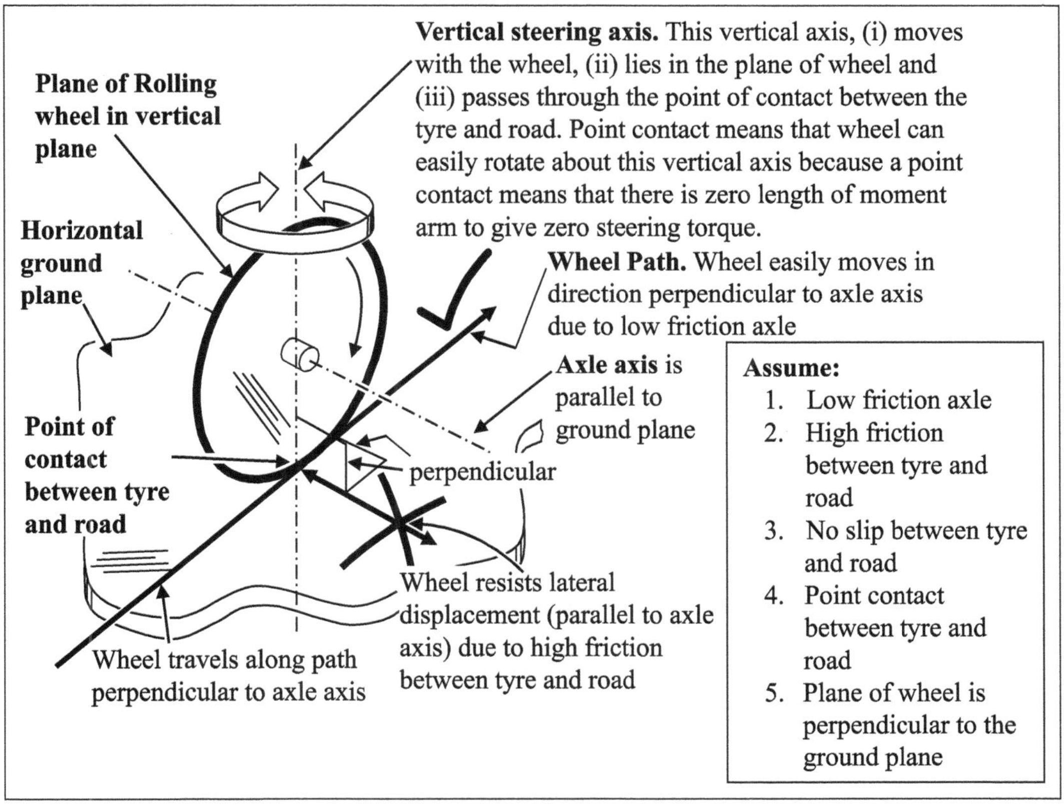

Figure 1.1 Motion of a single wheel in a ground plane.

wheel axle is always maintained parallel to the ground plane meaning that the plane of the rolling wheel is vertical, i.e. perpendicular to the ground plane. The rolling wheel will travel along a path that is perpendicular to the axle axis. Due to friction between the wheel contact point with the ground plane, the wheel will resist any motion that has a component in the direction of the axle axis.

If the wheel is turned about the vertical steering axis then it will take up a new path that remains perpendicular to the axle axis. This is what we know as 'steering the wheel'. At this point we realise that the wheel possesses similar properties to a particle that is travelling along the circumference of a circle. The particle represents the wheel position which is represented by the point contact between the wheel and the ground plane. One property of the particle is that the direction of its instantaneous velocity vector is always perpendicular to a line joining the particle, b and the centre of the circle, i, as shown in figure 1.2 below. As a corollary, the particle has no velocity component that is co-axial with the line ib.

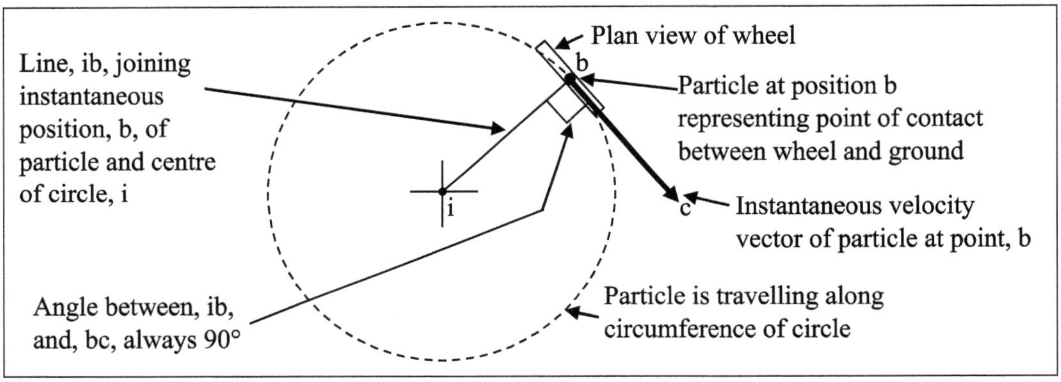

Figure 1.2 Plan view of wheel describing a circle with centre at, i on the ground plane. Particle at position, b represents point of contact between wheel and ground plane.

1.3 The Instantaneous Centre of Rotation, the ICofR

The point, i, which is the centre of the circle, is known as the instantaneous centre of rotation, (the ICofR), of the particle. In the case shown above, figure 1.2, the ICofR remains fixed with respect to the ground, which is a special case. Vehicles become useful when the ICofR position is able to move, for example, consider the following particle path in figure 1.3 below.

Figure 1.3 above shows a particle moving from left to right along the path, pqrs. The particle initially has its ICofR at, i_1, then later, i_2, and finally at infinity distance from the particle and perpendicular to the path direction. The concept of ICofR is regularly utilised by animals to negotiate terrain. For example, figure 1.4 below, shows an ant walking around a prey at a fixed distance whilst focussing on it. It leads to the question; how does an ant compute the joint angles of its six legs to enable such motion?

Figure 1.4 could equally represent a footballer circling an opponent who is in possession of the ball. The footballer is doing this as a strategy in anticipation of making a tackle against

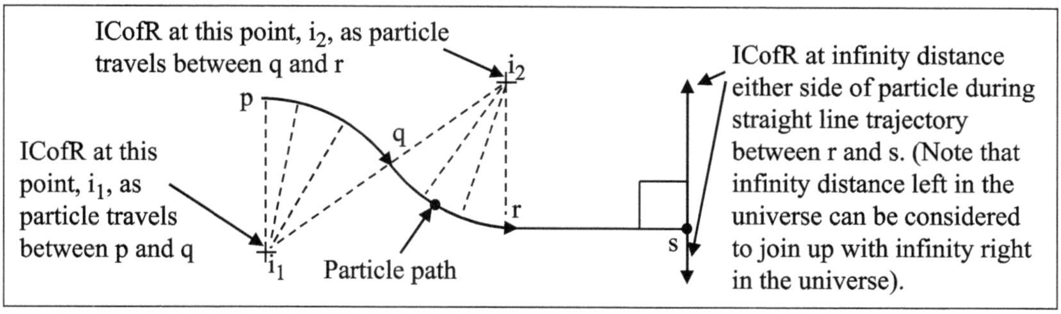

Figure 1.3 Moving of the Instantaneous Centre of Rotation, ICofR.

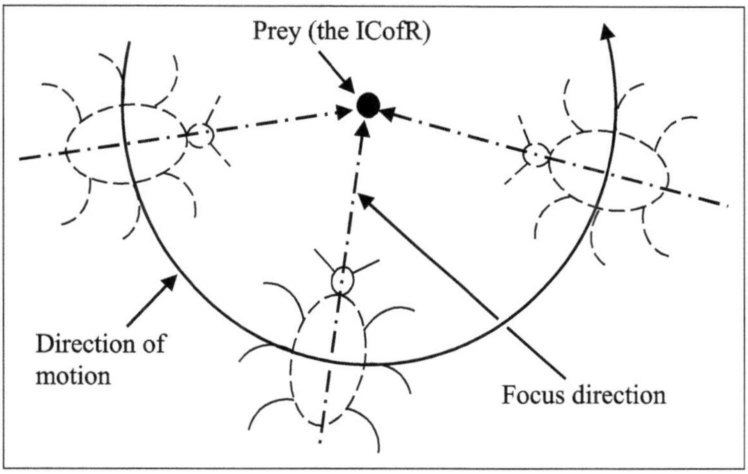

Figure 1.4 Ant encircling its prey which represents the instantaneous centre of rotation, ICofR of ant motion.

the opponent. The opponent, representing the ICofR could be travelling along a given path whilst the circling of the opponent by the footballer is going on. This presents very interesting real-time computation problems that will be discussed in a later chapter. These problems are common with swarm behaviour robotics and game strategy for collaborative and competitive robots.

Anyway, let's get back to wheels rather than walking ants. We have a problem regarding articulating the steering of the wheel. In order to get a wheel to turn about an instantaneous centre of rotation we need to steer the wheel. For example, to get a wheel to follow the same path as the particle in the figure 1.3 above, the plane of the wheel must always be tangential to the particle path or, alternatively, the axle must always be perpendicular to the tangent of the particle path. As a result, the axle axis passes through the ICofR. As an example you can see the wheel changing its direction in figure 1.5 below which is a plan view of the wheel as it travels along the path, p, to, q, to, r, to, s.c.

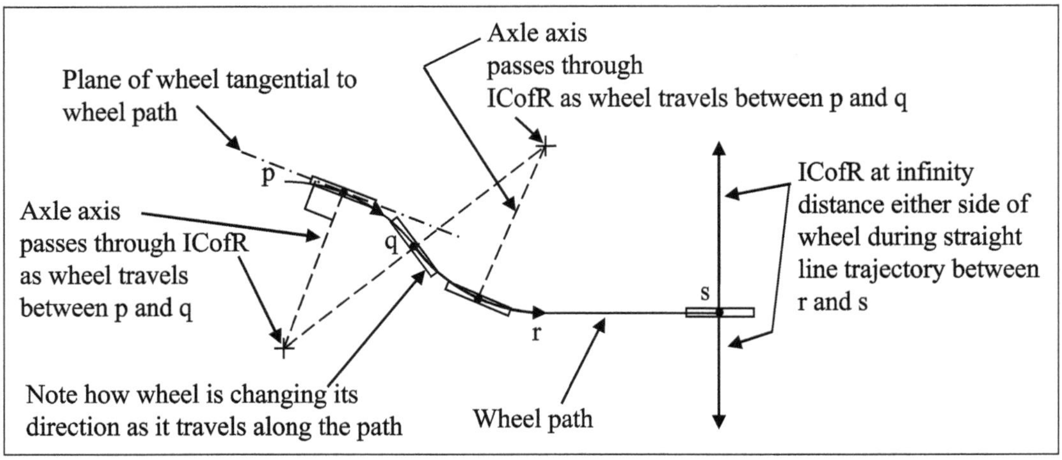

Figure 1.5 Wheel axis always passes through ICofR of wheel path.

We now stipulate a rule which is as follows:
The ICofR lies at a point anywhere along the axle axis, as shown in figure 1.6 below.

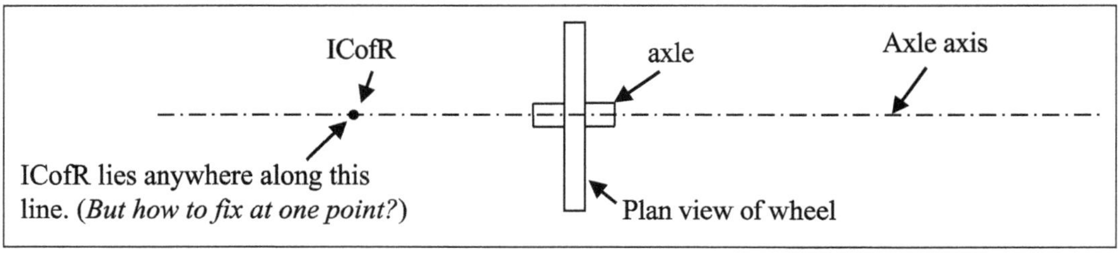

Figure 1.6 Plan view of wheel and the *ICofR which must lie on the axle axis.*

1.4 Fixing the ICofR at a set point

However, in order to control the wheel to follow a set path we need a method which can fix the ICofR at a set point with respect to the ground plane with the point lying along the axle axis. In other words, the single wheel needs to react against something with respect to the ground to be able to steer. (Note that a unicyclist experiences this problem). A simple, but nonetheless ingenious, solution to this problem is to mount the wheel on a vertical steering axis and then to mount this axis to a chassis and to add one or more wheels that are displaced from the steering wheel. For simplicity, we specify that the steering axis is perpendicular to the ground plane and the additional wheel axle or wheel axles are parallel to the ground plane. Now, how many wheels do we add? If we add just one wheel then we create a bicycle but this will require a stabilising system to keep it upright such that the wheel axles are maintained parallel to the ground plane. Also, a bicycle does not replicate a hexapod which is inherently stable since it walks on at least 3 legs. So, we will add two more wheels to create a tricycle, which is stable and will not fall over when at standstill and will maintain the wheel axles parallel to the ground plane. Such a tricycle arrangement is shown in figure 1.7 below.

Suppose now we want the tricycle to travel in a circle. Figure 1.8 below shows a plan view of the tricycle with the steered wheel, W_F, turned to the right. The tricycle ICofR must lie

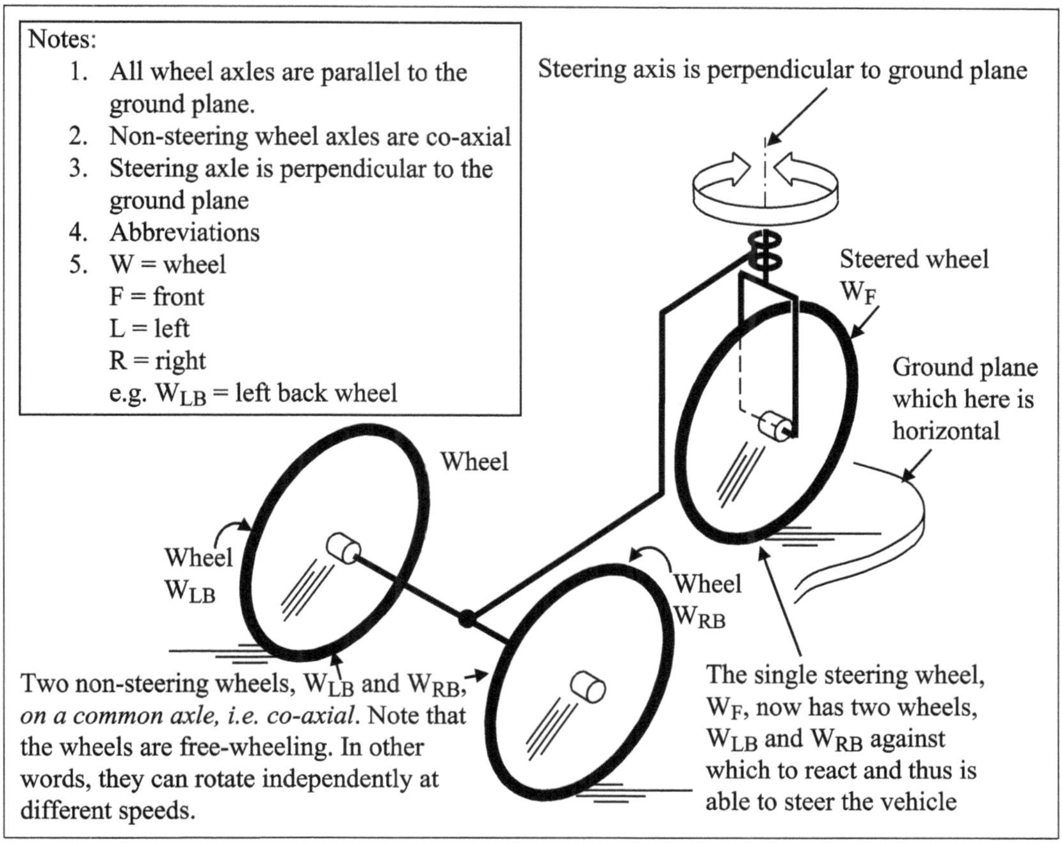

Figure 1.7 A tricycle with a steered wheel is a practical realisation of controlling the location of a vehicle's ICofR such that the vehicle's position on a ground plane can be controlled.

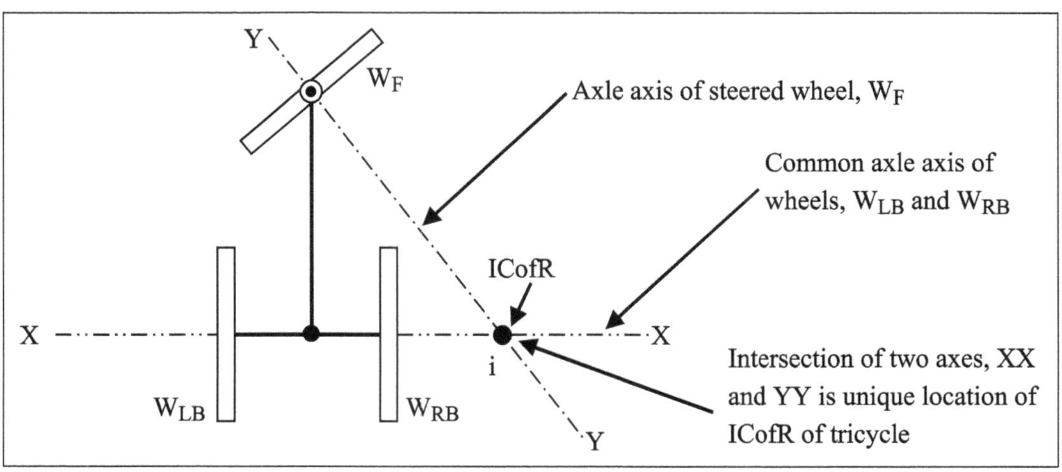

Figure 1.8 Plan view of tricycle showing example of fixing the location of the ICofR to enable the vehicle to travel in a circle.

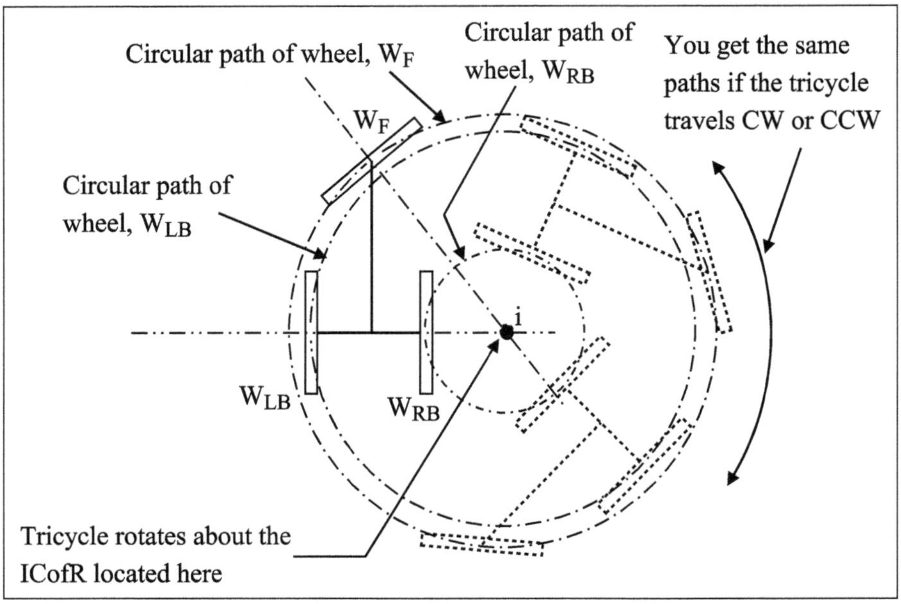

Figure 1.9 Plan view of the tricycle travelling CW and CCW in a circle with a fixed ICofR.

somewhere along the axle axis of each wheel, so where is the ICofR located? Well, two of the wheels, W_{LB} and W_{RB}, share a *common axle axis*, XX in figure 1.8, hence the ICofR will lie somewhere along this common axis. However, the ICofR must also lie somewhere along the steered wheel axis, YY. It follows, therefore, that the ICofR for the tricycle will lie at a unique point, i which is the intersection of XX and YY.

1.5 The Instantaneous *Axis* of Rotation, the IAofR

(To be compared with Instantaneous *Centre* of Rotation, ICofR)

In figure 1.9 above, the tricycle is moving in the plane of the paper which is considered the ground plane. The tricycle is rotating about the ICofR as if an imaginary stiff rod has been welded to the chassis as shown in figure 1.10 below, and this stiff rod is connected to a vertical post via a frictionless bearing. The vertical post is perpendicular to the ground plane and passes through the ICofR. Two pegs on the post exist to stop the bearing from sliding axially along the post, hence restricting motion of the tricycle body to one degree of freedom of motion which is that of rotating about the axis of the post. The pegs mean that you can remove the ground from beneath the wheels (but not from the post) and obtain the same motion of the tricycle chassis. The axis of the post raises an important new concept which is the Instantaneous Axis of Rotation, (the IAofR), i.e. 'axis' as distinct from 'centre'. Note that the IAofR and the ICofR are related in that the ICofR is an end-on view of the IAofR. The IAofR is an important concept because it means we can dispense with the ground and then think of a generalised body that rotates about the IAofR whose axis is an arbitrary vector in space relative to an inertial frame, figure 1.11, where, for example, the x-y plane could be the ground plane. Such a concept becomes useful later when programming the motion of a six degree-of-freedom hexapod robot.

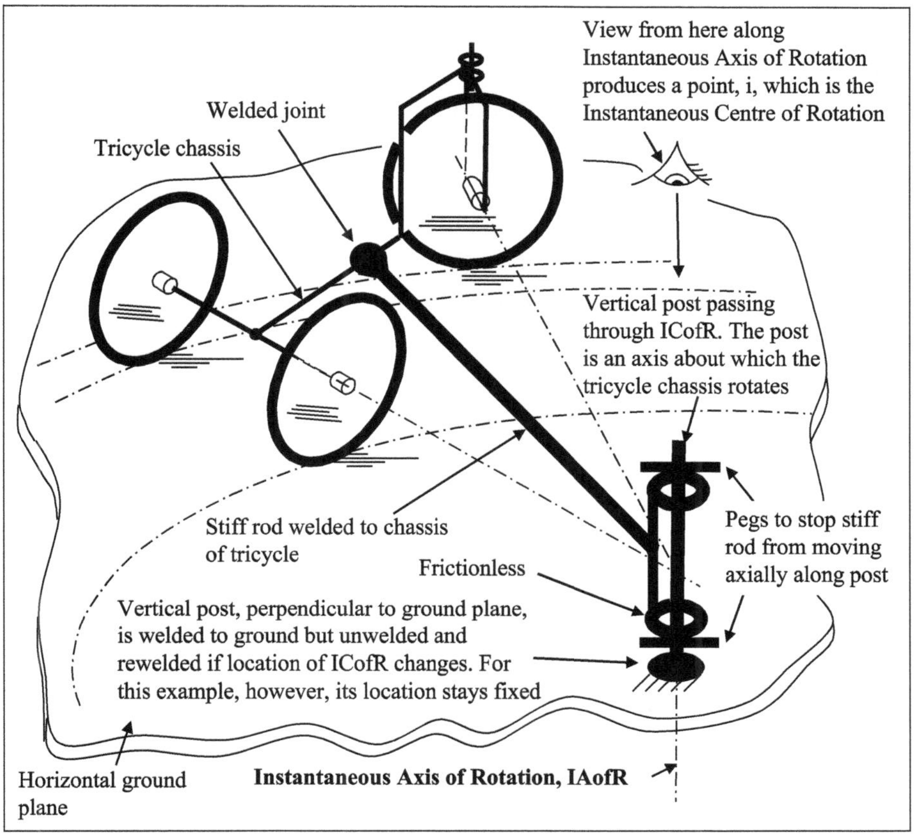

Figure 1.10 View of tricycle showing virtual reality stiff rod acting to rotate vehicle around an instantaneous axis of rotation.

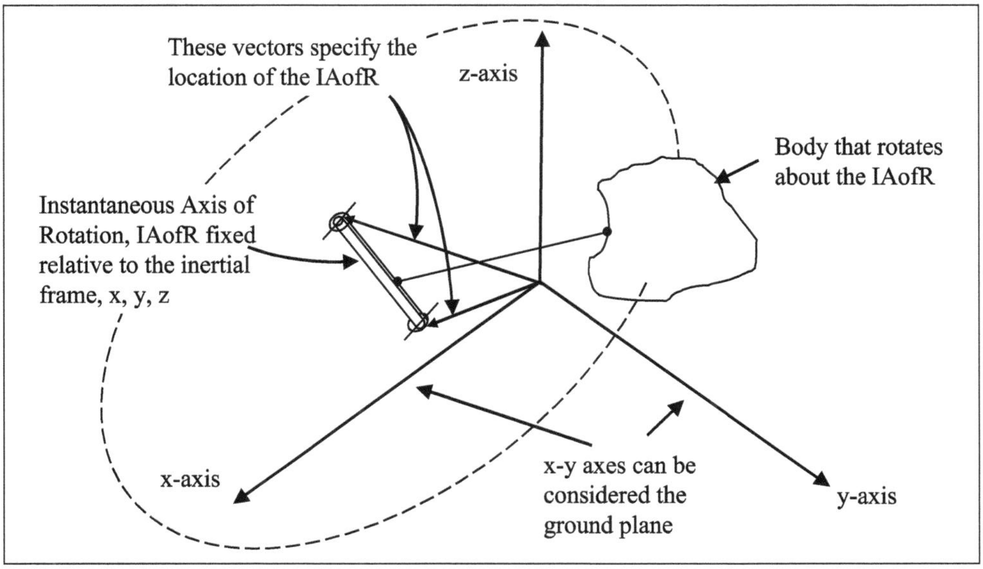

Figure 1.11 Diagram showing principle of Instantaneous Axis of Rotation, IAofR, placed in an inertial frame with axes, x, y and z. A body rotates about the IAofR which is placed in a given position and given direction with respect to the inertial frame.

1.6 Wheel angular velocity when vehicle turns corners

The next step is to bring attention to the angular velocity of the wheels as the tricycle turns corners. The figure 1.12 shows the distances from the ICofR, L_F, L_{LB} and L_{RB}, of each wheel W_F, W_{LB}, and W_{RB}, respectively. We assume that the tricycle is turning clockwise, CW, about the ICofR at a turning angular velocity of, Ωrad/sec. Thus, the tangential velocity, v, of the outer rim of each wheel, W, is given by, see figure 1.12 :-

Tangential velocity of W_F wheel, $v_F = \Omega \times L_F$
Tangential velocity of W_{LB} wheel, $v_{LB} = \Omega \times L_{LB}$
Tangential velocity of W_{RB} wheel, $v_{RB} = \Omega \times L_{RB}$
And the angular velocity, ω, of each wheel is given by, see figure 1.12 :-

Wheel angular velocity, $\omega_1 = \frac{v_F}{R_F} = \frac{\Omega.L_F}{R_F}$

Wheel angular velocity, $\omega_{LB} = \frac{v_{LB}}{R_{LB}} = \frac{\Omega.L_{LB}}{R_{LB}}$

Wheel angular velocity, $\omega_{RB} = \frac{v_{RB}}{R_{RB}} = \frac{\Omega.L_{RB}}{R_{RB}}$

To simplify the equations, if the radius, R, of each wheel is the same then the angular velocity, ω, of each wheel is proportional to its distance, L, from the ICofR. This poses interesting problems when an automobile has more than one wheel driven by just one engine such as, (i) the front two wheels driven or, (ii) the rear two wheels or, (iii) all four wheels. This

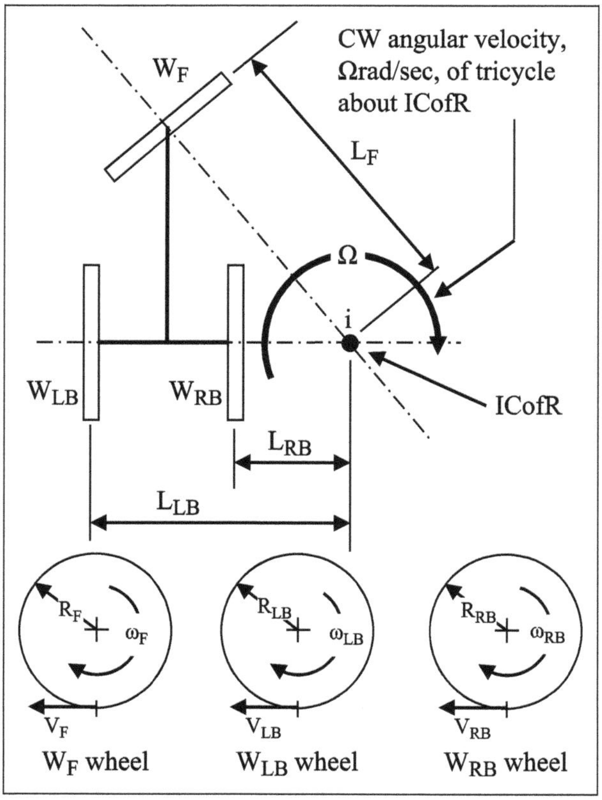

Figure 1.12 Diagrams showing that the wheels have their angular velocities, ω and tangential velocities, v proportional to their distances, L from the ICofR.

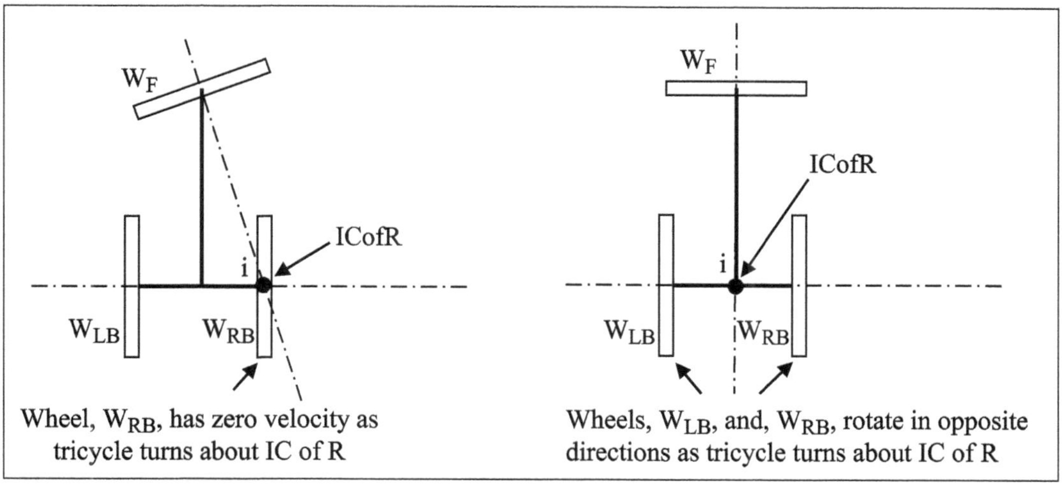

Figure 1.13 Two special cases of IC of R location.

problem is solved by using a mechanical differential which is a mechanical adding machine that readers can find out more on YouTube.

Suffice to say that a mechanical differential allows two wheels to be driven at different speeds with one input of mechanical power. It is also interesting to bring attention to two special cases where case one is where the ICofR lies at the point of contact with wheel, W_{RB} (or similarly, W_{LB},), or case two, where the ICofR lies mid-way between W_{LB} and W_{RB}. Figure 1.13 below shows how the steered wheel, W_F is positioned in these two situations.

In conclusion, wheeled vehicles must solve two engineering problems for locomotion; the first is how to set all wheel axles to point towards a common point, ICofR, the instantaneous centre of rotation and the second problem is how to drive all wheels at an angular velocity proportional to the distance that each wheel is displaced from ICofR. For the case of vehicles with three or four wheels, a mechanical differential with the remaining wheels being allowed to freewheel takes care of the second problem. However, the first problem associated with steering is now addressed.

1.7 Mechanical computation of steering angle

There are many combinations of wheeled vehicles that can be constructed that require all the wheel axles to pass through a common point, ICofR. For example, figure 1.14 shows a conventional 4-wheeled automobile with two front steered wheels. (In comparison, a lift truck puts the steered wheels at the rear). We can assume that the front wheels are freewheeling and the rear wheels are driven through a differential and thus the problem of the wheel angular velocity being proportional to their distance from the point, ICofR is solved. The problem remaining is how to set all the axles to pass through the point, ICofR?

The solution used on most automobiles is a four-bar steering linkage as in the figure 1.15. We generally think of computers using electronic technology but often not just one technology can be used to compute solutions to problems. In this case we use a mechanical system to compute the correct steering angles for each steered wheel. The 4-bar linkage is known as

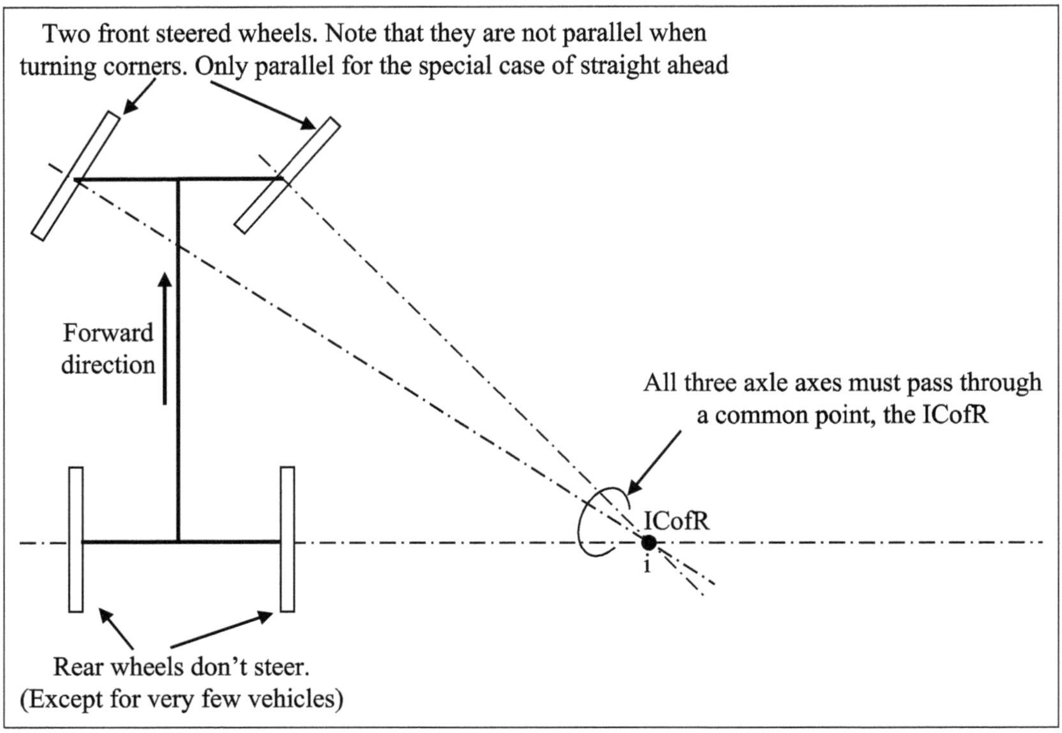

Figure 1.14 4- wheeled vehicle with two front steered wheels.

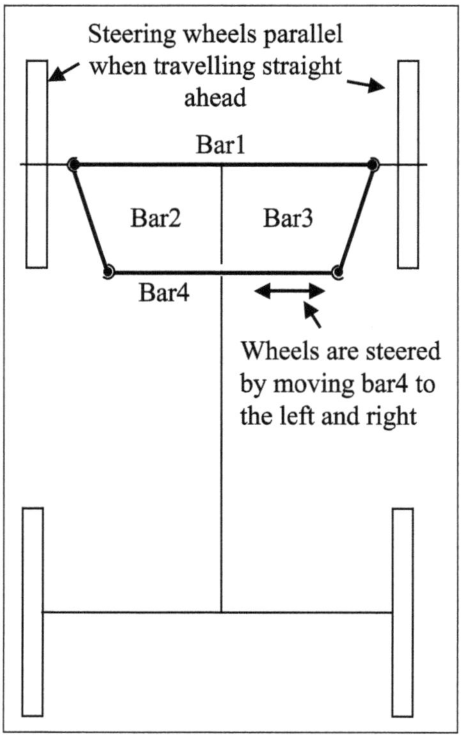

Figure 1.15 4-wheeled automobile with front wheel steering and 4-bar linkage mechanical computer designed for getting the axle axes to pass through common point.

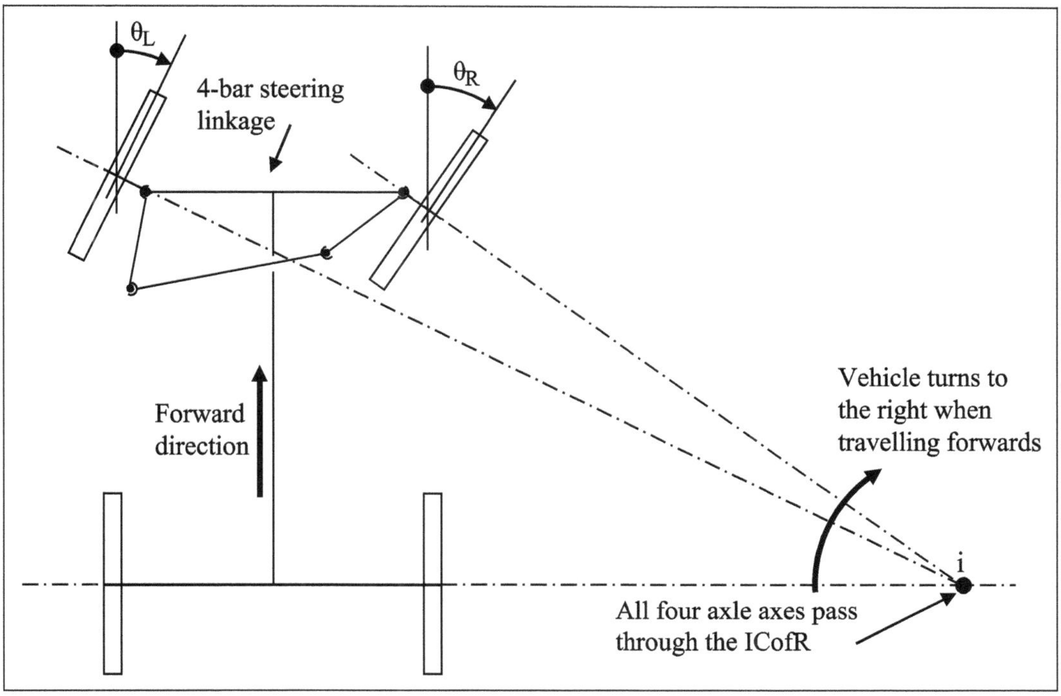

Figure 1.16 The 4-bar steering linkage displaced to arrange the vehicle axes to pass through the Instantaneous Centre of Rotation.

the Ackermann steering arrangement and readers can access more information on YouTube about its kinematic motion and design.

Figure 1.16, shows an example of the steered wheels turned to the right. The 4-bar linkage acts as a mechanical computer which computes the correct steering angles, θ_{LF}, and, θ_{RF}, so that the steered wheel axle axes pass through a common point located on the projected rear wheel axle axes.

Now we move on to considering a more advanced vehicle that is capable of omnidirectional motion. Such a vehicle comes close to replicating the locomotion principles of an omnidirectional hexapod.

1.8 An Omnidirectional tricycle

Figure 1.17 illustrates an omnidirectional vehicle in a tricycle arrangement and figure 1.18 shows its engineered manifestation designed and built by the author.

The omnidirectional tricycle, figure 1.17, is another interesting arrangement of a 3-wheeled vehicle with each wheel independently steered by an angular displacement servomechanism, called a servo motor and independently speed driven by a wheel servomotor. Each wheel system adds an orthogonal degree of freedom of motion in the ground plane such that the vehicle possesses three degrees of freedom, meaning that the vehicle can;

1. Translate in any heading direction in the x-y axes of a ground plane, (two degrees of freedom)

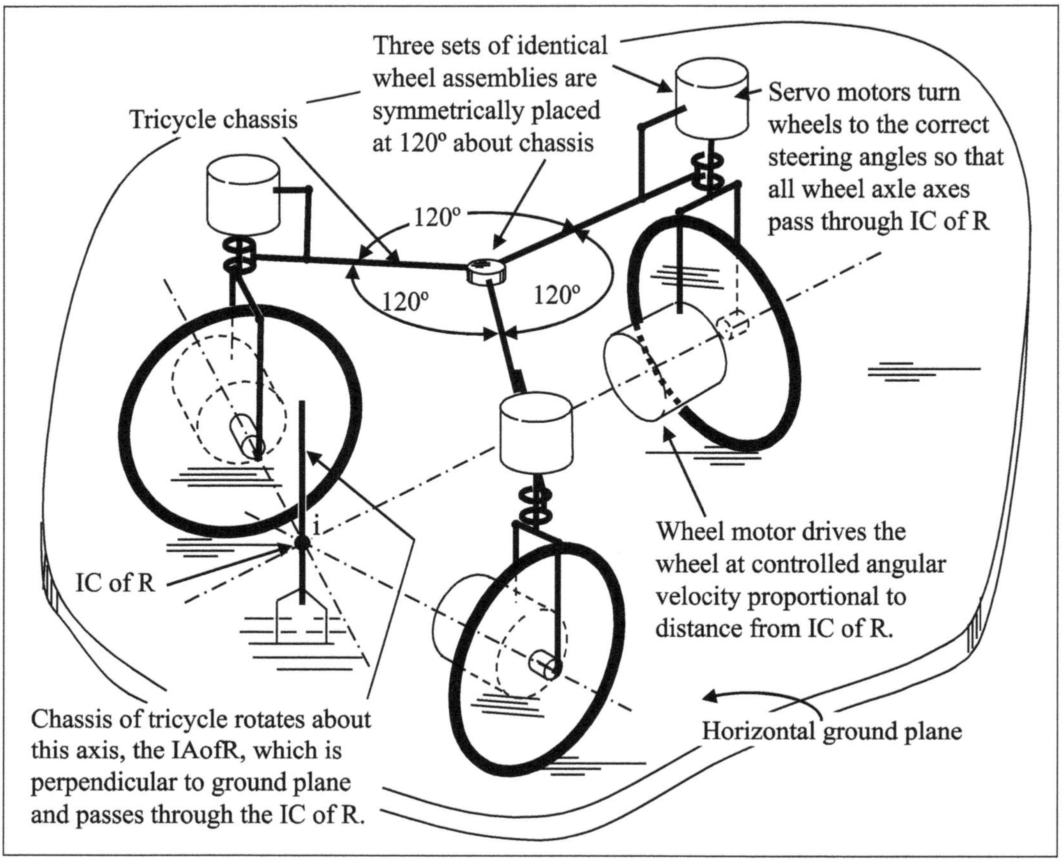

Figure 1.17 An omni-directional tricycle that replicates hexapod locomotion principles .

2. Rotate about an IAofR that is located anywhere on the ground plane and is perpendicular to the ground plane, (one degree of freedom).
3. Provide *any simultaneous combination of 1 and 2*, i.e. simultaneous translation and rotation, i.e. omnidirectional motion. (This is holonomic motion. The tricycle in figure 1.7 can do 1. and 2. but not simultaneously so is non-holonomic)

Such a vehicle, figure 1.17, is considered capable of omnidirectional motion in a ground plane, i.e. holonomic motion. So also, is a hexapod with 3 degree-of-freedom legs, capable of omnidirectional motion using three of its six legs in an alternating double tripod gait which is the most popular walking technique of insects. The double tripod gait works by supporting the robot body with three of its legs that are driven backwards with respect to the body to ambulate the body forward and the remaining 3 legs driven forwards in the air. When the supporting legs have finished their travel, they lift up and travel forwards in the air whilst the remaining legs become the ambulating legs and then the cycle repeats. Each leg of a hexapod can be driven in any heading direction at a controllable speed giving a leg the same attributes as a wheel in the omnidirectional tricycle. Thus, a hexapod using alternating double tripod gait shares a common

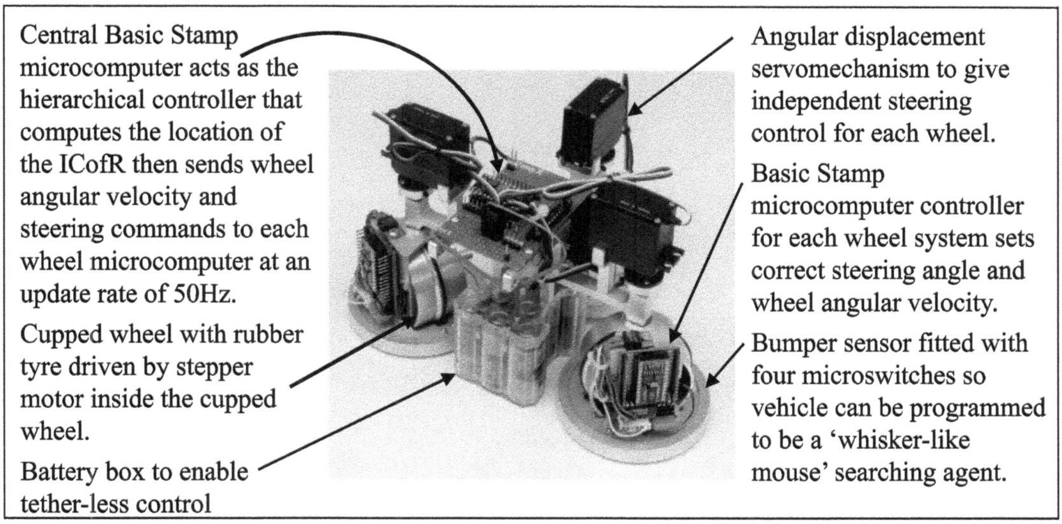

Figure 1.18 Omnidirectional wheeled vehicle. An engineered manifestation of the tricycle shown in figure 1.17.

locomotion technique with the omnidirectional tricycle. More precisely, an omnidirectional vehicle is considered a holonomic vehicle because its number of controllable degrees of freedom, i.e. three, is equal to its instantaneous degrees of freedom, i.e. three. In contrast, the tricycle in figure 1.7 is capable of three degrees of freedom in the ground plane but under certain constraints meaning that the tricycle can obtain any position, x, y in a ground plane and at any heading, but it can only do so with a specially prepared procedure, similar to when you park your car, which is a non-holonomic vehicle. Here you can park precisely in one manoeuvre but otherwise it requires multiple n-point turns. With a holonomic omnidirectional vehicle, the manoeuvre can be done in one manoeuvre.

1.9 Simultaneous body translation and rotation

As an example of simultaneous translation and rotation, figure 1.19 shows a pattern of motion that the omni-directional tricycle can accomplish. The pattern is similar to sun and planet wheel gear motion which involves rotation whilst translating. A virtual sun and planet gear is used to formalise the motion.

There are some interesting points concerning figure 1.19. These points will become important when we discuss the omni-directional tricycle next and program the omni-directional six-legged walking robot in a later chapter. The interesting points are illustrated and listed in figure 1.20 below.

The planet wheel can also be on the other side of the sun wheel as in figure 1.21, in which case the planet wheel will rotate in the opposite direction.

Sun and planet wheel motion is an important class of robot motion because it can be used as part of a behaviour pattern menu. For example, the motion can be used by a robot to back out of a door or retreat along a passage or all-round scanning and searching of terrain with an on-board computer vision system mounted on the chassis.

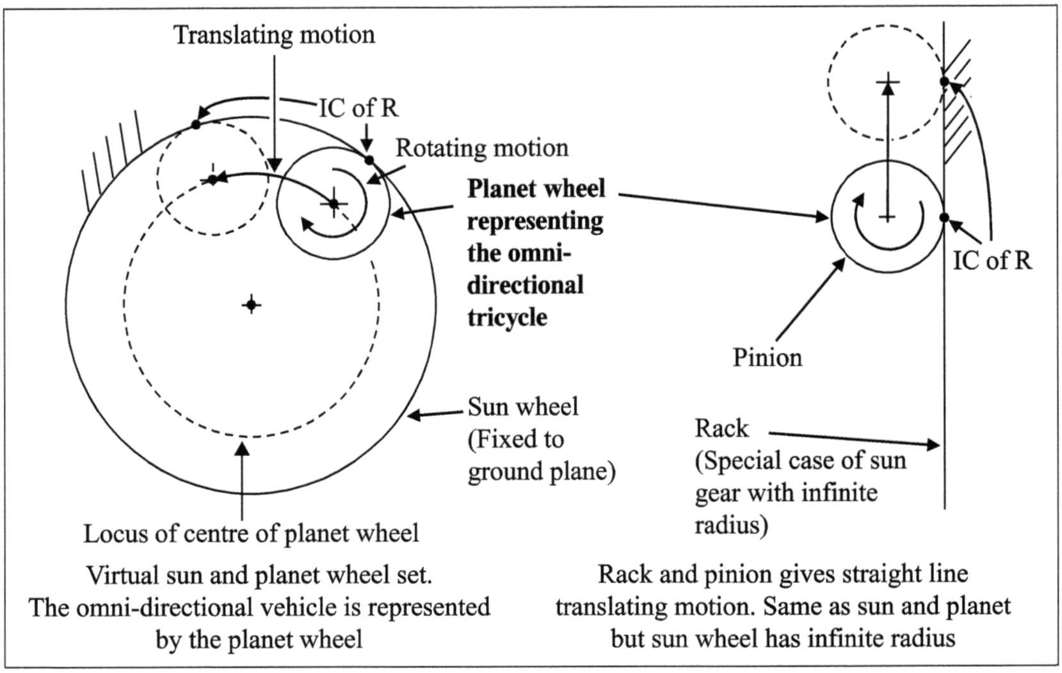

Figure 1.19 Illustrations of simultaneous translation and rotation related to the IC of R.

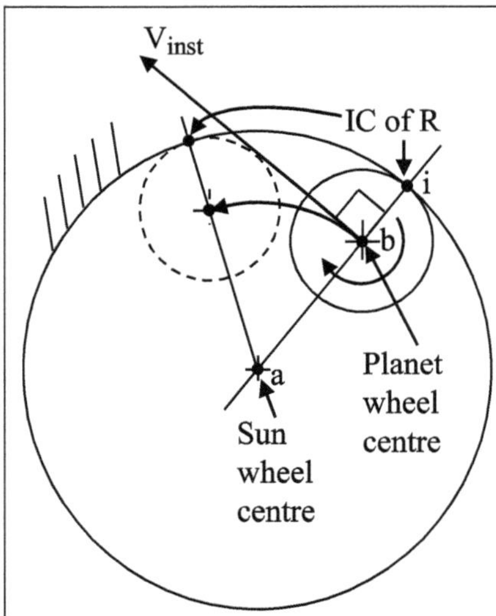

1. The sun wheel centre, planet wheel centre and IC of R all lie on a straight line, abi.
2. Instantaneous velocity, v_{inst}, of the planet wheel centre has a direction perpendicular to line, abi.
3. As already defined, the instantaneous rotation of the planet wheel is point, i, but the instantaneous rotation of the planet wheel centre, (cf. the wheel itself), is the sun wheel centre, i.e. point, a.

Figure 1.20 Some interesting points concerning sun and planet gear wheel behaviour.

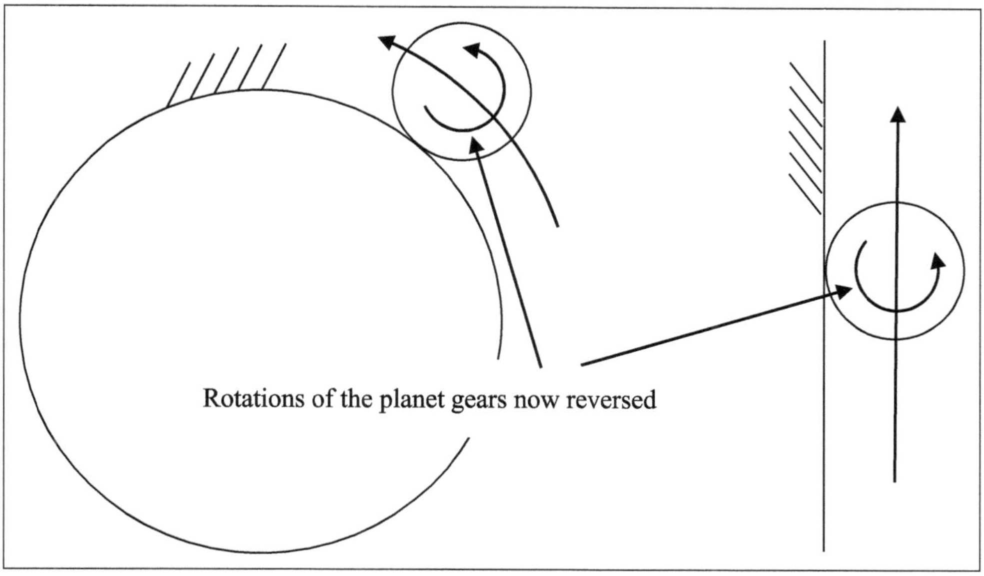

Figure 1.21 Planet wheels on the outside of sun wheel.

1.10 Achieving sun and planet motion

An omnidirectional tricycle can be programmed to carry out sun and planet motion by arranging the ICofR to rotate in a circle with respect to the ground and at the same time for the tricycle to rotate about the ICofR, figure 1.22.

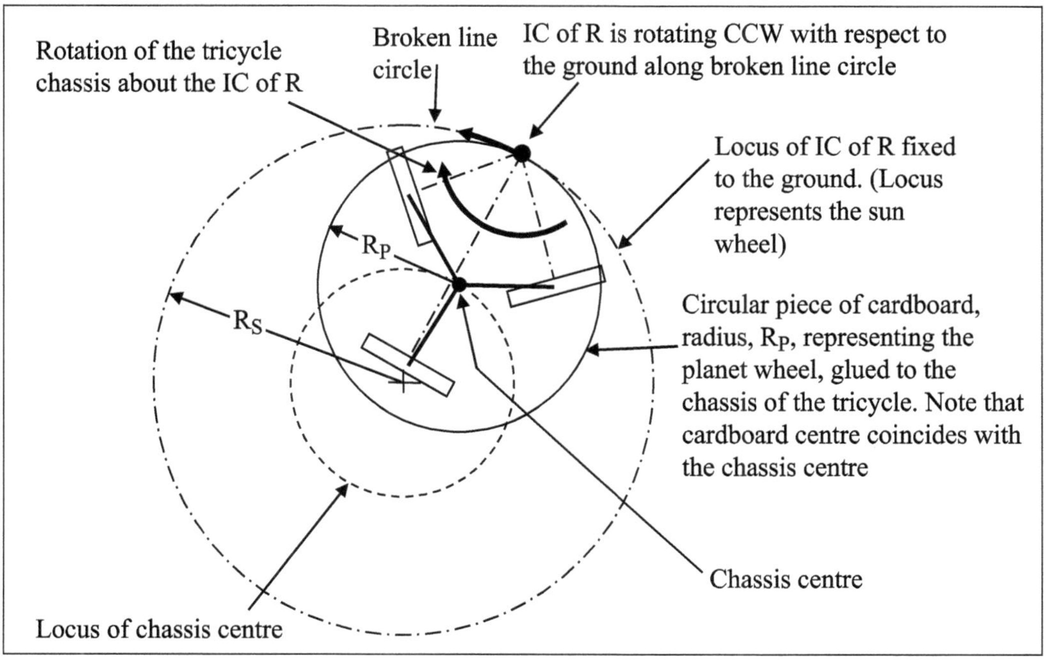

Figure 1.22 The omni-directional tricycle doing a 'Viennese waltz'.

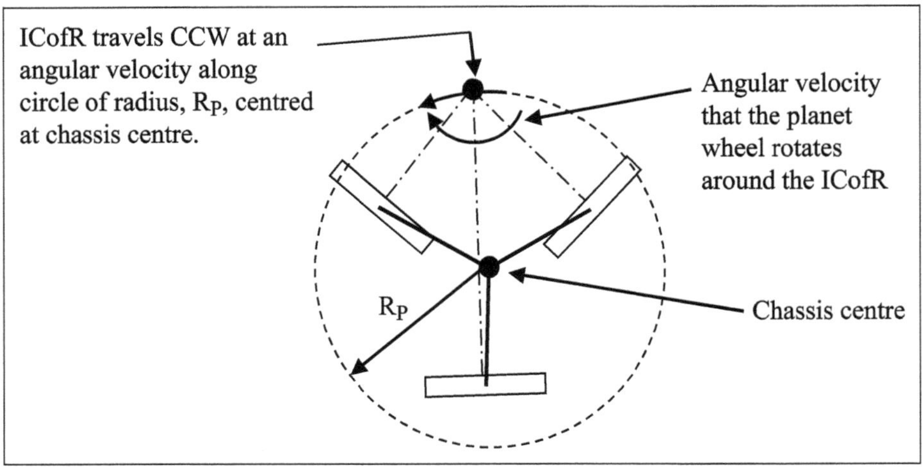

Figure 1.23 Position of IC of R during 'Viennese waltz' behaviour.

The radii of the sun and planet wheels are given the symbols, R_S, and, R_P, respectively. Now, if we imagine sitting still on the chassis centre, which becomes our frame of reference, then it can be seen that the ICofR rotates around us at radius, R_P, with centre positioned at the chassis centre, in a CCW direction, which is along the circumference of the piece of cardboard glued to the chassis, figure 1.23.

In order to program a robot to carry out sun and planet motion we need to know, (i) the radius of the planet wheel, (ii) the radius of the sun wheel (iii) the location of the sun centre and (iv) the angular velocity that planet wheel is rotating around the sun centre.

1.11 Viennese Waltz behaviour

The Viennese Waltz is another way of illustrating sun and plant motion with legs rather than wheels. It is a ballroom dance that requires a human dancing couple, joined as a coordinated four-legged paired unit, to rotate quickly whilst simultaneously translating, figure 1.24. Furthermore, the paired unit has to negotiate around the dance floor so, added to

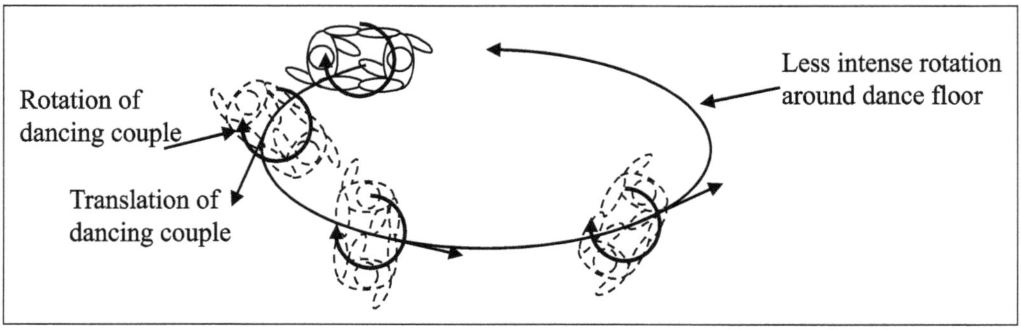

Figure 1.24 Bird's eye view of dancing couple doing the Viennese Waltz.

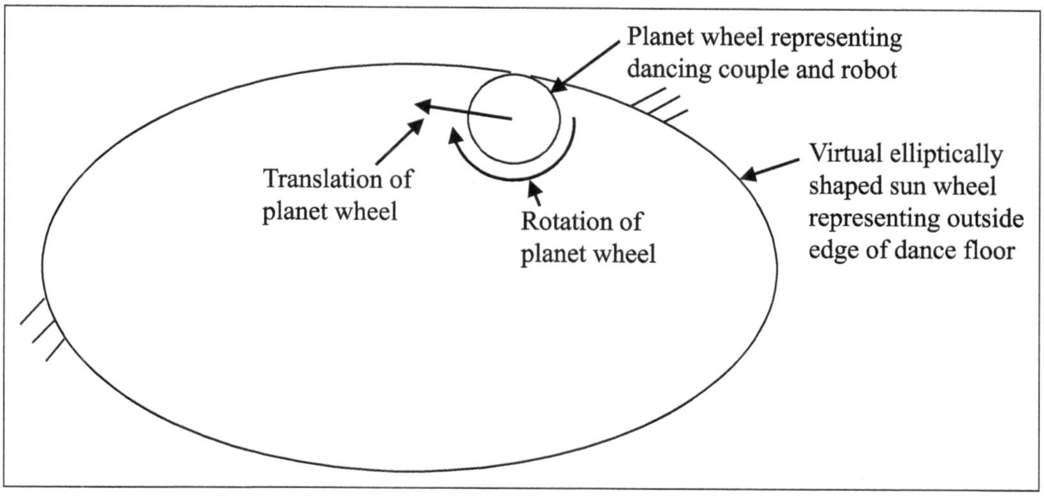

Figure 1.25 Sun and planet wheel modelling robot Viennese Waltz behaviour.

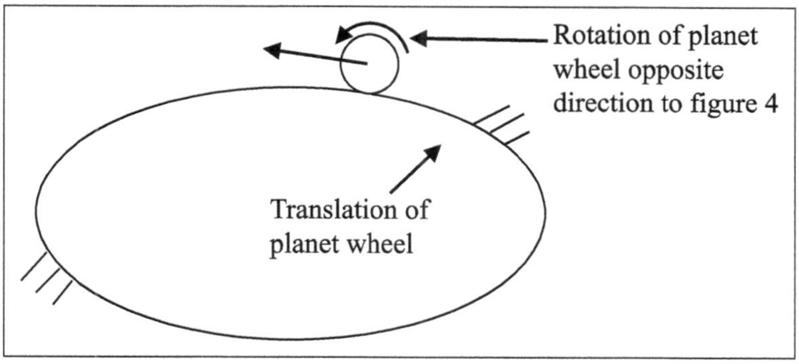

Figure 1.26 Planet wheel on outside of sun wheel.

the fast rotation and simultaneous translation, there is a less intense rotation around the dance floor.

In fact, ballroom dancing represents a very interesting study for the analysis and synthesis of legged and wheeled robot behaviour patterns. So also, interestingly, does the study of motion behaviour patterns and motion strategies of football players in a premier league match. Such a behaviour pattern is a fundamental requirement and a highly useful motion strategic property for omni-directional robots.

The classic "sun and planet wheel" is used to model Viennese Waltz behaviour where the planet wheel represents the robot and the sun wheel is a virtual wheel figure 1.25.

Also note that the planet wheel is a virtual wheel that is fixed to the robot and the diameter of the virtual planet wheel can be varied. The motion of the "planet wheel" is then programmed into the omni-directional walking robot such that the robot gives the required rotation whilst translating.

Furthermore, the planet wheel could be on the outside of the sun wheel, figure 1.26, in which case the rotation of the planet wheel is reversed.

1.12 Applications of Viennese waltz behaviour

1.12.1 Retreating motion in a corridor

The robot is shown, figure 1.27, backing out of a corridor representing retreating motion, i.e. translating, and simultaneously rotating such that the head of the robot is ready to face an opponent in the direction of its exit path.

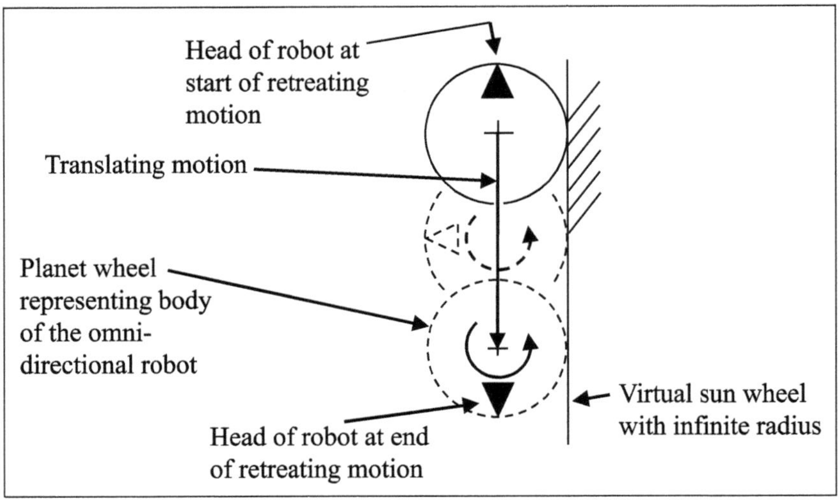

Figure 1.27 Application of Viennese Waltz behaviour. Retreating motion in a corridor.

1.12.2 Rotating on the spot

If the radius of the virtual planet wheel is set to zero then the robot will turn on the spot, figure 1.28.

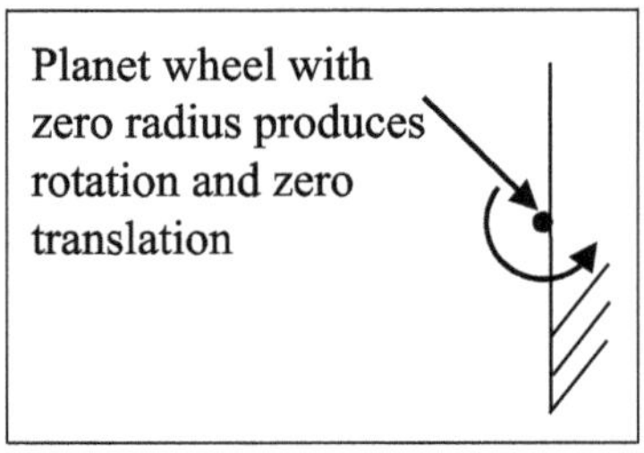

Figure 1.28 Application of Viennese Waltz behaviour. Rotating on the spot.

1.12.3 Rotating about a fixed point

If the radius of the sun wheel is set to zero and the planet wheel rotates on the outside of the sun wheel then the robot will rotate about a fixed point which is the sun wheel, figure 1.29.

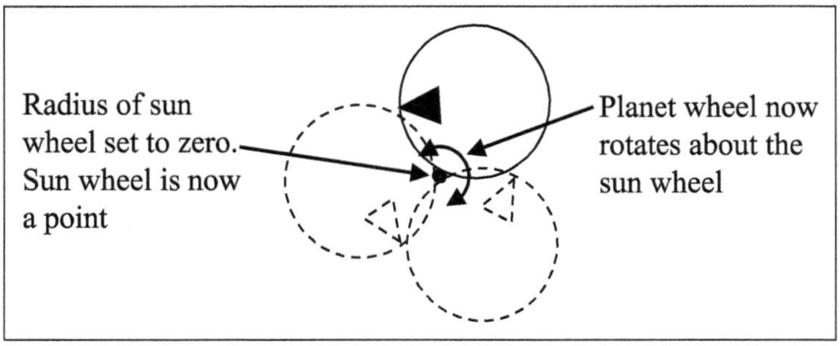

Figure 1.29 Rotating about a fixed point which is the sun wheel set to zero radius.

1.12.4 Motion in a straight line with no rotation, i.e. pure translation

This occurs when the radii of the sun and planet wheels are set to infinite radius, figure 1.30.

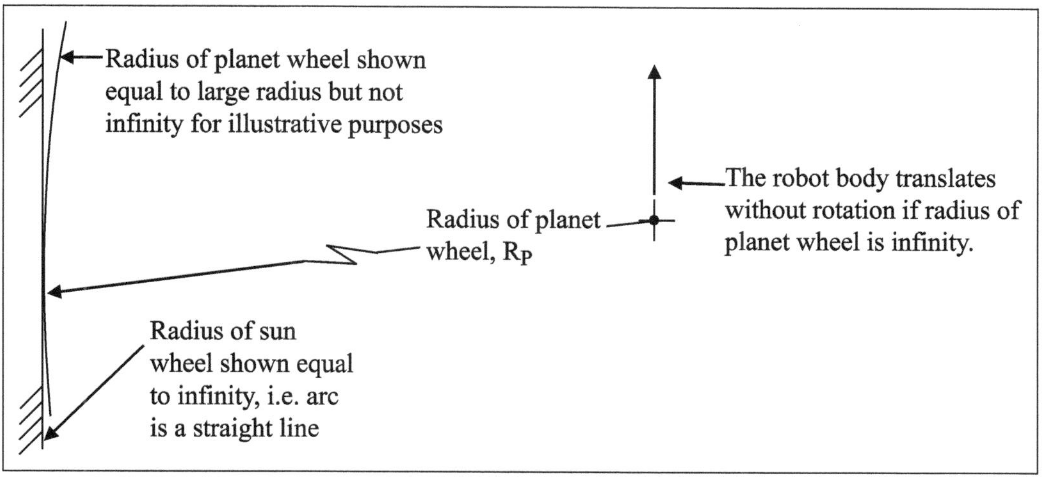

Figure 1.30 Motion in a straight line with sun and planet wheel radii set to infinity.

1.13 Analysis and programming of Viennese waltz behaviour into the robot

1.13.1 Analysis of a planet wheel circulating *inside* sun wheel

Viennese Waltz motion is programmed into an omnidirectional robot by assuming a virtual sun and planet wheel, figure 1.31. The radii of each wheel depend on the required behaviour pattern. The contact point between the sun and planet wheels is an instantaneous centre of rotation, ICofR.

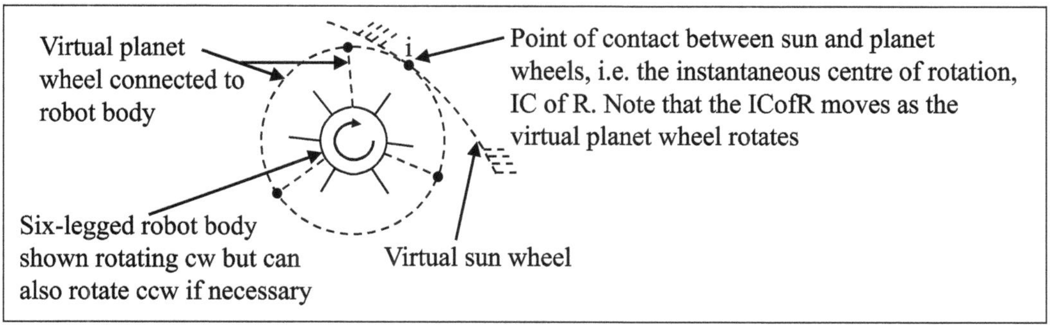

Figure 1.31 Virtual sun and planet wheels that are used as a model to produce Viennese Waltz behaviour in an omnidirectional robot.

In order to program Viennese waltz behaviour, it is only necessary to specify the following three variables; (see figure 1.32 part 1 and part 2 below for explanation):

1. Two variables which are the coordinates, θ_P, R_P, of the instantaneous centre of rotation, ICofR, with respect to the robot body.
2. Third variable which is the radius of the sun wheel, R_S.

part 1. ***Part 1 meaning of symbols***

Analysis of Viennese Waltz motion with a sun and planet wheel.

R_P = radius of planet wheel

R_S = radius of sun wheel

$\delta\theta_{planet}$ = incremental rotation of planet wheel about contact point between sun and planet wheels, Contact point is the IC of R of planet wheel.

$\delta\theta_{ICR}$ = incremental rotation of IC of R w.r.t. the robot body.

$\delta\theta_{sun}$ = incremental rotation of instantaneous centre of rotation about sun wheel centre.

θ_{P1}, θ_{P2} = angular positions of IC of R w.r.t robot body datum at positions 1 and 2 respectively.

Note that: $\theta_{P2} - \theta_{P1} = \delta\theta_{ICR}$

Part 2. ***Detailed kinematic analysis.***

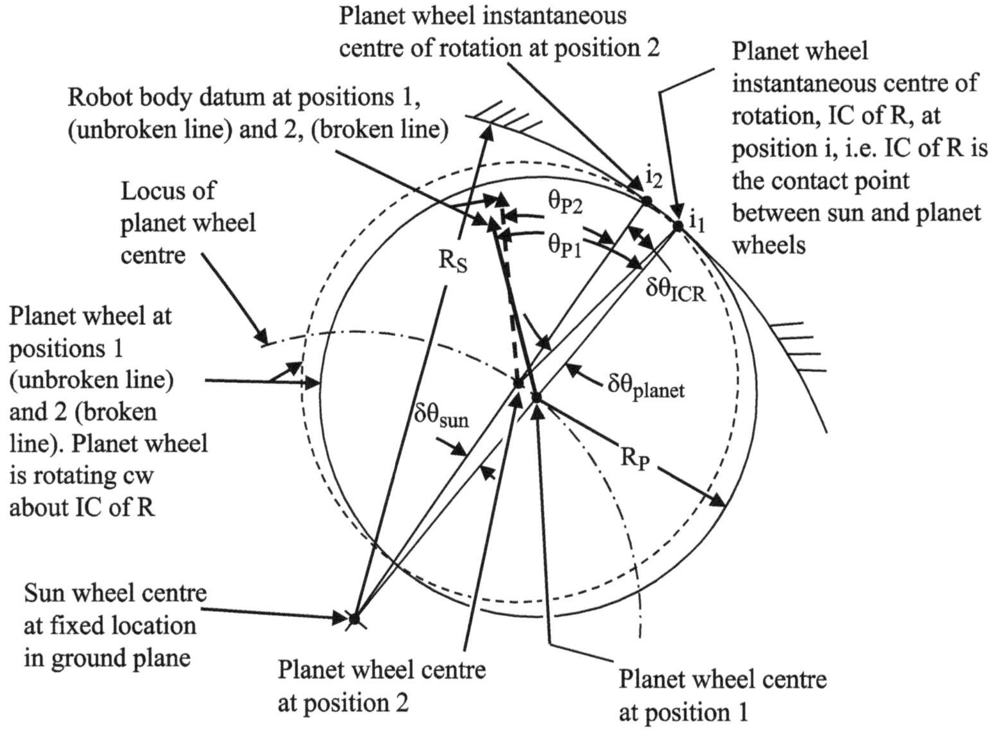

Figure 1.32 Analysis of Viennese Waltz motion with a sun and planet wheel shown with planet wheel in two positions, position 1 and position 2, separated by planet rotation discrete angle, $\delta\theta_{planet}$.

Figure 1.32 part 1 and part 2 above is a complete analytical overview picture and looks rather complicated so the important features are extracted, simplified and shown in figure 1.33.

Analysing figure 1.33, we obtain:-

$$\delta\theta_{ICR} = \frac{-\delta\theta_{sun} R_s}{R_p}$$

(negative sign because ICofR w.r.t. planet wheel moves CW which is opposite direction to the ICofR motion w.r.t. sun wheel)

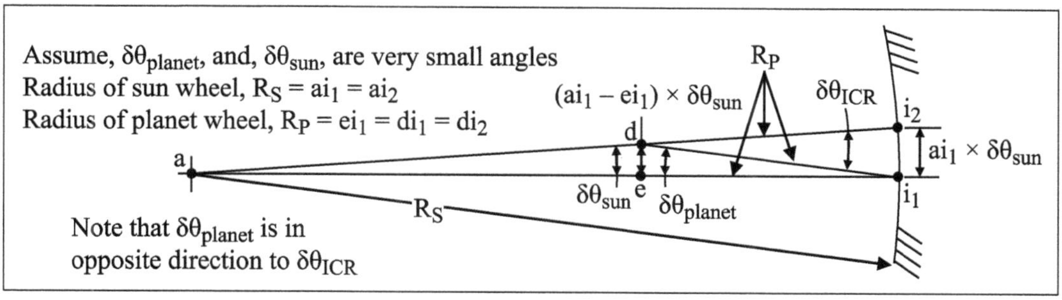

Figure 1.33 Discrete simplified analysis of the sun and planet wheels with the planet wheel shown in two positions as the planet wheel rotates by discrete angle, $\delta\theta_{planet}$.

$$\delta\theta_{planet} = \frac{\delta\theta_{sun}(R_s - R_P)}{R_p}$$

Eliminating, $\delta\theta_{sun}$, we get:-

$$\delta\theta_{ICR} = -\delta\theta_{planet} \times \left(\frac{1}{1 - \frac{R_P}{Rs}} \right) \qquad \text{Eq. 1.1}$$

or in dimensionless form

$$\frac{\delta\theta_{ICR}}{\delta\theta_{planet}} = -\frac{1}{1 - \frac{R_P}{R_S}} \qquad \text{Eq. 1.2}$$

....where, $\delta\theta_{planet}$ is a small suitable value chosen by the robot programmer

Eq. 1.1 is used to compute the coordinates of the new position of the ICofR, $new\theta_{ICR}$, with respect to the planet, i.e. the robot, like this:

$$new\theta_{ICR} = old\theta_{ICR} + \delta\theta_{ICR} \qquad \text{Eq. 1.3}$$

....where, $\delta\theta_{ICR}$, is given by eq. 1.1

R_P, remains the same unless a new trajectory is required. In practice the very small angles, $\delta\theta_{ICR}$, and, $\delta\theta_{planet}$, are not small. In fact, $\delta\theta_{ICR}$, enlarges to, $\Delta\theta_{ICR}$, and, $\delta\theta_{planet}$, enlarges to, $\Delta\theta_{planet}$. Hence eq. 1.1 leads to errors in the robot body following the required path. However, these errors are compensated by inserting a compensation coefficient gain factor, k where k>1, into eq.1.1 like this:-

Planet wheel circulating inside sun wheel

$$\Delta\theta_{ICR} = -\Delta\theta_{planet} \times k \left(\frac{1}{1 - \frac{R_P}{Rs}} \right) \qquad \text{Note, } k > 1 \qquad \text{Eq. 1.4}$$

...where k can be found by calculation or, empirically, by experiment.

The k value, being greater than unity, means that, $\Delta\theta_{ICR}$ increases as $\Delta\theta_{planet}$ becomes finite and furthermore, k increases as $\Delta\theta_{planet}$ increases. You can see a YouTube movie clip of a hexapod robot programmed by the author doing such a manoeuvre, i.e. a planet wheel circulating inside sun wheel where the planet wheel is represented by the hexapod robot; *https://www.youtube.com/watch?v=IOX8aw2l0_c*

Now let's look at the complementary situation where the planet wheel is on the outside of the sun wheel.

1.13.2 Analysis of a planet wheel circulating *outside* sun wheel

Figure 1.34 illustrates the case with the planet wheel outside the sun wheel using very small discrete values of angle changes, $\delta\theta_{ICR}$, $\delta\theta_{sun}$, and $\delta\theta_{planet}$

Analysing figure 1.34:

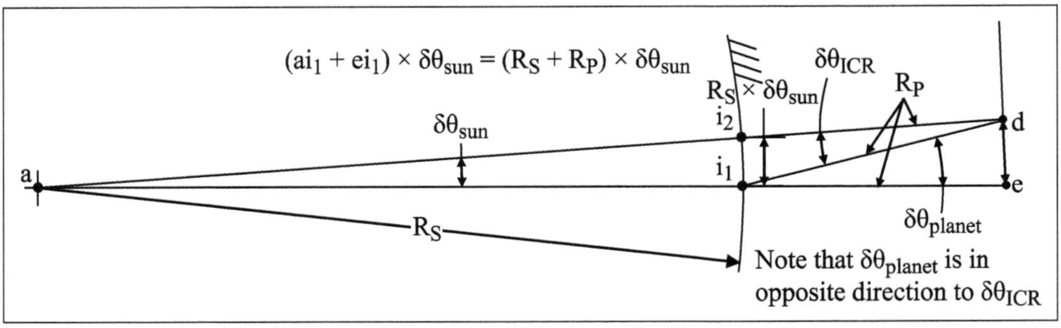

Figure 1.34 Discrete simplified analysis of the sun and planet wheels with the planet wheel circulating outside the sun wheel.

$$\delta\theta_{sun} = -\frac{R_P}{R_S} \times \delta\theta_{planet} \qquad \text{Eq. 1.5}$$

$$\text{Also}: \delta\theta_{ICR} = \frac{de}{R_P} = \frac{(R_S + R_P)}{R_P} \text{x } \delta\theta_{sun} \qquad \text{Eq. 1.6}$$

Substituting $\delta\theta_{sun}$,(Eq. 1.5), into Eq. 1.6, we obtain;

$$\delta\theta_{ICR} = -\delta\theta_{planet} \times \frac{(R_S + R_P)}{R_S} \qquad \text{Eq. 1.7}$$

The very small angle values are set to practical achievable values as follows: $\delta\theta_{ICR} = \Delta\theta_{ICR}$ and $\delta\theta_{planet} = \Delta\theta_{planet}$, thus with a compensation coefficient, k, where k>1. Thus, Eq. 1.7 becomes:

Planet wheel circulating *outside* sun wheel

$$\Delta\theta_{ICR} = -\Delta\theta_{planet} \times \frac{1}{k}\left(1 + \frac{R_P}{Rs}\right) \qquad \text{Note, } k > 1 \qquad \text{Eq. 1.8}$$

This time, $\Delta\theta_{ICR}$ decreases as $\Delta\theta_{planet}$ becomes finite. You can see a YouTube movie clip of a hexapod robot programmed by the author doing this manoeuvre, i.e. a planet wheel circulating outside sun wheel where the hexapod robot is represented by a planet wheel; *https://www.youtube.com/watch?v=cAm4uO_Pcp0*

1.14 Practical examples of a robot acting as the planet wheel

We now show examples of how the quantities, $\Delta\theta_{ICR}$ and $\Delta\theta_{planet}$ are used in a robot.

1.14.1 Robot rotating as a planet wheel on the *inside* of a sun wheel

Figure 1.35 shows a plan view of a hexapod robot that could equally represent an omnidirectional 3-wheeled robot. The planet wheel is on the inside of the sun wheel. As an example, $R_P/R_S= 4$ and $\Delta\theta_{planet} = 45°$.

Thus, the relationship, eq. 1.4 applies, i.e. $\Delta\theta_{ICR} = - \Delta\theta_{planet} \times 1 \times (1/(1-R_P/R_S))$ Substituting values with k = 1,

$\Delta\theta_{ICR} = - 21° \times 1 \times k(1/(1-1/4)) = - k \times 28°$

...and at present k is unknown. We only know that k>1 for practical discrete values of $\Delta\theta_{planet}$

The diagram on the right-hand side of figure 1.35 shows the robot rotated CW about the initial ICofR, shown on the left-hand side, i_n, by 21°, i.e. $\Delta\theta_{planet} = 21°$. The location of the next ICofR, i_{n+1}, is given by $\Delta\theta_{ICR} = -k \times 28°$. If k=1 then the location is underestimated, thus k>1 and found by experiment or calculation.

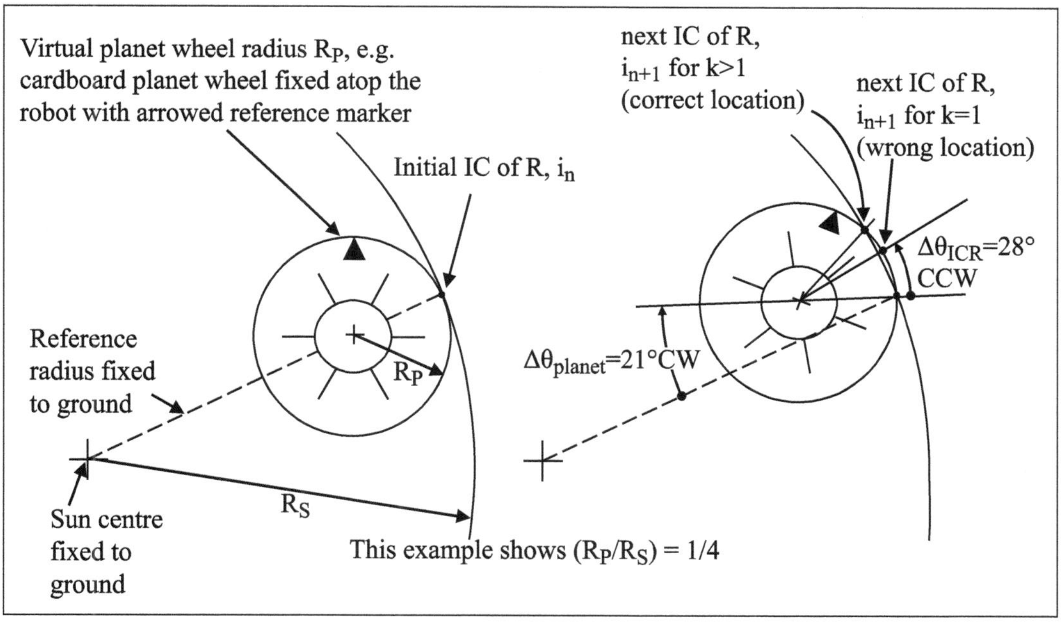

Figure 1.35 Practical example of robot acting as the planet wheel rotating inside a virtual sun wheel.

1.14.2 Robot rotating as a planet wheel on the *outside* of a sun wheel

Figure 1.36 shows the planet wheel on the outside of the sun wheel. As an example, $R_P = R_S/2$ and $\Delta\theta_{planet} = 30°$.

Thus, the relationship, eq. 1.8 applies, i.e. $\Delta\theta_{ICR} = - \Delta\theta_{planet} \times (1/k) \times (1+R_P/R_S))$ Substituting $R_P/R_S = 1/2$ and $\Delta\theta_{planet} = 30°$, we obtain: $\Delta\theta_{ICR} = -30° \times 1/k \times (1+1/2)) = -1/k \times 45°$

...and at present k is unknown. We only know that k>1

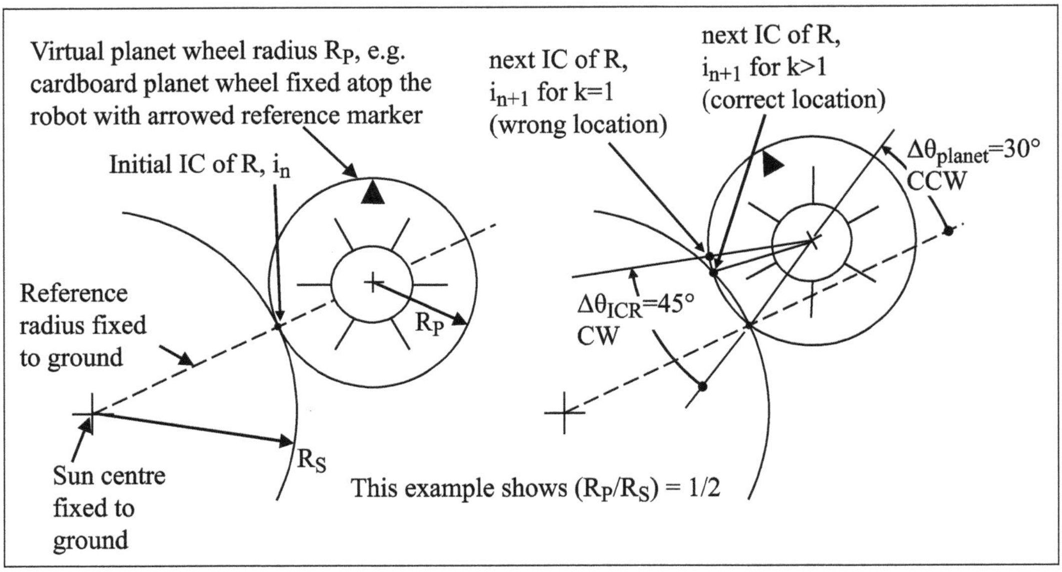

Figure 1.36 Practical example of robot acting as the planet wheel rotating inside a virtual sun wheel.

The diagram on the right-hand side of figure 1.36 shows the robot rotated CW about the initial ICofR, shown on the left-hand side, i_n, by 30°, i.e. $\Delta\theta_{planet} = 30°$. The location of the next ICofR, i_{n+1}, is given by $\Delta\theta_{ICR} = -(1/k) \times 45°$. If k = 1 then the location is underestimated, thus k>1 and found by experiment or calculation.

1.15 More interesting ideas for wheeled vehicles

1.15.1 Tricycle with 3 degree-of-freedom wheels

Let us further develop our omni-directional tricycle by adding an extra degree of freedom to each wheel assembly. The extra degree of freedom changes the height of the wheel along the z-axis. Each wheel now becomes a 3dof wheel assembly as shown below, figure 1.37. Each wheel has a vertical z-axis that is positioned by a drive motor. The system is similar to an automobile with active suspension which means each wheel is independently displaced by a hydraulic jack or otherwise to keep the car level or to tilt when the car turns a corner in the same way that an aeroplane banks when turning so your cup of coffee does not spill.

3dof wheels will enable the tricycle to travel over undulating terrain whilst the chassis stays level thus becoming a stabilised level platform. Such a platform will give a smooth comfortable ride for passengers or will enable the stabilised and accurate deployment of missiles from moving military vehicles or it will enable an on-board computer vision system to take images that have no need of being rotated to be referenced to the horizon.

In fact, our own eyeballs are stabilised platforms about the three rotation axes, i.e. roll, pitch and yaw, as shown in figure 1.38.

It is significant to point out that the eyeball has limited angle of rotation in each of its axes; roll (approx. +/−20°), pitch (approx. +/−45°), and yaw (approx. +/−45°).

The tricycle chassis, fitted with 3dof wheel assemblies, can go one step further than the eyeball because it can be stabilised in six degrees of freedom as shown next.

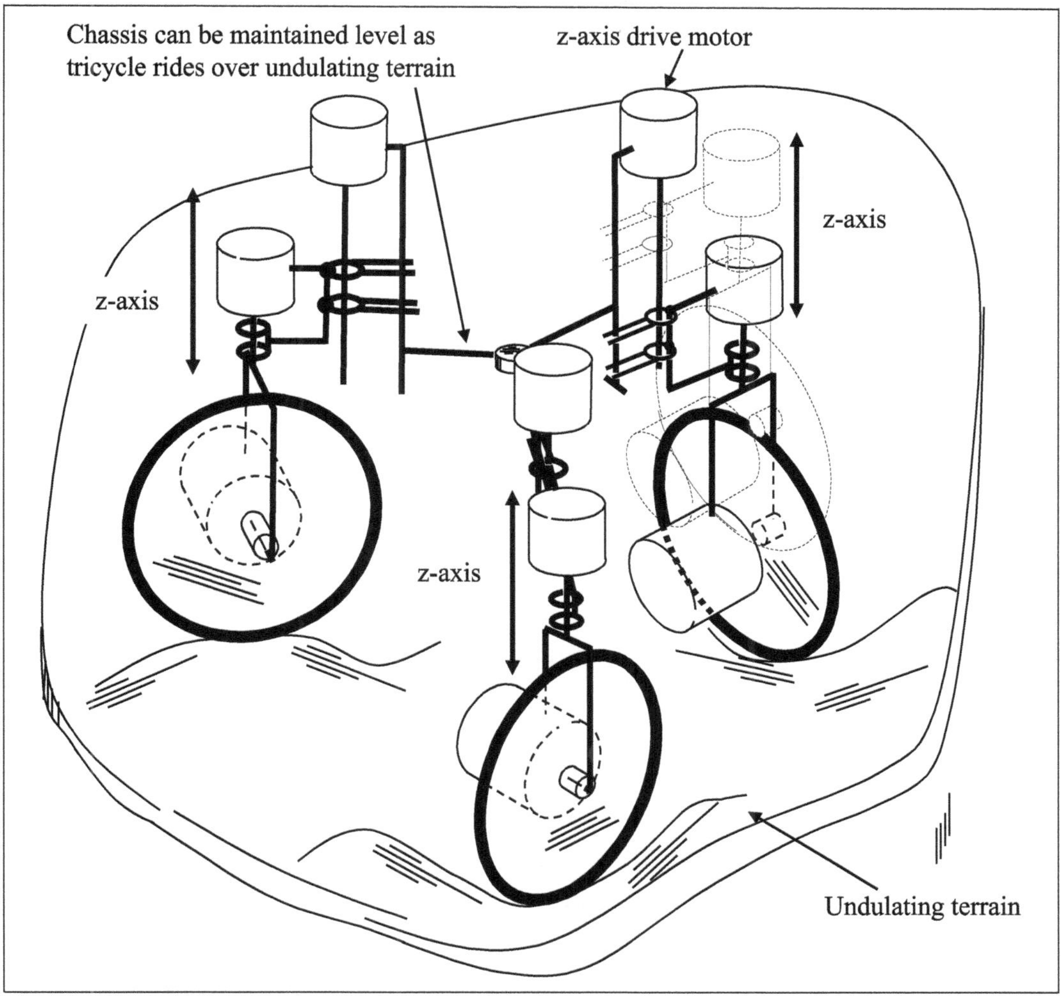

Figure 1.37 Tricycle with 3dof wheels. The degrees of freedom are
1. individual angular velocity control of each wheel
2. individual steering angle of each wheel
3. individual height control of each wheel along the z-axis.

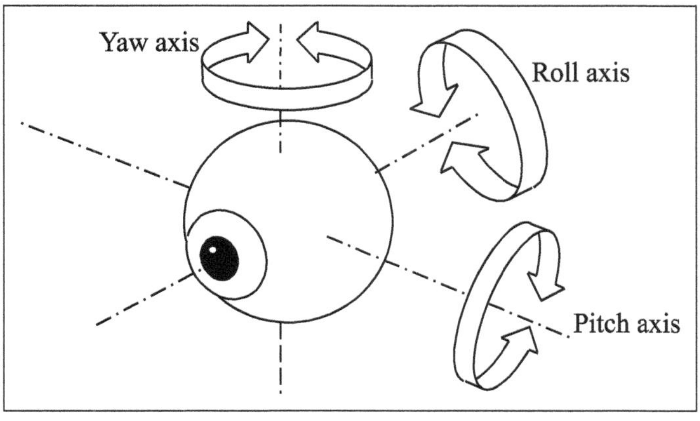

Figure 1.38 3 dof eyeball. Our own eyeballs are stabilised in 3 rotation degrees-of freedom.

1.15.2 Programming the tricycle to emulate a 4-wheeled automobile

It is possible to program our omnidirectional tricycle, figure 1.17 or figure 1.37, to emulate vehicles, with which we are familiar such as a 4-wheel automobile with a fixed back axle, figure 1.7 and Figure 1.14. For example, consider the following example, figure 1.39.

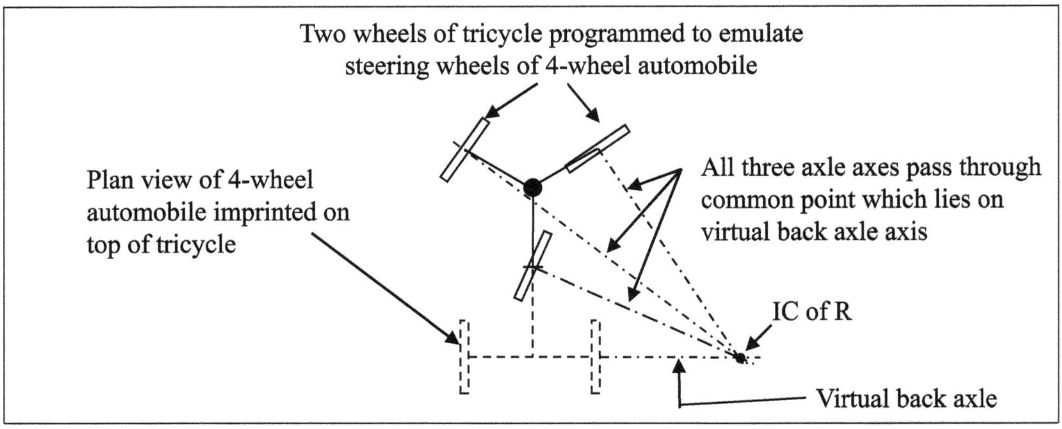

Figure 1.39 Simulating a 4-wheel automobile with an omnidirectional tricycle.

1.15.3 Multi-segmented trailer motion

Multi-segmented towed trailers pulled by a tractor, figure 1.40 are popular at airports where they are used to transport passenger baggage to and from parked aircraft. The

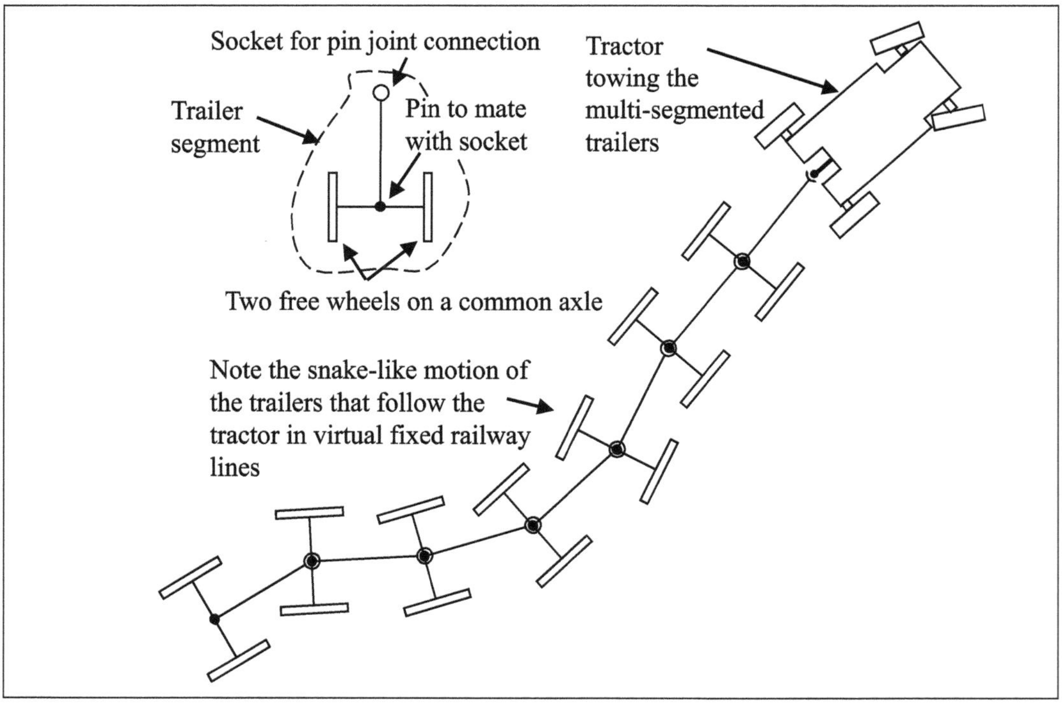

Figure 1.40 Multi-segmented trailer wheeled vehicle pulled by a tractor is used by airport baggage handlers when loading and unloading aeroplanes.

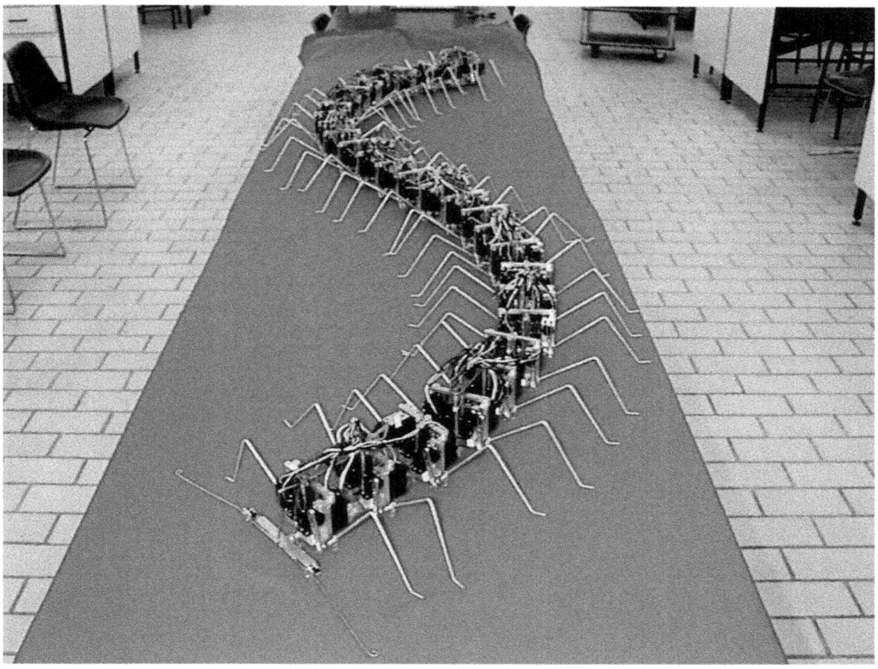

Figure 1.41 60-legged centipede-like robot that emulates the snake-like motion of the multi-segmented wheeled trailer.

author built a 10-segmented walking 60-legged, centipede-like robot, figure 1.41 to emulate such a multi-segmented trailer system. Each 'trailer' segment was a six-legged robot with 2 dof legs and each segment was joined with a sensor that measured the angle between adjoining segments. An electrical spine sent angle information down the spine at the same speed as the robot was walking forward. These angle values were picked up by each segment as they appeared and the legs of that segment produced that angle between segments. Such programming emulated the same snake-like motion of the multi-segmented towed trailer.

We now conclude this chapter which has served as an introduction to vehicle wheeled motion together with the relation between wheels and legs. Chapter 3 deals with a real practical vehicle, which is a tricycle with two steered wheels and one dual-servo driven wheel. The tricycle raises mathematical real-time computing problems and provides their solutions plus it reinforces theoretical concepts that have been raised in this chapter. Furthermore, the tricycle control algorithms serve as a forerunner to the programming of legged robots. However, before proceeding to the Tricycle Robot, the next chapter, chapter 2, deals with a Servo Teach module which is a fundamental tool for controlling and calibrating servomechanisms which are used for actuating all the robots in this book.

2. The ServoTeach module

2.1 Introduction

The ServoTeach module is a manually adjustable electronic circuit board, figure 2.1, that derives three independent, variable-angle, control signals for Futaba angular displacement servomechanisms, figure 2.2 that are used as artificial muscles for the robots described in this book. These servomechanisms or 'servos' actuate the rotation of an output shaft that pushes or pulls mechanical cranks, levers, links and limbs of the robot that produce locomotion. The ServoTeach module manually positions the servo output shaft to a required angle and the angle value integer number can be read at a computer screen. The integer number is derived from a Basic Stamp2 microcomputer. The number represents fractions of degree units that are calibrated in experiments described in following chapters.

The ServoTeach module is used to produce a series of numbers that represents the desired trajectory of a robot limb. From this series a mathematical function is derived, usually a Taylor sum of integer polynomial series to represent the numbers as a continuous function.

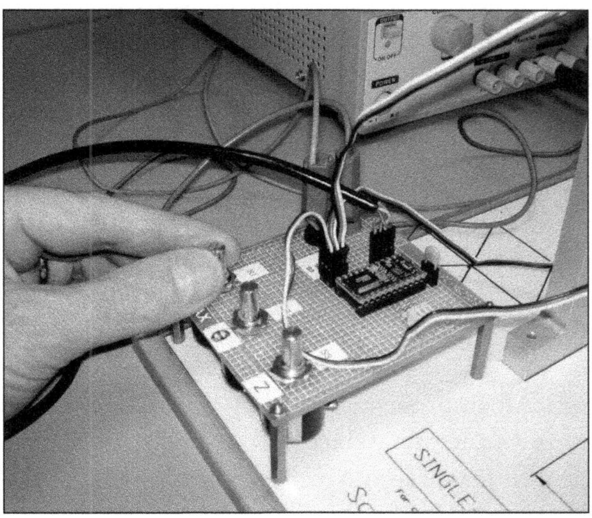

Figure 2.1 The ServoTeach Module.

2.2 The Futaba integrated servomechanism, the 'Servo'

The Futaba integrated servomechanism, or just abbreviated to 'servo', is wonderful compact little device used to actuate the steering wheels and drive wheel of the Tricycle Robot in the next chapter. Servos were originally designed as actuators for the aileron, elevator and rudder control surfaces of radio-controlled model aircraft as well as the speed control of the diesel/glowplug engine/electric motor powered propeller that provided the aircraft propulsion force. Hence, sometimes, servos are referred to as r/c servos, i.e. radio-controlled servos. Servos have an output shaft that rotates $+/-70°$, so attaching a lever to the shaft enables it to actuate the position of, for example, an engine throttle or a rudder.

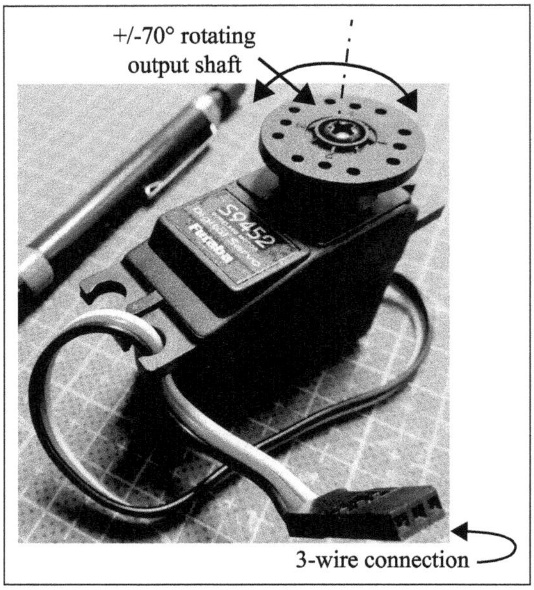

Figure 2.2 The Futaba servo.

The beauty of the servo is its ease of control through a 3-wire connection. The three wires are (i) 0V ground, (ii) 4.5V to7.4V power and (iii) signal, (1ms to 2ms pulse length repeated at 40Hz to 80Hz but nominally 50Hz) that controls the angle of the output shaft between +/−45° where the shaft angle is proportional to the pulse length. Note that all these values are approximate and dependent on the model number and brand. The earliest recollection that the author has of these small black boxes is in the mid 1960's when he was a youngster. They originated some decades before that time but in the form of solenoid-actuated incremental bidirectional motion control similar to a clock 'escapement'. Otherwise, the size, outward construction and mounting arrangement of servos have changed little in the intervening years but the insides have improved significantly with the use of double ball bearing races, metal gears, brushless DC motors, digital control, increased stall torque and increased no-load slew speed to name probably the most significant improvements. They are still relatively lightweight and now manufacturers are producing many sizes, e.g. fat high-power ones that can be used for robots, or thin ones to fit inside slim wings or micro servos for the amazing tiny insect-sized quadcopters that are available now. There are also variants of the servo output power, for example, you can have high no-load slew speed at lowered stall torque, or conversely, lower no-load speed at higher stall torque depending on your application. A visit to www.towerhobbies.com will show the extent of the range of servos that can be purchased.

Furthermore, Futaba is just one brand; there are many others, e.g. HiTec, and now Chinese manufacturers are producing very good low priced servos, e,g, JX Servo. Other servos have been invented in the last few years, e.g. Dynamixel, which differ in their more sophisticated control protocol and that they can be controlled as a standard angle position servo or as a continuous rotation velocity servo.

However, all servos are similar in that an internal sensor measures the output shaft angle and internal electrical signal and power circuitry then automatically drives the angle of the output shaft, either in the clockwise or counter clockwise direction, using negative feedback control to make the shaft angle equal to the demanded angle. For Futaba servos, the demanded angle is represented by a pulse length that varies between 0.8ms and 2.2ms that positions the servo to +/−70° from its central position approximately subject to calibration, or in the case of the Dynamixel servo, by a specialised serial communication protocol. All these measurement and control functions are carried out inside the little black box of the servo, hence its name given by the author of, 'integrated servomechanism'. Also, there is a mass of information on YouTube concerning these servos, just search for 'rc servo'.

2.3 The servo teach module

The ServoTeach module can independently actuate three servos with three manually adjustable 3-terminal potentiometers that are wired as 2-terminal variable resistors as shown in the ServoTeach module circuit diagram, figure 2.3 below. A Basic Stamp2px microcomputer is central to the module and it measures the resistance of each of the variable resistors using the ingenious RCtime instruction which is an analogue-to-digital converter, i.e. an ADC. The RCtime instruction measures the resistance of each variable resistor that is connected in charge time units (0.81μs for the BS2px) that are not ohm units but nonetheless represent ohm units via a scale factor that can be found by experiment if necessary.

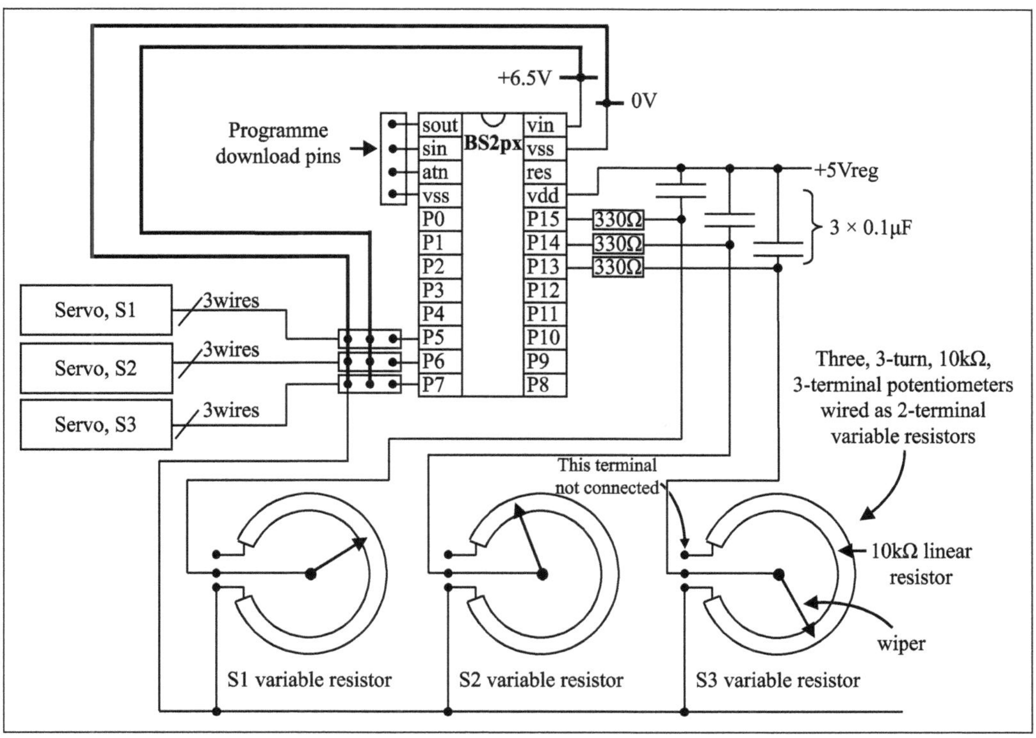

Figure 2.3 Circuit diagram of the ServoTeach module.

A detailed explanation of the how the RCtime instruction works can be found by an internet search and by using the Parallax editor help facility that is located in the Basic Stamp editor downloaded from the www.parallax.com website. Adequate explanation of the RCtime instruction is freely available and thus it is not necessary to repeat such explanation in this book.

The ServoTeach module measures S1, S2 and S3 variable resistances, figure 2.3 at pins p15, p14 and p13 respectively via 0.1μF capacitors wired to each of these variable resistors. The RCtime instruction then measures each of the resistances and these values are scaled to change them into pulse widths that vary from 0.8ms to 2.2ms that are used, via the PULSOUT instruction, to position three servos, S1, S2 and S3 to required angular positions. The programme code that carries out the ADC measurement and sends scaled pulse widths to the servos is given below. The values of these pulse widths that represent servo angle are sent to the computer screen using the DEBUG instruction, figure 2.4. These values are then recorded by students who then devise mathematical functions that pass through these values. The programme code, described with comments, has a special hysteretic filter to extract noise from the RCtime measurements so that the servo is not fed with signals that cause jitter and that can provide steady values on the computer screen. The hysteretic filter is not an averaging filter so there is no delay in measuring numbers.

After running the code, a debug window will open and show you the three numbers, S1, S2 and S3 being sent to servos S1, S2 and S3, respectively, similar to this:

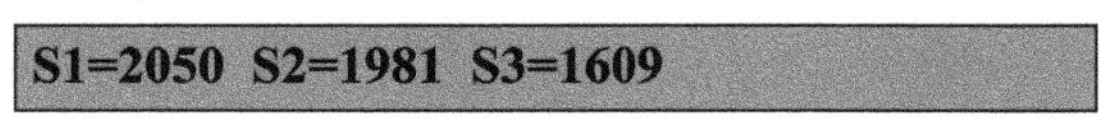

Figure 2.4 debug window showing S1, S2 and S3 numbers that are displayed on the computer screen.

2.4 Servo teach module programme code

```
'{$STAMP BS2px}
'ServoTeach with backlash filtering
'Program for the ServoTeach/programming module with backlash noise filtering
'The program manually controls the position of the three servos, S1, S2 and S3
'Use this program to manually rotate a servo by 'adjusting the three variable resistor
'knobs marked S1, S2 and S3.
'The debug window will show the values of the servo angles as S1, S2 and S3
'The program has backlash filtering to cut out noise so that the servos don't jitter.
a     VAR    Word    'three variables, a, b and c for
                     measuring raw RCtime values
b     VAR    Word
c     VAR    Word
S1    VAR    Word    'three variables, S1 , S2 and S3
                     that have been scaled from raw
                     'values of a, b and c respectively
S2    VAR    Word
S3    VAR    Word
```

```
backlash     CON            3          'the backlash value that represents
                                       the peak to peak noise 'elimination
                                       value
again:       HIGH 15                   'discharge all three 0.1μF capacitors
             HIGH 14
             HIGH 13
             PAUSE 1                   'wait 1ms to complete discharge
             RCTIME 15,1,a             'measure values of the three variable
                                       resistors, S1, S2 'and S3 variable
                                       resistor. These values represent servo
                                       'demand angles
             RCTIME 14,1,b
             RCTIME 13,1,c
             a=a-1    MIN    0         'set RCtime starting values to zero
             MAX 1444                  and set 'maximum values
             b=b-1    MIN    0
             MAX 1444
             c=c-1 MIN 0 MAX
             1444
             a=a*8/7 +1050             'scale the three variable resistor values
             b=b*8/7+1050
             c=c*8/7+1050
BacklashfilterS1:           IF a<(S1-backlash) THEN skip1
             GOTO SKIP2
skip1:       S1=a+backlash
skip2:       IF a>S1+backlash THEN SKIP3
             GOTO SKIP4
SKIP3:       S1=a-backlash
BacklashfilterS2:
SKIP4:       IF b<S2-backlash THEN SKIP5
             GOTO SKIP6
SKIP5:       S2=b+backlash
SKIP6:       IF b>S2+backlash THEN SKIP7
             GOTO SKIP8
SKIP7:       S2=b-backlash
```

```
'BacklashfilterS3
SKIP8:      IF c<S3-backlash THEN SKIP9
            GOTO SKIP10
SKIP9:      S3=c+backlash
SKIP10:     IF c>S3+backlash THEN SKIP11
            GOTO pout
SKIP11:     S3=c-backlash
'All filters done so output the angle demand values to the servos
pout:       DEBUG  "S1=",DEC
            S1,"  S2=",DEC  S2,"          'show'values on screen
            S3=",DEC S3,CR
            PULSOUT 5,S1                  'output values to servo S1, S2
                                          and S3
            PULSOUT 6,S2
            PULSOUT 7,S3
            GOTO again                    'repeat the process again to
                                          update angle demand values
```

2.5 Hysteresis noise filter

The noise filter is a hysteresis type filter and is shown in figure 2.5 where it is modelled by two intermeshing gears with backlash between the gear teeth. The backlash allows the noisy RCtime values, represented by the jittering input gear, to jitter inside the backlash angle without being fed to the output gear. The advantage of such a filter is that there is no time lag produced as there would be with an averaging filter

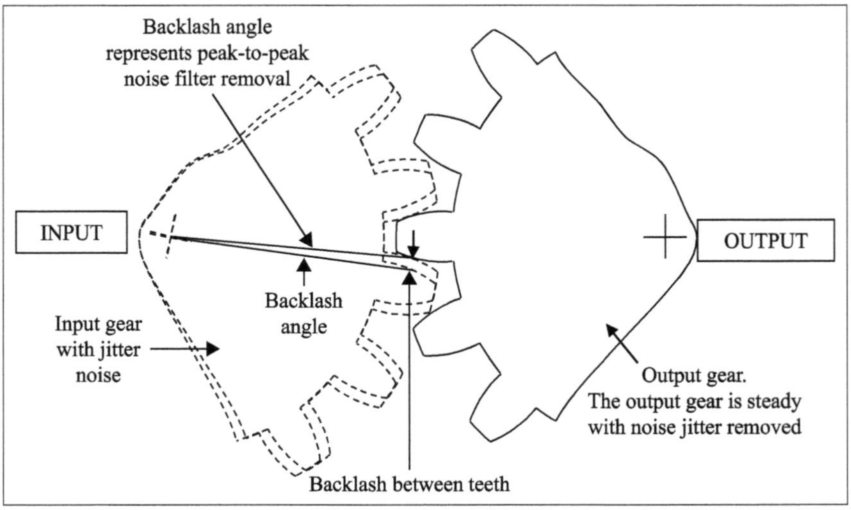

Figure 2.5 Backlash filter modeled as a mechanical gear. The filter is a hysteretic method for eliminating noise without producing a time lag.

Figure 2.6 illustrates an example of measuring a noisy signal and elimination the noise component. The backlash envelope is programmable and the figure shows the envelope set at +/−2 units. How does one decide on this value? Well, by trial and error so if the peak to peak noise component is, say, 5 units then the envelope should be set to +/−3 units.

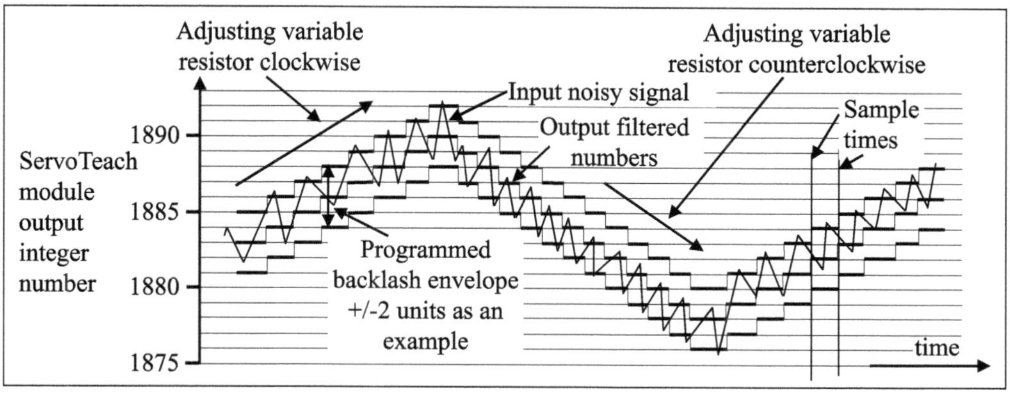

Figure 2.6 Typical response of the hysteretic filter showing input noisy signal, output filtered numbers and the programmed backlash envelope.

3. The Tricycle Robot: a mobile programmable 'drive-by-wire' electric vehicle

A forerunner to (i) hexapod and vehicle motion behaviour patterns in a ground plane and (ii) the use of piecewise truncated Taylor series to solve complicated equations using integer numbers

3.1 Introduction

In this chapter we discuss the programming of a 3-wheeled electric vehicle called the Tricycle Robot, figure 3.1, to undertake basic motion in a ground plane that utilises the concept of the instantaneous centre of rotation in relation to vehicle steering theory. The concept is transferrable to, and lays the foundation for, robot legged motion described in later chapters. The tricycle has two front wheels for steering and one rear driven wheel for locomotion. The driven wheel is rotated by a pair of synchronously-acting angular displacement servomechanisms (Futaba brand) that actuate the wheel via push-rods like a railway steam engine providing variable speed locomotion in both forward and reverse directions.

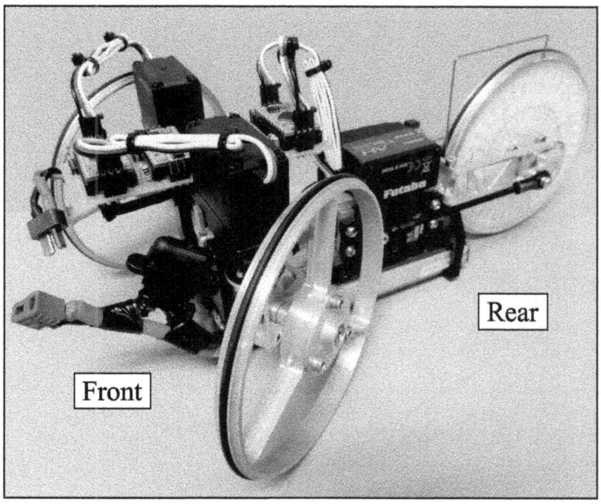

Figure 3.1 The Tricycle Robot. Two free-wheeling steering wheels at the front and one drive wheel at the rear.

The steering wheels do not provide for locomotion and are thus considered free wheels. They are steered individually and coordinately with a second pair of servomechanisms. Thus, there are four servos that enable controlled motion in a ground plane. An on-board computer system, comprised of four hierarchically connected Basic Stamp microcomputers, acting as a multicore, parallel processing computer architecture system, is responsible for the real-time computation of mathematical functions that enable precise motion. The mathematical functions are accurately represented by a number of truncated Taylor series which facilitates the tricycle turning about a specified instantaneous centre of rotation, ICofR, by a specified distance in the ground plane. One specified ICofR coupled with a matching specified distance represents one behaviour pattern. The computer system manages vehicle locomotion by compiling a string of behaviour patterns manifested by electrical signals sent via wires to the 4 servos. Thus, the push-rod tricycle, carrying its own battery, becomes a mobile programmable, 'drive-by-wire' electric vehicle.

3.2 Description of tricycle components

Figure 3.2 shows a side view of the tricycle.
Note:

(i) The two steering wheels that are steered independently by two steering servos.
(ii) The driven wheel and its fiducial marker used for driven wheel angle calibration
(iii) The under slung lithium polymer, LiPo, battery that lowers the centre of gravity of the vehicle as well as enabling tether-less mobile operation

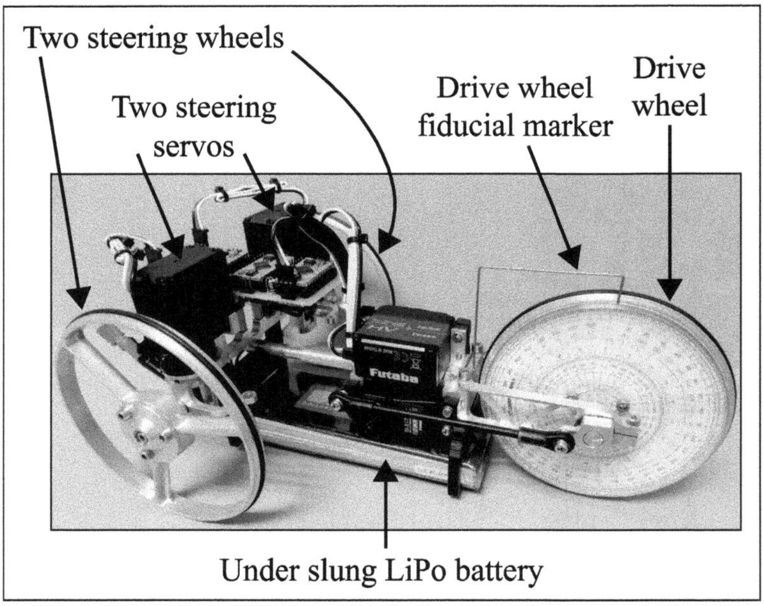

Figure 3.2. Left side view of tricycle.

Figure 3.3 shows a top side view of the tricycle.
Note:

(i) The two push-rods that articulate the driven wheel.
(ii) The on-board Basic Stamp microcomputer system (4 × BS2px) that manages locomotion via the steering wheels and the push-rod actuation of the driven wheel

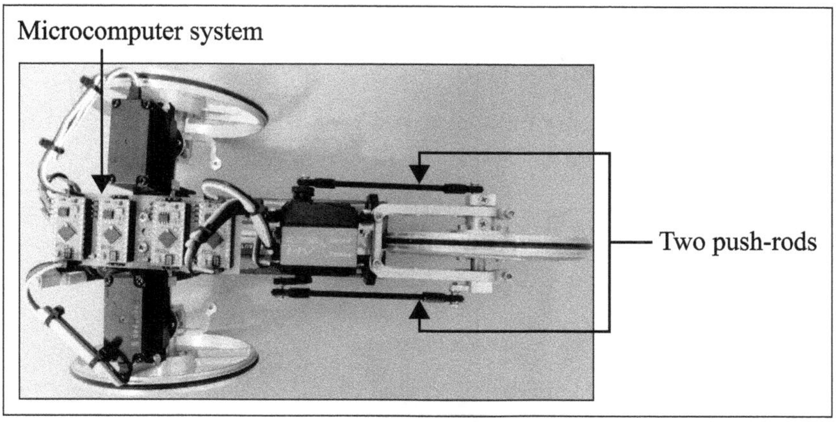

Figure 3.3 Top view of tricycle.

Figure 3.4 shows an underside view of the tricycle.
Note:

(i) The lithium polymer battery with 5.2 amp-hr capacity. Also note the in-line fuse (25amp) that is necessary for battery short circuit safety protection. The battery is under slung to lower the vehicle centre of gravity for increased locomotion stability.
(ii) Ackermann steering arms are disconnected because we use servo computed steering actuation. Ackermann steering, a mechanical computation of steering actuation was briefly described in chapter 1.

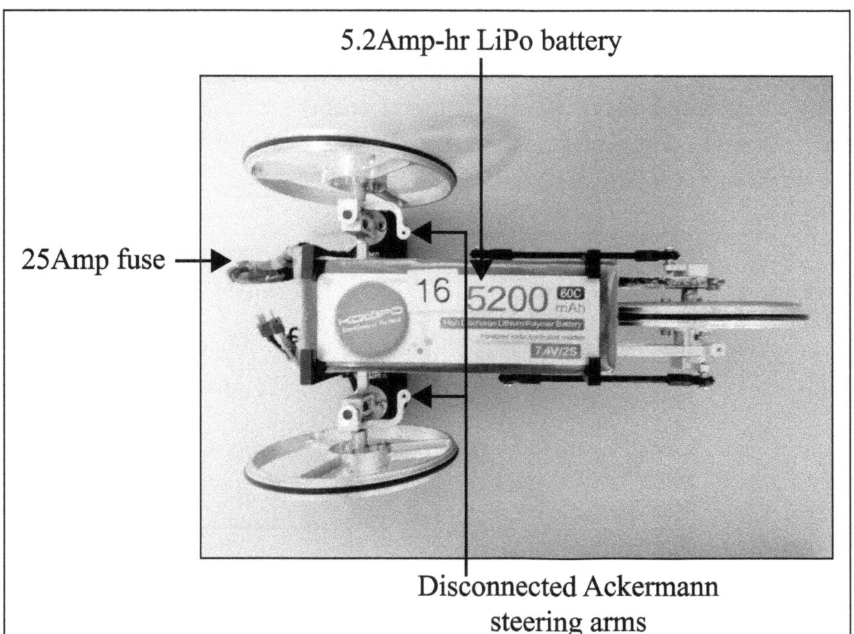

Figure 3.4 Under side view of tricycle.

Figure 3.5(a) shows the wheel drive servos
Note:

(i) The pair of oppositely facing Futaba servos, type BLS 172SV (BLS = brushless DC motor, 3.7N-m stall torque at 7.4V, no-load slew speed 0.11sec/60°)

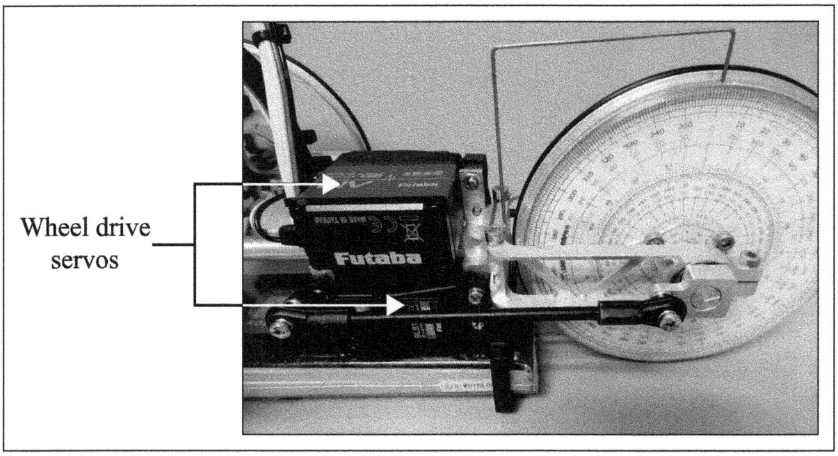

Figure 3.5(a) The pair of wheel drive servos.

Figure 3.5(b) shows the driven wheel that provides for vehicle locomotion.
Note:

(i) A protractor, mounted on the wheel, is used for 1° angle resolution calibration of the wheel.
(ii) The wheel has an O-ring rubber tyre mounted at its rim. The tyre diameter is 97mm giving a circumference of 305mm (nearest 1mm)
(iii) The driven wheel angle fiducial marker that is used for driven wheel calibration
(iv) The crank mounted on the axle of the driven wheel

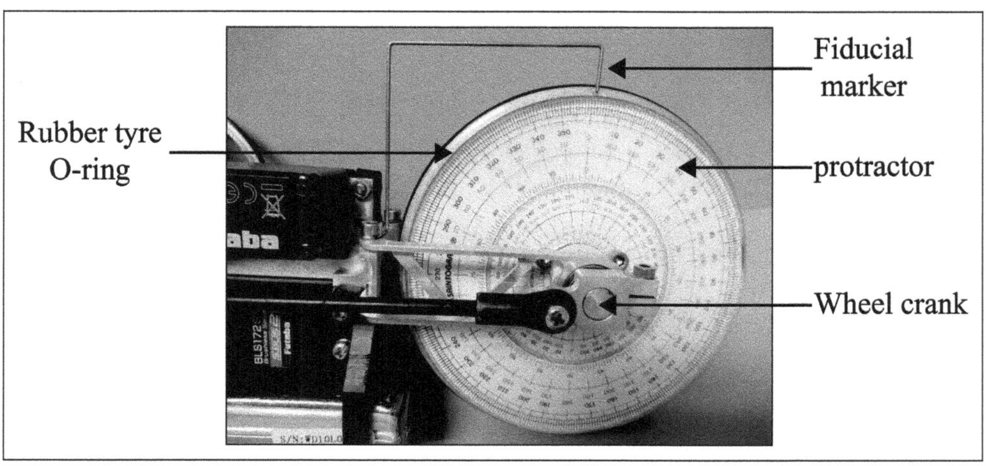

Figure 3.5(b) The driven wheel that provides vehicle locomotion and its fiducial marker.

Figure 3.6 shows a detail of the steering mechanism.
Note:

(i) Two independently driven BLS471SV direct drive servos that actuate the steering axles
(ii) Specification of the BLS471SV is 1.4N-m stall torque at 7.4V, no-load slew rate 0.07sec/60°

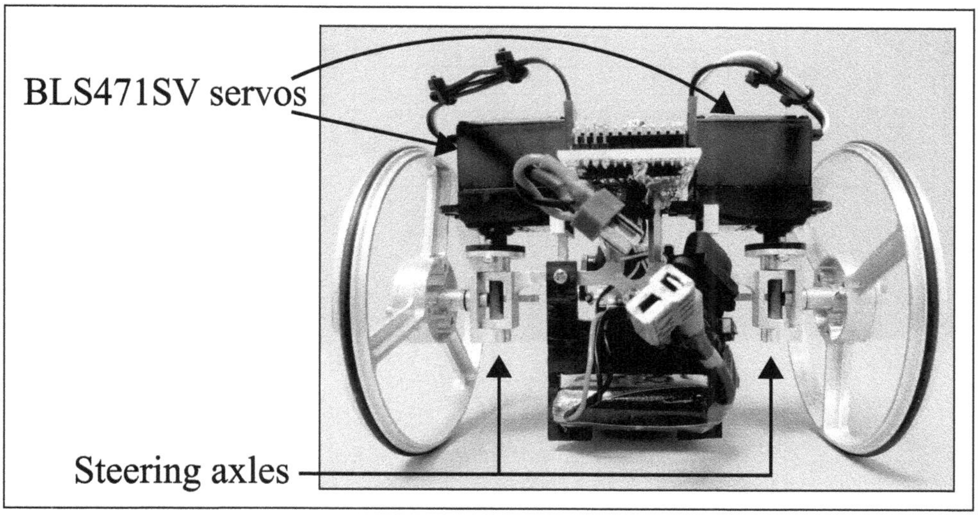

Figure 3.6 Direct drive steering mechanism.

3.3 The tricycle microcomputer system

The microcomputer system, figure 3.7, is comprised of four Basic Stamp2px, (BS2px), microcomputers. Figure 3.7 also shows the electrical wiring of system and the software table showing pin input/output, i/o, connections to the servos. The microcomputers are annotated as BS2px#1 to #4. Microcomputer, BS2px#1 is used as a clock timer which sends signals to the remaining microcomputers every 20ms, i.e. at a rate of 50Hz. These timer signals are primarily used as digital system update rate signals that initiate polled interrupt algorithms in each of the microcomputers.

The signals can be simple constant-width pulses, e.g. 5μs pulse width, or can be serial communication pulses that transmit control information such as steering and distance requirements. The microcomputer, BS2px#4 is used to control all four servos, i.e. two servos for steering control and two for the drive wheel control. Microcomputer, BS2px#3 is used as an odometer that computes the x, y and ϕ local positioning coordinates in the ground plane that represent, respectively, the 2-dimensional location and heading of the tricycle. Their values are computed from the scalar distance travelled of the tricycle, L together with the value of the ICofR and whether it is to the left or the right of the tricycle. These values are transmitted using serial communication pulses from BS2px#4 to BS2px#3. The remaining BS2px#2 is a spare microcomputer that is to be used innovatively by students to compute, for example, strategic level commands based on sensed information such as from line-following sensors. Two more points to

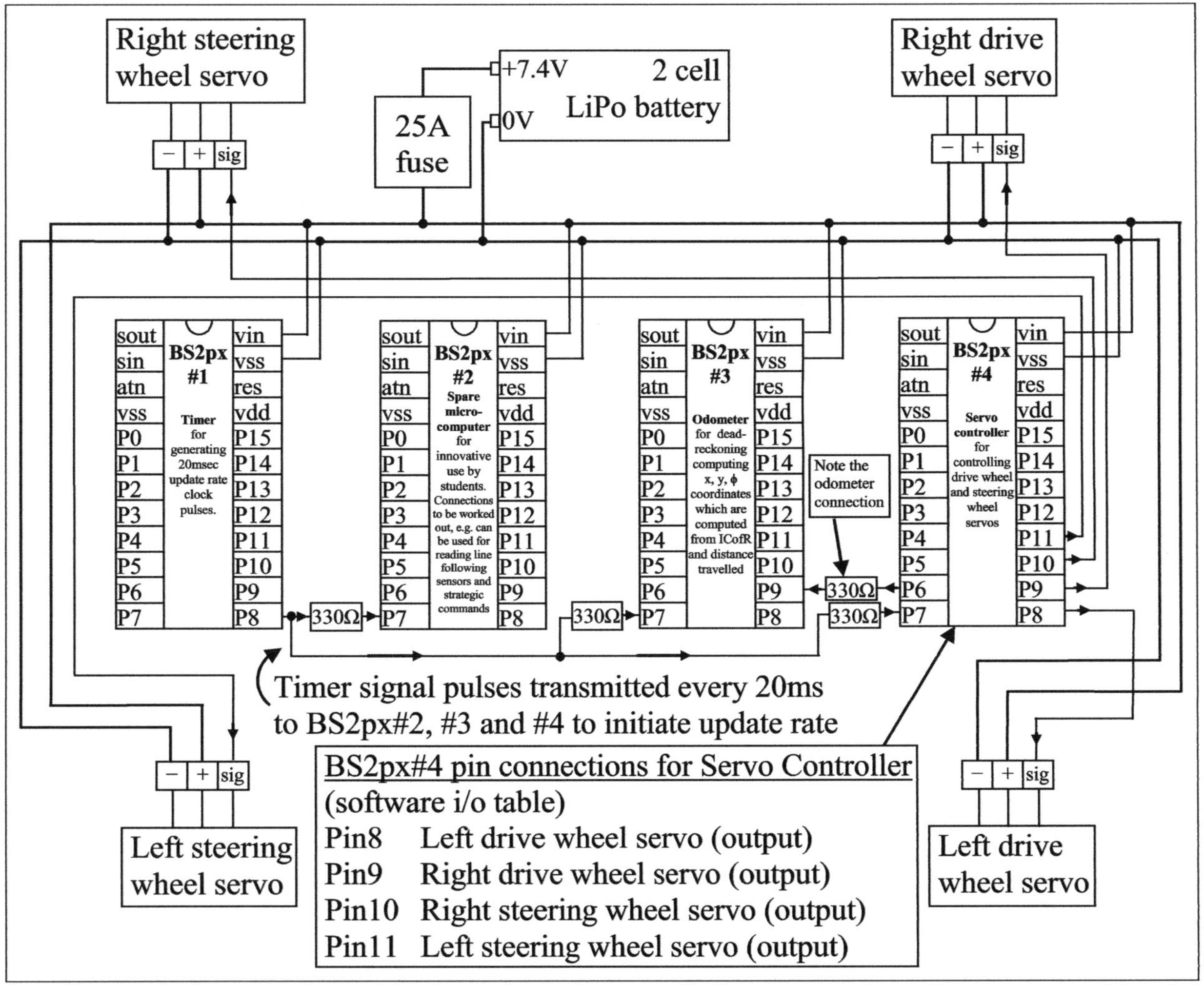

Figure 3.7 Electrical hardware circuit diagram of tricycle that shows the multi-core parallel processing microcomputer system. Note the software pin connections to the four servos and the odometer connection between BS2px#3 and #4.

note concerning figure 3.7 are (i) the 330Ω protection resistors that interconnect the microcomputers and (ii) the Lithium Polymer, LiPo, 2-cell, 7.4V battery plus its 25A fuse that provides electrical energy to the system.

Futaba servo angles are generated from the BS2px PULSOUT instruction that produces a pulse width in integer multiples of 0.81μs. As an example, the instruction, PULSOUT 8, 1235 will produce a pulse width of 1235 × 0.81μs = 1000μs = 1ms that will be output from pin 8 that will rotate the right drive wheel servo, (see software i/o table, figure 3.7), to an extreme end of its rotation angle, e.g. 45° CCW from its mid-range position. Similarly, the instruction, PULSOUT 8,2469 will produce a 2ms pulse width that will rotate the servo to 45° CW from its mid-range position. The integers, 1235 and 2469 are called 'pulsout values' and their dimensionless units are identified as pulsout units, 'pu units'.

3.4 The dual push-rod mechanism that actuates the tricycle drive wheel

The drive wheel is rotated in a counter clockwise, CCW, or clockwise, CW, direction in order to locomote the tricycle either forwards or backwards respectively. The mechanism converts reciprocating motion of a pair of servo cranks into continuous rotating motion of the drive wheel. Reciprocating motion is provided by two servos that actuate push-rods that are, in turn, connected to crank arms on the drive wheel, figure 3.8. The crank arms are rotationally displaced by 90° about the drive wheel axle. The push-rods are displaced forwards and backwards by the two synergistically acting servos, one being driven approximately sinusoidally and the other driven approximately cosinusoidally such that the drive wheel rotates CCW to give forward motion. If the driving phases of the servos are exchanged then the drive wheel will rotate CW giving backwards motion.

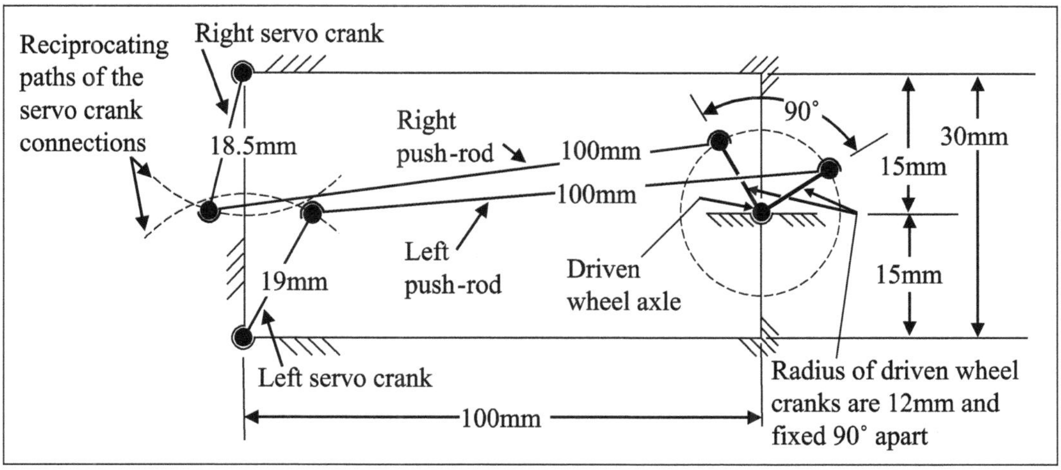

Figure 3.8 Layout and dimensions of the dual push-rod mechanism that actuates the driven wheel to rotate CW or CCW.

3.5 Kinematic analysis of the push-rod mechanism

Figure 3.9 illustrates the left push-rod mechanism that is shown with the left wheel crank at angle, θ_{LW}, and left wheel drive servo angle, θ_{LS}, where both angles are referred to vertical datum angle references.

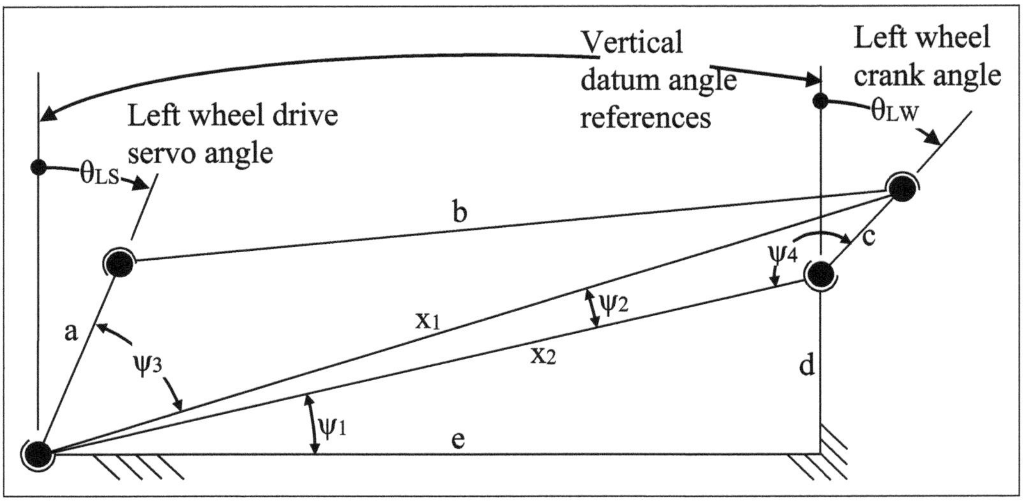

Figure 3.9 Inverse kinematics analysis of the push rod mechanism. The diagram shows the left side mechanism.

The task is to calculate the following inverse kinematics relationship:

Left wheel drive servo angle, $\theta_{LS} = f(\theta_{LW}, a, b, c, d, e)$ Eq. 3.1

where; θ_{LW} is the left wheel crank angle,
and a, b, c, d, e are fixed length constants equal to the following values,
a= 18.5mm, b = 100mm, c = 12mm, d = 15mm, e = 100mm

Formulating θ_{LS} as a function of θ_{LW}, (eq. 3.1), means that we are deriving the *inverse kinematics* relationship of the mechanism, i.e., what is the angle, θ_{LS} of the left wheel drive servo crank given that we wish to position the left wheel crank angle, θ_{LW} from 0° to 359° in steps of 1°? In contrast, formulating the *forward kinematics* relationship would be the other way around, i.e., what is the left wheel crank angle given the left wheel drive servo crank angle.

Thus, we can write;

$\theta_{LS} = f(\theta_{LW}, a, b, c, d, e)$ Inverse kinematics relationship
and $\theta_{LW} = f(\theta_{LS}, a, b, c, d, e)$ Forward kinematics relationship

When programming robots, the inverse kinematics relationship is the most important relationship since we wish to program the end-effector, in this case, the wheel angle, to be positioned to the correct location by positioning the two wheel drive servos to the correct angles. End-effectors are devices such as grippers at the end of robot arms but here it is the drive wheel.

Now let us derive, $\theta_{LS} = f(\theta_{LW}, a, b, c, d, e)$

Referring to figure 3.9, we carry out the following analysis to derive, θ_{LS},

$$\psi_1 = \tan^{-1}(d/e) \quad \text{Eq. 3.2}$$

$$\psi_4 = 90^\circ + \theta_{LW} + \psi_1 \quad \text{Eq. 3.3}$$

Substituting eq. 3.2 into eq. 3.3;

$$\psi_4 = 90^\circ + \theta_{LW} + \tan^{-1}(d/e) \quad \text{Eq. 3.4}$$

$$x_2 = \text{sqrt}(d^2 + e^2) \quad \text{Eq. 3.5}$$

Cosine rule,

$$x_1 = \text{sqrt}(x_2{}^2 + c^2 - 2cx_2\cos(\psi_4)) \quad \text{Eq. 3.6}$$

Substituting eq. 3.4 and eq. 3.5 into eq. 3.6;

$$x_1 = \text{sqrt}(d^2 + e^2 + c^2 - 2c\text{sqrt}(d^2 + e^2)\cos(90^\circ + \theta_{LW} + \tan^{-1}(d/e)) \quad \text{Eq. 3.7}$$

Sine rule,

$$(\sin\psi_2)/c = (\sin\psi_4)/x_1$$

thus,

$$\psi_2 = \sin^{-1}((c/x_1)\sin(\psi_4)) \quad \text{Eq. 3.8}$$

Substituting eq. 3.4 and eq. 3.6 into eq. 3.8,

$$\psi_2 = \sin^{-1}(\ (c/(\text{sqrt}(x_2{}^2 + c^2 - 2cx_2\cos(90^\circ + \theta_{LW} + \tan^{-1}(d/e)\,)))\ \sin(90^\circ + \theta_{LW} + \tan^{-1}(d/e))\ ) \quad \text{Eq. 3.9}$$

Cosine rule,

$$b^2 = a^2 + x_1{}^2 - 2ax_1\cos(\psi_3) \text{ .....and rearranging...}$$

$$\psi_3 = \cos^{-1}((a^2 + x_1{}^2 - b^2)/2ax_1) \quad \text{Eq. 3.10}$$

Substituting eq. 3.7 into eq. 3.10,

$$\psi_3 = \cos^{-1}((a^2 + d^2 + e^2 + c^2 - 2c\text{sqrt}(d^2 + e^2)\cos(90^\circ + \theta_{LW} + \tan^{-1}(d/e)) - b^2)/ (2a(\text{sqrt}(d^2 + e^2 + c^2 - 2c\text{sqrt}(d^2 + e^2)\cos(90^\circ + \theta_{LW} + \tan^{-1}(d/e))))\,) \quad \text{Eq. 3.11}$$

And finally, we have the result for, θ_{LS} given in four angle functions, which are;

$$\theta_{LS} = 90^\circ - \psi_1 - \psi_2 - \psi_3 \quad \text{Eq. 3.12}$$

Writing out eq. 3.12 fully as the sum of the four angle functions, we obtain the final inverse kinematics equation, eq. 2.13, for the left wheel drive servo angle, θ_{LS}

$$\theta_{LS} = +90^\circ \left.\begin{array}{ll} - \tan^{-1}(d/e) & \psi_1 \\ - \sin^{-1}\left\{\dfrac{c.\sin\left(90^\circ + \theta_W + \tan^{-1}(d/e)\right)}{\text{sqrt}\left[c^2 + d^2 + e^2 - 2c.\text{sqrt}\left(d^2 + e^2\right)\cos\left(90^\circ + \theta_W + \tan^{-1}(d/e)\right)\right]}\right\} & \psi_2 \\ - \cos^{-1}\left\{\dfrac{\left(a^2 - b^2 + c^2 + d^2 + e^2 - 2c.\text{sqrt}\left(d^2 + e^2\right)\cos\left(90^\circ + \theta_W + \tan^{-1}(d/e)\right)\right)}{2a.\text{sqrt}\left[c^2 + d^2 + e^2 - 2c.\text{sqrt}\left(d^2 + e^2\right)\cos\left(90^\circ + \theta_W + \tan^{-1}(d/e)\right)\right]}\right\} & \psi_3 \end{array}\right\}$$

Eq. 3.13

The first two functions, i.e. 90° and ψ_1 (=$\tan^{-1}(d/e)$), are straightforward constants. The remaining two functions, ψ_2 and ψ_3, however, are complicated equations which are a common occurrence for revolute mechanisms. Revolute mechanisms have rotary joints which are recognised by the black circles in figure 3.9. It can be said that, eq. 3.13 is not so complicated since it can be simplified due to the commonly occurring sub-functions such as in the denominators of ψ_2 and ψ_3, e.g. sqrt($d^2 + e^2$). The reader may see eq. 3.13 as easily computed in real-time with a microcomputer equipped with a maths package. However, for the Basic Stamp 2px (BS2px) microcomputer it is a realtime impossibility because (i) the BS2px does not have a trigonometrical instruction set, (ii) it cannot process decimal point numbers, and (iii) it is a 16-bit microcomputer so it cannot handle integer numbers greater than 65535. Soon we will show that eq. 3.13 can, in actual fact, be computed accurately by the BS2px. The lack of BS2px computing powers, like the whole series of BS2 microcomputers, should *not* be seen as a severe shortcoming, but rather, as a major *advantage*. Why is this so? Well it is because many engineering students are unaware of how mathematical functions are computed and this may limit their inventiveness in devising computational solutions in many areas of technology. For example, students would do well to ask themselves, how is arctan, sine, cosine and square root functions computed and to what precision are they computed by computers and calculators? Thus, Basic Stamp microcomputers serve as valuable teaching tools, not only for science and engineering students but also for maths and computer science students. Furthermore, multiple Basic Stamps can be linked together, using serial communication connections according to your own creative design, to form a specialised multicore, parallel processing computer system, i.e., a dedicated computer architecture with an operating system to your own design. This idea will be implemented progressively as we progress to the hexapod robot.

Two further unlikely significant advantages of the Basic Stamp microcomputer are:

(i) Its low memory capacity;
   (a) RAM volatile: only thirteen 16-bit words which can be split in combinations of 8-bit bytes, 4-bit nibs and 1-bit bits.
   (b) RAM scratchpad of 128 bytes.
   (c) EEPROM programme memory of 8 pages of 2k bytes

(ii) Its integer number computation. (No decimal point numbers!).

Once again, why are these two 'significant advantages' not seen as severe limitations? Well, the low memory capacity is an advantage because it is concerned with drilling discipline into students to be economical and ingenious with their programme variables and algorithms. As regards integer number computation, this advantage is concerned with the discipline of engineering precision, resolution and accuracy. Without exception, engineers are concerned with ingenuity and economics when they design and build machines and processes. In fact when doing so, engineers should first ask the question,

'What computation accuracy do I need and to what resolution should my computation be carried out to achieve that needed accuracy?'

(Remember, computing to a given resolution does not guarantee precision and accuracy to the same value as the resolution). If you are unsure of the meaning of resolution, precision and accuracy then just Google these terms.

If not already done so, students should extricate themselves from their obsession with decimal places because it is the number of significant figures, *not* the number of decimal places that dictates the precision of a number, e.g. 1.234E3 has the same precision as 1234. The first number has 3 decimal places and the second has none but they are both stated to a precision and accuracy of +/−0.5, the resolution being 1. The answer to the question above concerning what is required for computing the location of a mobile robot that is roaming in the ground plane in the author's Mechatronics Studio could be, 'I wish a location resolution of computation of 1mm, which will give me, I hope, a positioning precision of less than +/−10mm and hopefully an accuracy of +/−50mm'. We use the verb, 'hope' because, until you carry out experimental work and/or a detailed analysis, you don't know what precision and accuracy you'll actually get. However, you have to start with assumptions and fine tune later. Why is our positioning precision less than the resolution? Well, that is because errors build up during computation of mathematical functions and as the mobile robot travels in the ground plane due to for example, slippage between the wheel and the ground plane. The idea is that computing to 10 times finer than the required precision will give me the required accuracy following the computation. Don't forget that you must also be capable of measuring to 1mm resolution; otherwise it's not possible to obtain the precision that you want. Furthermore, your accuracy is likely to be less than your precision due to errors in your measurement. Once you have settled on a resolution, 0.1mm in this case, then every integer represents 0.1mm. For example, the Basic Stamp integer, 1234 represents 1234 × 0.1mm = 123.4mm or 0.1234m.

In conclusion, there is no need for decimal point numbers because the resolution of your measurement has been chosen to be sufficiently high, (high resolution means small measurement increment), such that the computation any truncated numbers with their lost decimal places does not erode the computational accuracy more than the required precision.

When students get hold of this idea, through its practise, it is a major step forward in their engineering aptitude.

3.6 So, what can Basic Stamp microcomputers compute?

Virtually without exception, all digital computers can only add. Innately, they cannot subtract, multiply, divide or carry out any mathematical functions. In order to do subtraction, computers add the 2's complement and require complex algorithms for all other formulation such as decimal point number manipulation and trig functions. So, what can Basic Stamps do? Well, Basic Stamps are equipped with a low level basic maths computation instruction set that can multiply and divide positive integers but cannot derive decimal point solutions. (In fact, most sensors and actuators that form mechatronic and machine systems work with integers. Sensors that send information to a microcomputer and actuators that receive numbers from a microcomputer, have no comprehension of decimal point numbers). The Basic Stamp, for example, will compute 5/2 to equal 2 and not 2.5 and 2/5 will equal 0 and not 0.4 meaning that the decimal place is not computed. (To produce decimal point computation for, say, 5/2 requires additional formulation via an algorithm that subtracts 2 from 5 twice (= 2.) giving a remainder of 1, which is less than 2 so the remainder is multiplied by 10 giving $1 \times 10 = 10$ from which, 2 is subtracted 5 times thus giving the result of 2.5). However, trouble lies ahead because −5/2 gives 32765 which is not the expected result of −2. The reason is that −5 is computed as the 2's complement for a 16-bit system as 65531 which is then divided by 2 which gives 32765 and not the expected −2. Also, 5/−2 gives 0 which is also not the expected result of −2. The reason is that 5 is divided by the 2's complement of 2 which is 65534 thus the computation is 5/65534 giving a result of 0. The conclusion is that an algorithm is required to compute division calculations correctly.

Multiplication is more convenient but negative number results are given in 2's complement format. For example, −5*2 is computed as 65526 which is 2's complement for −10 which is the correct result. Also −5*−2 is computed correctly as 10.

We will now show how eq.3.13 is computed by the Basic Stamp in order to rotate the drive wheel.

3.7 Wheel servo computation algorithm to rotate drive wheel

Figure 3.10 shows the inverse kinematic relationships based on eq. 3.13. The graphs show the left and right wheel drive servo angles in degree units that will position the drive wheel at a given angle, θ_W where this angle is referenced to the fiducial marker, figure 5(b). The relationships are very close to sine and cosine shapes, (shown with dashed lines in figure 3.10), despite the complicated looking equations of eq. 3.13.

It is the responsibility of BS2px#4 to compute the two mathematical relationships in figure 3.10 so a plan of action is required. The plan is first to convert the wheel drive servo angles to pulse widths, called 'pu units', using the PULSOUT instruction and secondly to formulate a method for computing the two mathematical relationships. In fact we will circumvent the conversion of wheel drive servo angles to PULSOUT pu units by, instead, directly measuring the pu units with the experimental system in figure 3.11 that uses the Servo Teach module described in chapter 2. This method also serves to calibrate the BS2px#4 microcomputer matched together with the two wheel drive servos. The calibration will change if you interchange BS2 microcomputers since the on-chip clock speed can vary significantly from BS2 chip to chip. See the important note in figure 3.11.

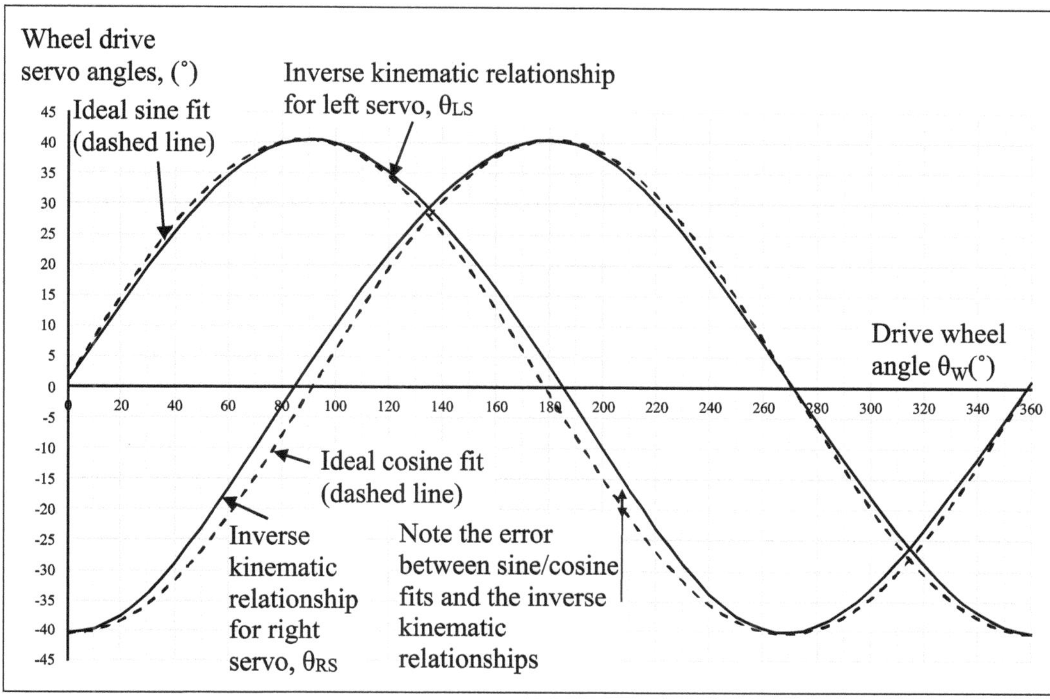

Figure 3.10 Inverse kinematic relationships, (eq.3.13), showing relationships between servo drive wheel angles and the wheel angle. Note the comparison with ideal sine and cosine relationships.

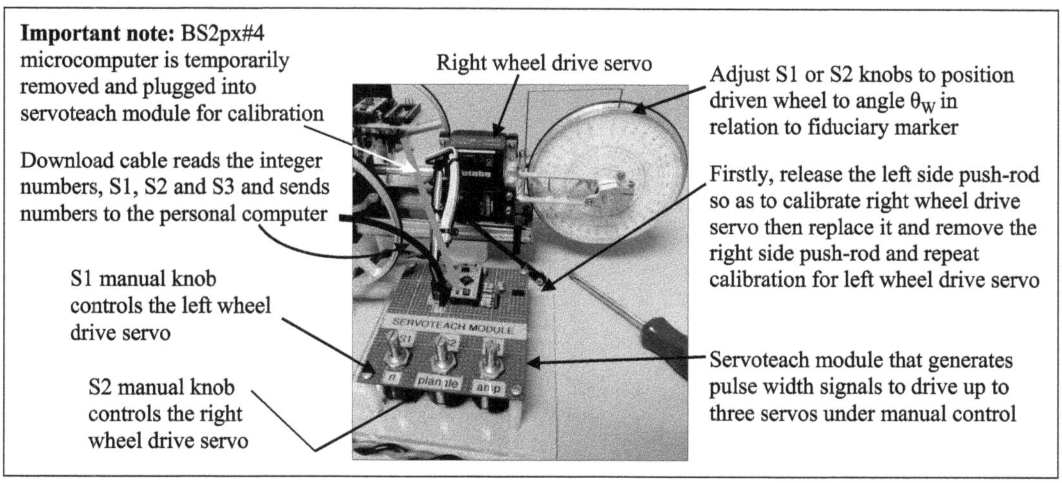

Figure 3.11 Experimental apparatus for calibrating the two items of wheel drive servo angle, θ_{LS} and θ_{RS} as a function of the wheel angle, θ_W.

The Servo Teach module uses a manually adjustable knob that varies the pulsout number to rotate either the left or right wheel drive servo to position the drive wheel to an angle θ_W The pulsout number is sent to the computer screen in order to record its value as a function of θ_W. Following the experiment, figure 3.11, the calibrated drive wheel angles in pulsout units are shown in figure 3.12. The dimensional units of the pulsout units are shown as (–), meaning that they are dimensionless units; they are digital numbers that represent servo angles. The approximate calibration for the BS2px microcomputer is 12.5 pulsout units per degree.

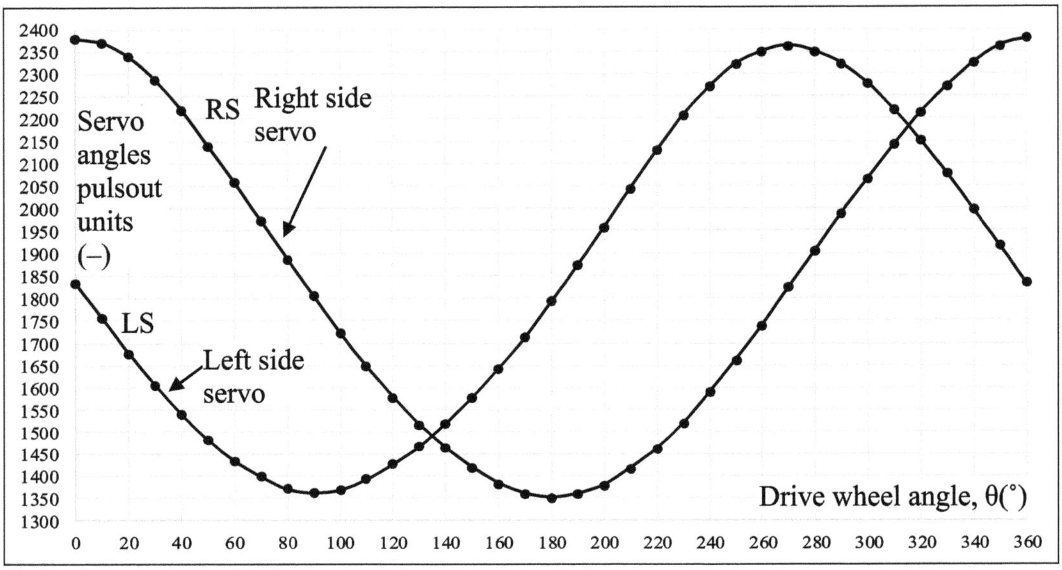

Figure 3.12 Calibrated values for drive wheel servos, LS and RS to position drive wheel angle, see experiment, figure 3.11.

3.8 How to compute servo angles, LS and RS using the Basic Stamp microcomputer

The technique is to compute the LS and RS values, figure 3.12 by firstly dividing the drive wheel angle into 6 sections identified as,

section 0 which is drive wheel angle from 0° to 60°, figure 3.13
section 1 which is drive wheel angle from 60° to 120°, figure 3.14
section 2 which is drive wheel angle from 120° to 180°, figure 3.15
section 3 which is drive wheel angle from 180° to 240°, figure 3.16
section 4 which is drive wheel angle from 240° to 300°, figure 3.17
section 5 which is drive wheel angle from 300° to 360°, figure 3.18
....and secondly to fit accurately 3rd order Taylor series equations to each of these sections.

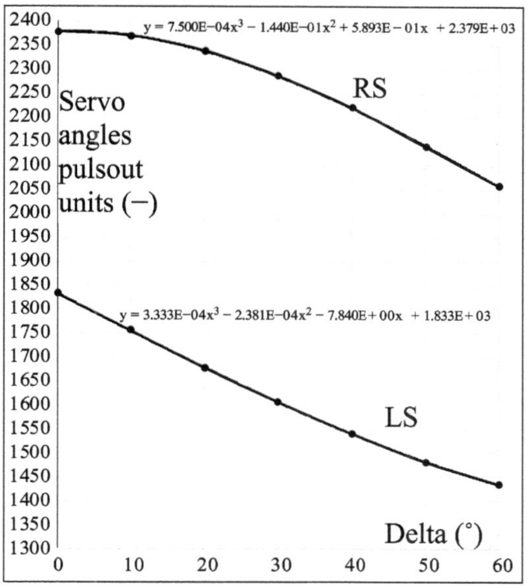

Basic Stamp equations

Section 0

0° to 60°

Left wheel drive servo, LS

```
LS =  1833
        -(784*delta/100)
        -(  0*delta/100*delta/10)
        +( 33*delta/100*delta/50*delta/20)
```

Right wheel drive servo, RS

```
RS = 2379
        +( 59*delta/100)
        -(144*delta/100*delta/ 10)
        +( 75*delta/100*delta/50*delta/20)
```

Figure 3.13 Equations for Section 0 for drive wheel angle 0° to 60°.

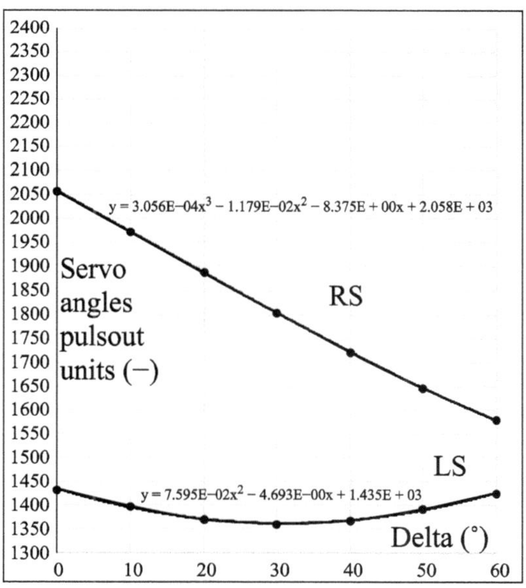

Basic Stamp equations

Section 1

60° to 120°

Left wheel drive servo, LS

```
LS =  1435
        -(469*delta/100)
        +( 76*delta/100*delta/10)
        +(  0*delta/100*delta/50*delta/20)
```

Right wheel drive servo, RS

```
RS =  2058
        -(838*delta/100)
        -( 12*delta/100*delta/10)
        +( 31*delta/100*delta/50*delta/20)
```

Figure 3.14 Excel spreadsheet equations Section 1, 60°to 120°.

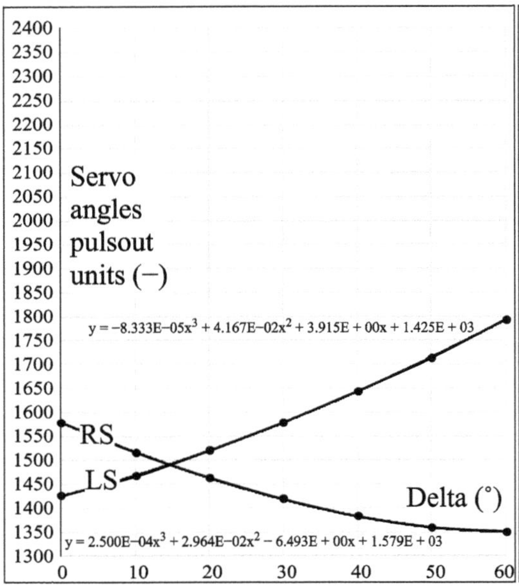

Basic Stamp equations
Section 2
120° to 180°

Left wheel drive servo, LS
LS = 1425
+(392*delta/100)
+(42*delta/100*delta/10)
–(8*delta/100*delta/50*delta/20)

Right wheel drive servo, RS
RS = 1579
–(649*delta/100)
+(30*delta/100*delta/10)
+(25*delta/100*delta/50*delta/20)

Figure 3.15 Excel spreadsheet equations Section 2, 120°to180°.

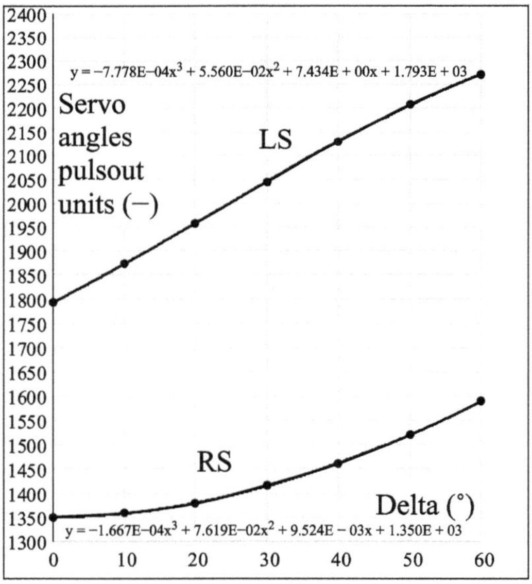

Basic Stamp equations
Section 3
180° to 240°

Left wheel drive servo, LS
LS = 1793
+(743*delta/100)
+(56*delta/100*delta/10)
–(78*delta/100*delta/50*delta/20)

Right wheel drive servo, RS
RS = 1350
+(1*delta/100)
+(76*delta/100*delta/10)
–(17*delta/100*delta/50*delta/20)

Figure 3.16 Excel spreadsheet equations Section 3, 180°to 240°.

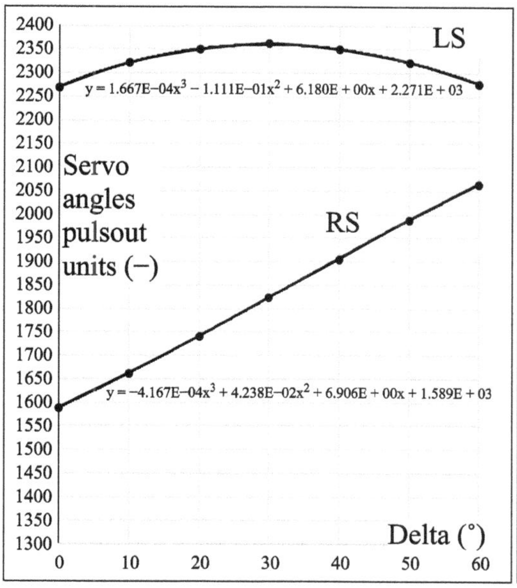

Basic Stamp equations
Section 4
240° to 300°

Left wheel drive servo, LS

```
LS =  2271
      +(618*delta/100)
      -(111*delta/100*delta/ 10)
      +( 17*delta/100*delta/50*delta/20)
```

Right wheel drive servo, RS

```
RS =  1589
      +(691*delta/100)
      +( 42*delta/100*delta/10)
      -( 42*delta/100*delta/50*delta/20)
```

Figure 3.17 Excel spreadsheet equations Section 4, 240° to 300°.

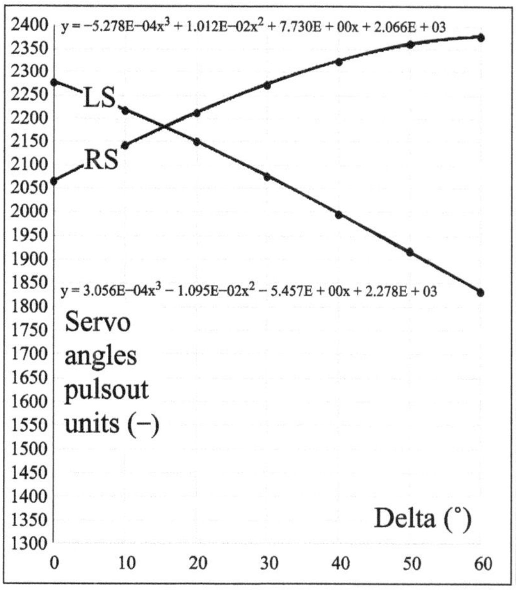

Basic Stamp equations
Section 5
300° to 360°

Left wheel drive servo, LS

```
LS =  2278
      -(546*delta/100)
      -( 51*delta/100*delta/10)
      +( 31*delta/100*delta/50*delta/20)
```

Right wheel drive servo, RS

```
RS =  2066
      +(773*delta/100)
      -( 10*delta/100*delta/10)
      -( 53*delta/100*delta/50*delta/20)
```

Figure 3.18 Excel spreadsheet equations Section 5, 300° to 360°.

The wheel angle resolution is 1° and we use the variable, 'delta' to represent the drive wheel angle displacement from the start of each section, for example, delta = 10° for a drive wheel angle of 310° in section 5. Another example is delta =0° for a drive wheel angle of 60° in section 1. Note that the maximum value of delta is 59°, (not 60°). As soon as delta reaches 60° then it shifts into the next section and becomes 0°.

The left side of figure 3.13 shows the LS and RS values for the drive wheel angle from 0° to 60°). The right side shows the Basic Stamp 16-bit integer computation whose methodology will now be explained.

3.9 Basic Stamp 16-bit integer accurate computational method

First look at the 3rd order Taylor series equation for the LS servo, (Note that y = LS and x= delta), y= 3.333E−04x³−2.381E−4x²−7.840E+00x+1.833E+03

Adjusting the equation to map LS as a function of delta and setting the coefficients to 3 significant integers except for the intercept which is set to the nearest integer; thus we make a table of the coefficients for each section according to table 3.1 below.

Table 3-1 Left side drive wheel servo angle coefficients showing first stage of computational simplification.

LS coefficients	delta⁰	delta¹	delta²	delta³
section 0	1833	-7.84	-0.000~~238~~	0.000333
section 1	1435	-4.69	0.076~~0~~	0
section 2	1425	3.92	0.041~~7~~ [2]	-0.000083~~3~~
section 3	1793	7.43	0.055~~6~~ [6]	-0.000778
section 4	2271	6.18	-0.111	0.000167
section 5	2278	-5.46	-0.051~~0~~	0.000306
	Nearest integer	3 sig figs	3 sig figs	3 sig figs

Note rounding up and rounding down based on largest magnitude of coefficient of delta² of -0.111 and delta³ of -0.000778

Three sig. figs. are used because this number gives the required accuracy for most engineering applications but the number should always be checked to see if it's adequate or not. Note the rounding up and rounding down that is done to simplify the computation. We are assuming that only 3 sig. figs are required to give accurate computation of the LS values. Let us see if this is simplification is fulfilled. Table 3.2 shows the rounded up and rounded down values together with the errors associated with the simplification.

Table 3-2 Left side, LS, drive wheel servo angle coefficients 'cleaned up' into 3 sig. fig format and showing computational errors

LS coefficients	delta⁰		delta¹	delta²	delta³	computed delta=60
section 0	1833	← no error	-7.84	0.000	0.000333	1435
section 1	1435	← no error	-4.69	0.076	0.000000	1427
section 2	1425	← +2 error	3.92	0.042	-0.000083	1793
section 3	1793	← no error	7.43	0.056	-0.000778	2272
section 4	2271	← +1 error	6.18	-0.111	0.000167	2278
section 5	2278	← no error	-5.46	-0.051	0.000306	1833

Table 3.2 computes the errors from the number simplifications as follows as an example;

For section 0 and with delta set to a value of 60;

LS= $1833 \times (60)^0 - 7.84 \times (60)^1 - 0.000 \times (60)^2 + 0.000333 \times (60)^3 = 1435$ (nearest integer)

The result of 1435 is exactly equal to the starting value for delta = 0 for the next section which is section 1. Thus there is no error. Doing the same calculation for the remaining sections gives a maximum error of 2, (section 1). If we assume that the maximum error that can be tolerated is +/−3 which represents approximately ¼° of servo angle then the simplified computation is "accurate" and is not "approximate". If ¼° error is not acceptable then the coefficient resolution should be increased to 4 sig. figs. If you do this then it comes with a cost which is the computational complexity increase (code length increase) together with the increase of real time it will take to execute the computation.

There is a method to determine whether 3 sig. figs. will give sufficient accuracy. The method is as follows;

1. If the maximum error of computation is set at +/− 3 then set the maximum error of computation for each of the 3 terms in the series also equal to +/−3 on the basis that the errors do not summate on the assumption that statistically the coefficients of $(\text{delta})^1$, $(\text{delta})^2$ and $(\text{delta})^3$ generally oscillate between positive and negative values, e.g. for section 0,the coefficients are, −7.84,+0.000 and +0.000333. Thus errors for each term are unlikely to add; they are more likely to cancel out but not necessarily completely, especially since there is an odd number of terms. You may be unlucky and all 3 terms have the same sign. However, each section series does have at least one sign reversal.
2. Now find the lowest resolution required for each term in the series like this;
   (i) For $(\text{delta})^1$; lowest resolution= error allowable/$(60)^1$=3/$(60)^1$=0.01 (specified to 1 sig. fig. rounded down)
   (ii) For $(\text{delta})^2$ lowest resolution= error allowable/$(60)^2$=3/$(60)^2$=0.001 (specified to 1 sig. fig. rounded up)
   (iii) For $(\text{delta})^3$; lowest resolution= error allowable/$(60)^3$=3/$(60)^3$=0.00001 (specified to 1 sig. fig.)
3. Now look at each column of coefficients in table 3.2 and check to see if there are only 3 numbers to the left of the specified number of decimal places. If there are more than 3 then, e.g. 4 numbers, then 4 sig. figs. are required. If there are less than 3, e.g. 2 numbers, then only 2 sig. figs. are required.

Hence the conclusion from point 2. above is that the coefficient of;

(i) $(\text{delta})^1$ should be specified to at least 2 decimal place resolution
(ii) $(\text{delta})^2$ should be specified to at least 3 decimal place resolution
(iii) $(\text{delta})^3$ should be specified to at least 5 decimal place resolution

Table 3.2 shows, in actual fact, that the resolution for each coefficient, as indicated in the conclusions from point 2. above, is adhered to. In fact, the resolution for the $(\text{delta})^3$ coefficient is specified to a greater than necessary resolution of 6 decimal places rather than 5 decimal places, e.g. for section 5 the coefficient is 0.000306 where 0.00031 would have been adequate. In this case it would have only been necessary to have specified the coefficients to 2 significant figures. Otherwise, the conclusion from point 3. above is that the coefficients for $(\text{delta})^1$ and $(\text{delta})^2$ been specified to a precision of 3 sig. figs., 3 sig. figs. respectively, so supporting the assumption that 3 sig. figs. precision is usually adequate for most engineering calculations.

3.10 Making the equations in table 3.2 suitable for the Basic Stamp microcomputer

A key feature of taking the trouble of preparing coefficients in table 3.2 is so that the computation numbers can be arranged in a highly compact and efficient computation format. Remember that the Basic Stamp is a 16-bit integer microcomputer so any number can neither have a decimal point nor exceed $2^{16}-1 = 65535$. Let us assume we have a set of integer coefficients, each with 3 sig. figs. and each set at their maximum value as follows; we use 1833 for the first coefficient just as an example,

$LS=1833+9.99*\text{delta}+0.999*\text{delta}^2+0.00999*\text{delta}^3$

We compute this equation with the Basic Stamp by adhering to a set of rules as follows;

(i) do all multiplication first and
(ii) all division last but with the caveat that
(iii) any multiplication does not exceed 65535 and that
(iv) any division uses the smallest value that does not impede the multiplication caveat and that is integer divisible into 100000.

Ok, let's show this technique. Remember that the maximum value of delta is 59 but we will simplify by saying the maximum value is 60. Also remember that the Stamp does all calculations from the left to the right and completes any calculation before moving right. However, any calculations in brackets are done separately.

We compute as follows;

LS=1833 + (999*delta/100) + (999*delta/100*delta/10) + (999*delta/100*delta/50*delta/200)

Now set delta=60 and check that all rules as stated above are adhered to;

LS=1833 + (999*60/100) + (999*60/100*60/10) + (999*60/100*60/50*60/20) Eq. 3.14

No problem here since 1833 does not exceed 65535	999*60<65535 so no problem	999*60*60=3596400 which is approximately 55 times more than 65535 so 999*60 must be divided by 100 which is smallest value greater than 55 and that is integer divisible into 100000 before multiplying by the next 60	999*60/100*60*60=2157840 which is approximately 33 times more than 65535 so 999*60/100*60 must be divided by 50 which is smallest value greater than 33 and that is integer divisible into 100000 before multiplying by the next 60

Now, all coefficients will not be greater than 999 so you can safely replace the 999 with the correct 3 sig. fig. coefficient without any impediment to the computation being corrupted by overflow of numbers greater than 65535. Furthermore a very convenient computational table, (or matrix), can be set up. For example, for the computation of LS for each section is like this;

LS values for each section in tabular format compatible with the Basic Stamp

```
section0: 1833 – (784*delta/100) – ( 0*delta/100*delta/10) + ( 33*delta/100* delta/50*delta/20)
section1: 1435 – (469*delta/100) + ( 76*delta/100*delta/10) + ( 0*delta/100* delta/50*delta/20)
section2: 1425 + (392*delta/100) + ( 42*delta/100*delta/10) – ( 8*delta/100* delta/50*delta/20)
section3: 1793 + (743*delta/100) + ( 56*delta/100*delta/10) – ( 78*delta/100* delta/50*delta/20)
section4: 2271 + (618*delta/100) – (111*delta/100*delta/ 10) +( 17*delta/100* delta/50*delta/20)
section5: 2278 – (546*delta/100) – ( 51*delta/100*delta/10) +( 31*delta/100* delta/50*delta/20)
```

Only 2 sig. figs. used of the 3 sig. figs. available because coefficients for $(\text{delta})^3$ need only to be specified to 5 decimal places

...and a similar method produces the following tabular format for the RS wheel drive servo angles based on Excel spreadsheet trendline equations in figures 3.13 to 3.18 inclusive.

RS values for each section in tabular format compatible with the Basic Stamp

section0: 1833 – (784*delta/100) – (0*delta/100*delta/10) + (33*delta/100* delta/50*delta/20)
section1: 1435 – (469*delta/100) + (76*delta/100*delta/10) + (0*delta/100* delta/50*delta/20)
section2: 1425 + (392*delta/100) + (42*delta/100*delta/10) – (8*delta/100* delta/50*delta/20)
section3: 1793 + (743*delta/100) + (56*delta/100*delta/10) – (78*delta/100* delta/50*delta/20)
section4: 2271 + (618*delta/100) – (111*delta/100*delta/ 10)+ (17*delta/100* delta/50*delta/20)
section5: 2278 – (546*delta/100) – (51*delta/100*delta/10) +(31*delta/100* delta/50*delta/20)

The subroutine that computes the LS and RS wheel drive servo angles is named '**servodrivewheelangles:**' and can be seen in the programming code at the end of this chapter. The beginning of the subroutine determines how many degrees, variable 'drivewheelangle', transposes into a new complete rotation from the 0° degree rotation starting position of the wheel displacement. Following that computation, the section is determined, using the 'branch' instruction, from the value of 'drivewheelangle' which will have a value between 0° and 359° inclusive. The programming code has been written with sufficient comments such that the algorithm can be understood just from these comments. It is also necessary to point out that the subroutine, '**sectiontravel:**' is a lengthy subroutine that does the following:

(i) Converts demanded tricycle distance, mm units, into drive wheel angle displacement in ° units. Conversion based on; the drive wheel circumference is 305mm (figure 3.6) so the calibration converting mm to wheel angle ° is 360°/305mm = 1.18°/mm (3 sig. figs.) so the wheel angle (°) is given by the following basic Stamp computation: drivewheelangle = drivewheeldistance + (18 × (drivewheeldistance/100))
(ii) Sets a limit of tricycle distance of 32,767mm forward and reverse displacement.
(iii) Computes acceleration and deceleration of the drive wheel if the demanded tricycle distance is greater than 180mm, otherwise the speed is set constant.

We now conclude the description of the drive wheel rotation and move on to the engineering of the steering system.

3.11 The Tricycle steering system

Figure 3.19 is a plan view of the tricycle outlining the steering theory. The diagram shows key geometrical dimensions together with the axles of the three wheels all projected to pass through a common point known as the Instantaneous Centre of Rotation, ICofR, which has already been well described in chapter 1.

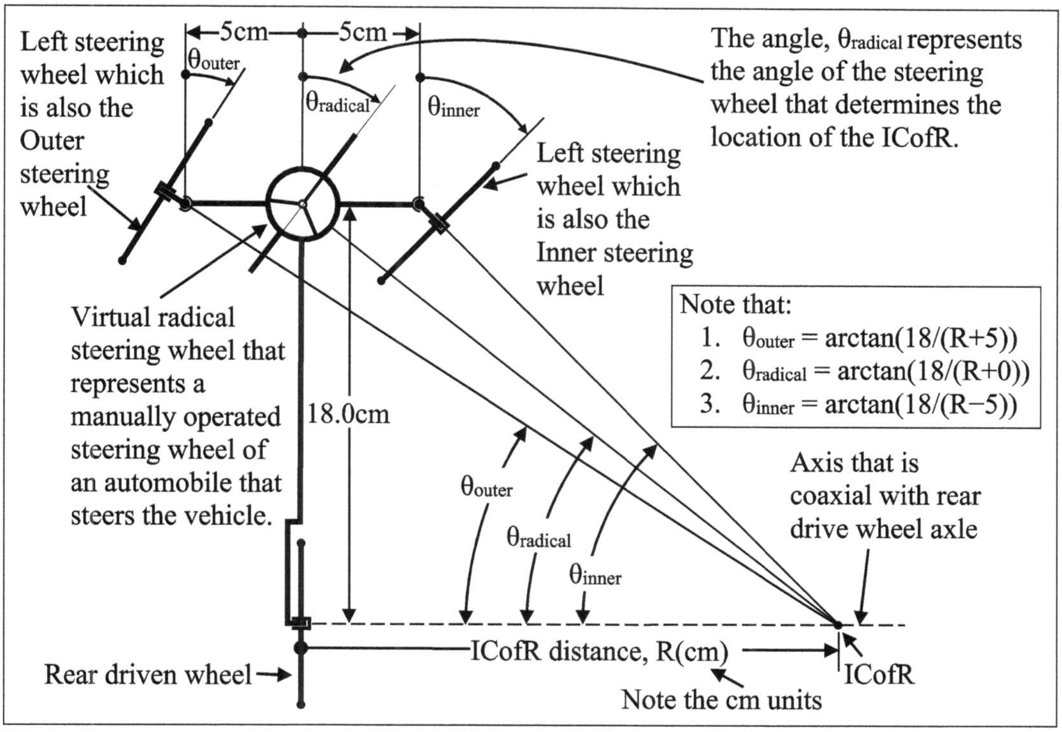

Figure 3.19 Plan view of the tricycle showing the steering theory. The radical steering wheel does not exist. It is virtual and represents the steering wheel that is manually operated by the driver of an automobile, (see chapter 1, figure 1.7 and figure 1.8).

Figure 3.19 also introduces the concept of a 'virtual radical steering wheel' which is to be thought of as the manually operated steering wheel that we all use when driving an automobile. A manual steering wheel, approximately speaking, rotates the virtual radical steering wheel directly such that it fixes the ICofR by the intersection of the virtual radical steering wheel axle projected onto the axis that is coaxial with the rear drive wheel axle. The virtual radical steering wheel serves as a master that fixes the location of the ICofR and that the left and right steering wheels must follow. In other words, once the ICofR has been fixed then the left and right steering wheel axles must align with the ICofR. Inspecting figure 3.19, it can be seen that, when the tricycle is turning right, the outer steering wheel, θ_{outer}, i.e. the left steering wheel angle, is less than the virtual radical steering wheel angle, $\theta_{radical}$. Conversely, the inner steering wheel, θ_{inner}, i.e. the right steering wheel angle, is greater than the virtual radical steering wheel angle, $\theta_{radical}$. The steering angles, θ_{outer}, and θ_{inner} and $\theta_{radical}$, all become equal when the vehicle is travelling straight ahead, which is a special case of the ICofR being set at infinity to the left or right of the rear drive wheel.

The tricycle can not only turn right, figure 3.19, but can turn left also, figure 3.20. The left steering wheel is called the outer steering wheel and the right steering wheel is called the inner steering wheel when the tricycle is turning right in which case the ICofR is on the right side of the vehicle. Figure 3.20 shows that roles of the steering wheels and location of the ICofR are all exchanged when turning left.

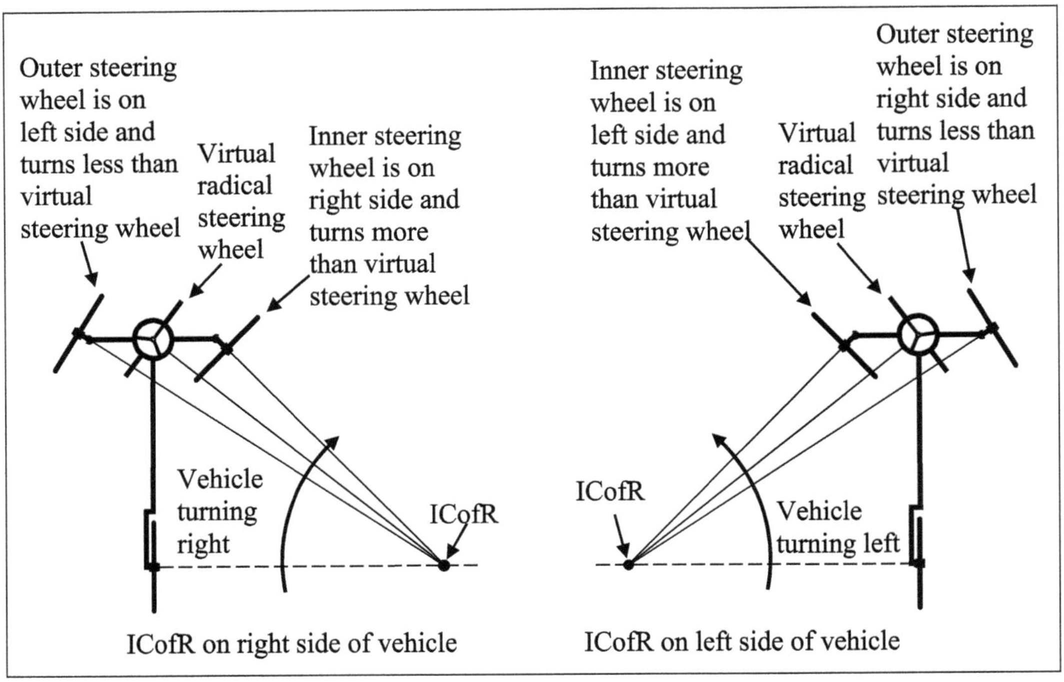

Figure 3.20 Note that when turning right, the ICofR is on the right side of the vehicle, and the left steering wheel is the outer steering wheel that turns less than the inner steering wheel on the right side. This situation is exchanged when the ICofR is on the right side of the vehicle.

So to summarise, what we have done is to set the virtual radical steering wheel to be the master locator of the ICofR and thus the next step is to compute the angle values of the left and right steering wheels as a function of the location of the ICofR. This is done using the following steps.

Step 1

Plot the radical virtual steering wheel angle, θ_{radical} (°) versus the value of the ICofR, (R cm), figure 3.21, (left side graph). Using figure 3.19, the relationship is given by;

$$\theta_{\text{radical}} = \arctan\left\{\frac{18(\text{cm})}{R(\text{cm})}\right\} \quad \text{Eq. 3.15}$$

The ICof R units are cm units since this will be the chosen resolution because mm units were considered too resolute and unobtainable due to the backlash (i.e. the 'looseness') in the gears of the steering servos. Figure 3.21 right side graph plots the relationship with servo pulsout units using the calibration factor of 12.2 pulsout units/degree. Note that the relationship, being an arctan function, is extremely non-linear and will be difficult to model using the Basic Stamp that is without trigonometrical functions. However, there is a solution, which is step 2.

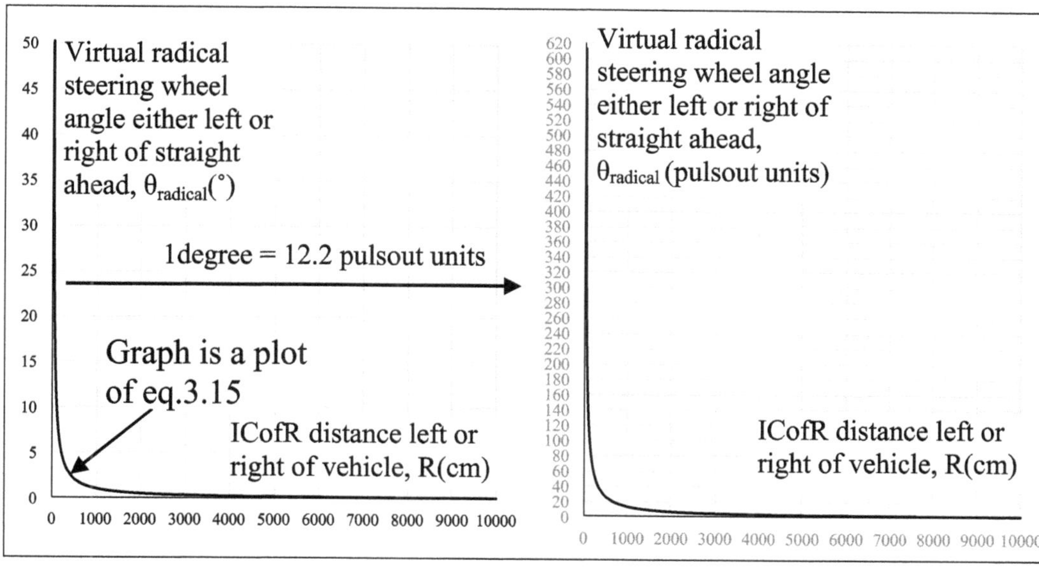

Figure 3.21 Graphs showing the highly non-linear relationship between radical steering wheel angle and the distance of the ICofR. The right-hand side graph converts the steering wheel angle to servo pulsout units.

Step 2

Sometimes it is more convenient to describe the amount of turning that a vehicle is undertaking as 'curvature' rather than referring to its ICofR value. 'Curvature' is defined as the inverse of ICofR, i.e. 1/R(cm). So, for example, if the ICofR is very large then the vehicle turning curvature is very small and conversely, if the ICofR is very small then the turning curvature is very big. If we now plot the virtual radical steering wheel servo angle against turning curvature, figure 3.22, then the relationship suddenly becomes more manageable by being far more linear.

It will be noticed in figure 3.22 that turning curvature is given as 10000/R(cm) and not 1/R(cm), because (i) we must work with integers and (ii) because the maximum value of ICofR will be 10000cm which is a value of R that accurately indicates travelling straight ahead. Why is R=10000cm chosen? Well the first reason is that the value of R for a vehicle travelling straight ahead is infinity, i.e. an extremely large number that is impossible to work with so there has to be a maximum, but nonetheless, large number that represents the vehicle travelling straight ahead. The value of 10000cm is chosen because that value represents a steering angle of $\theta_{radical}$ =arctan(18/10000) = 0.1° which is related to the highest precision that the steering servos can be controlled, (one pulsout integer unit is approximately equal to 0.1°, (12.2 pulsout units = 1°), notwithstanding the greater inaccuracy of servo backlash. If curvature is the variable chosen with which to work then an R value of infinity becomes equal to a workable value of zero curvature. If the maximum value of R is 10000cm then the question arises as to what is the minimum value of R, i.e. the maximum value of curvature. The answer is given by the maximum turning angle of the steering servos that is 50° for the inner steering wheel and in turn relates to a minimum R value of $\tan 50° = 18/(R_{min}-5)$, figure 3.19, thus R_{min}=20cm that relates to an R value for the inner steering wheel of 15cm.

So to summarise the range of R is from 15cm up to 10000cm and so turning curvature, figure 3.22, ranges from 10000/10000=1cm^{-1} to 10000/15=667cm^{-1} which is the maximum value of the curvature axis in figure 3.22.

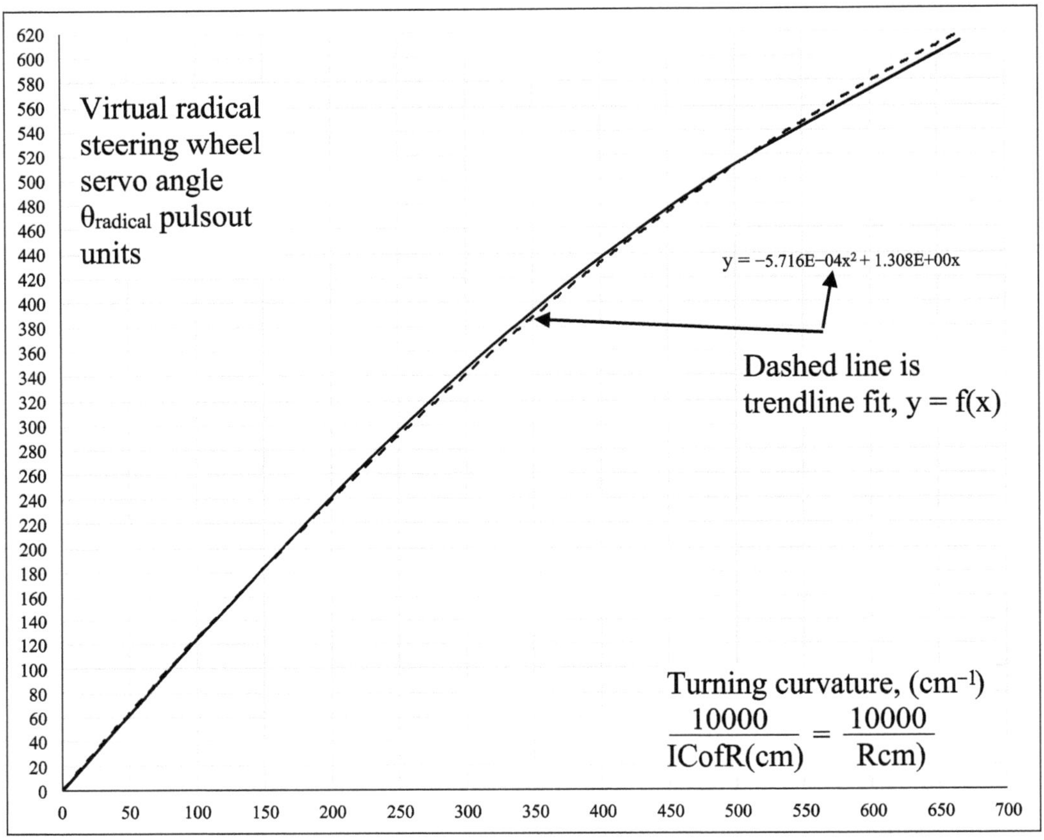

Figure 3.22 Graph showing manageable relationship between virtual radical steering servo angle and the vehicle turning curvature (=1/R). Relationship can be represented by a 2nd order polynomial series that gives a steering error of less than ½ degree that occurs at large turning curvatures. Otherwise, at low turning curvatures the steering error is less than 1/10 degree.

Finally, we use a 2nd order polynomial series to model accurately the relationship in figure 3.22, thus;

$$\theta_{\text{radical}}(\text{pulsout units}) = 1.31\left\{\frac{10000}{\text{ICofR}(\text{cm})}\right\} - 0.000572\left\{\frac{10000}{\text{ICofR}(\text{cm})}\right\}^2 \qquad \text{Eq. 3.16}$$

This equation, eq. 3.16, is used, given the value of the ICofR, to compute the value of the $\theta_{radical}$ angle in pulsout units. The equation can be seen in the Basic Stamp subroutine listed at the end of this chapter; it is entitled, **setsteeringangle:** and $\theta_{radical}$ is renamed steeringwheelpuanglenew and computed as follows;

steeringwheelpuanglenew=((10000/ICofR/2*131/50) – (10000/ICofR/ICofR*572/100))

Now we move to step 3 which shows how to compute the values of the left and right steering wheel servos.

Step 3

From a given ICofR, R value, we use eq. 3.16 to compute the virtual radical steering wheel angle, $\theta_{radical}$ and the next step is to compute θ_{outer} and θ_{inner} from this value, $\theta_{radical}$. Let us now calculate the relationships, θ_{outer} and θ_{inner} as functions of $\theta_{radical}$ as follows.

Given the following equations; (figure 3.19)

$\theta_{outer} = \arctan(18/(R+5))$ Eq. 3.17a

$\theta_{radical} = \arctan(18/(R+0))$ Eq. 3.17b

$\theta_{inner} = \arctan(18/(R-5))$ Eq. 3.17c

From eq. 3.17b

$$\frac{18}{R} = \tan\theta_{radical} \text{ thus, } R = \frac{18}{\tan\theta_{radical}} \quad \text{Eq. 3.18}$$

Substitute eq.3.18 into eq. 3.17a we obtain;

$$\theta_{outer} = \arctan\left\{\frac{18}{\frac{18}{\tan\theta_{radical}}+5}\right\}$$ Now multiply through arctan function by, $\frac{\tan\theta_{radical}}{18}$

$$\theta_{outer} = \arctan\left\{\frac{\tan\theta_{radical}}{1+\frac{5}{18}\tan\theta_{radical}}\right\} \quad \text{eq. 3.19}$$

...and by a similar analysis using eq. 3.17c above, we obtain the inner servo steering angle

$$\theta_{inner} = \arctan\left\{\frac{\tan\theta_{radical}}{1-\frac{5}{18}\tan\theta_{radical}}\right\} \quad \text{Eq. 3.20}$$

The relationships, eq. 3.19 and eq. 3.20 are shown in figure 3.23 with angles converted to steering servo pulsout units using the conversion calibration factor of 12.2pulsout units per degree. Trendline equations for these relationships are given as follows; (3 sig. fig. coefficient accuracy and $\theta_{radical}$, θ_{outer} and θ_{inner} all in pulsout units)

$$\theta_{outer} = 0.983*\theta_{radical} - 0.000253*(\theta_{radical})^2$$

$$\theta_{inner} = 1.01*\theta_{radical} + 0.000348*(\theta_{radical})^2$$

These above equations are computed, using the Basic Stamp, as follows; (note, (i) that k= $\theta_{radical}$ and (ii) k varies between 0 and 667 so the first multiplying factor for the coefficient must not exceed 65535/667=98)

θ_{outer} = (98*k/100) – (98*k/1000*k/387)

(–2 units full-scale error with k=667, which is inside the tolerance of +/–6 (½ °))

θ_{inner} = (98*k/97) + (98*k/1000*k/282) (zero full-scale error)

These equations are utilised in the subroutine '**poutsteering**' in the code listed at the end of this chapter.

We now conclude the three steps that compute the left and right steering wheel servo angles. Some readers may be thinking why not compute the steering wheel servo angles directly from the ICofR; in other words, why go to the trouble of computing the 'middleman' $\theta_{radical}$. The answer is that when you want to steer the vehicle from one ICofR to another then you need an 'impartial' steering wheel that is programmed to rotate linearly from one angle to another thus giving a smooth motion of the steering wheels.

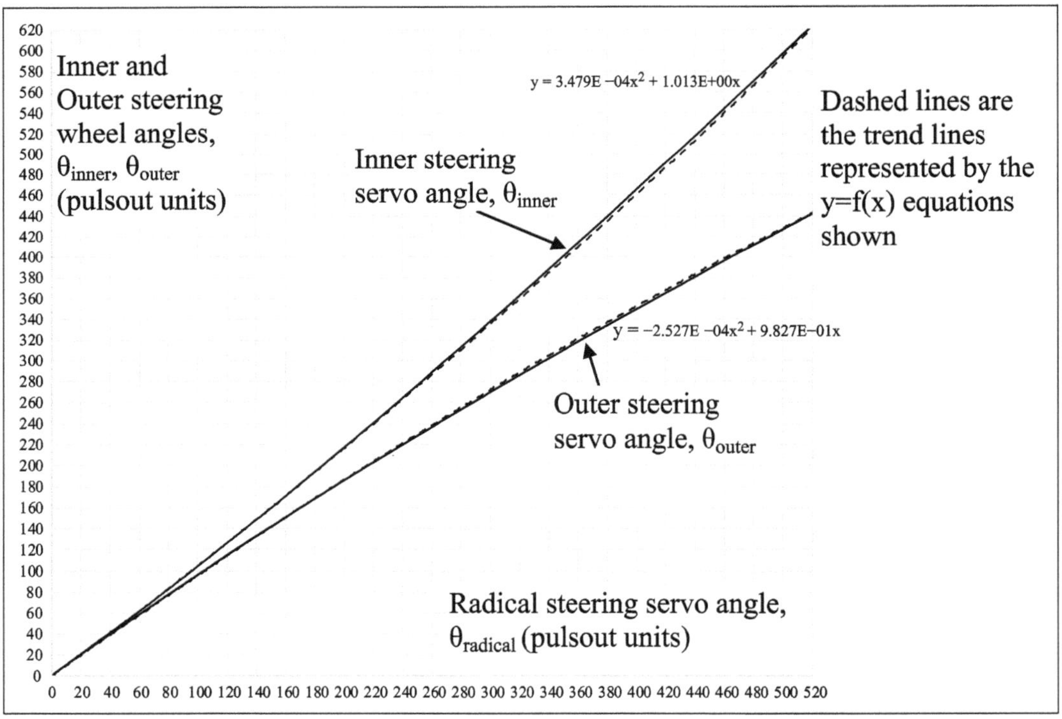

Figure 3.23 Relationship between radical steering wheel servo angle and the inner and outer steering wheel servo angles.

3.12 A planar odometer that computes vehicle location coordinates, x, y, ϕ

Those of us who drive a vehicle are aware of the odometer on the dashboard that reports on the distance travelled by the vehicle, usually with a precision of 0.1km. How is this value calculated? Well, the circumference of a car tyre can be measured or calibrated as, C, and is independent of its air pressure, (you can investigate this assumption to check its validity), so if we measure the number of wheel revolutions, n, then the distance travelled is simply the product of the two values, C and n. However, which wheel do we measure for its number of revolutions because if a car is travelling up hill and constantly turning left then the right side wheels will rotate more than the left side wheels. We don't want the extra cost and complication of measuring left side and right side wheels and averaging their revolutions so a neat solution is to measure the gearbox output shaft revolutions. So, if we know the fixed ratios of any mechanical gearboxes in between the gearbox and the drive wheels, e.g. the differential gearbox ratio, then it is a straightforward matter to compute the distance travelled by the centre of the vehicle. Note that the distance travelled by the vehicle is a scalar value, meaning that there is no indication of the vehicle vector displacement in the ground plane whilst it travelled that distance.

However, we can provide vector information if the radical virtual steering wheel angle is measured at the same time the distance travelled is measured, and, assuming no slip between the tyres and the road, then we can compute not only scalar distance travelled but also the vector path of the vehicle.

So, the idea behind the planar odometer is, assuming (i) no slip between the tricycle wheels and the table on which it is travelling, and (ii) knowing the history of the ICofR and distance travelled values, then it is possible to compute the tricycle x,y coordinates on the table together with the vehicle heading angle, figure 3.24.

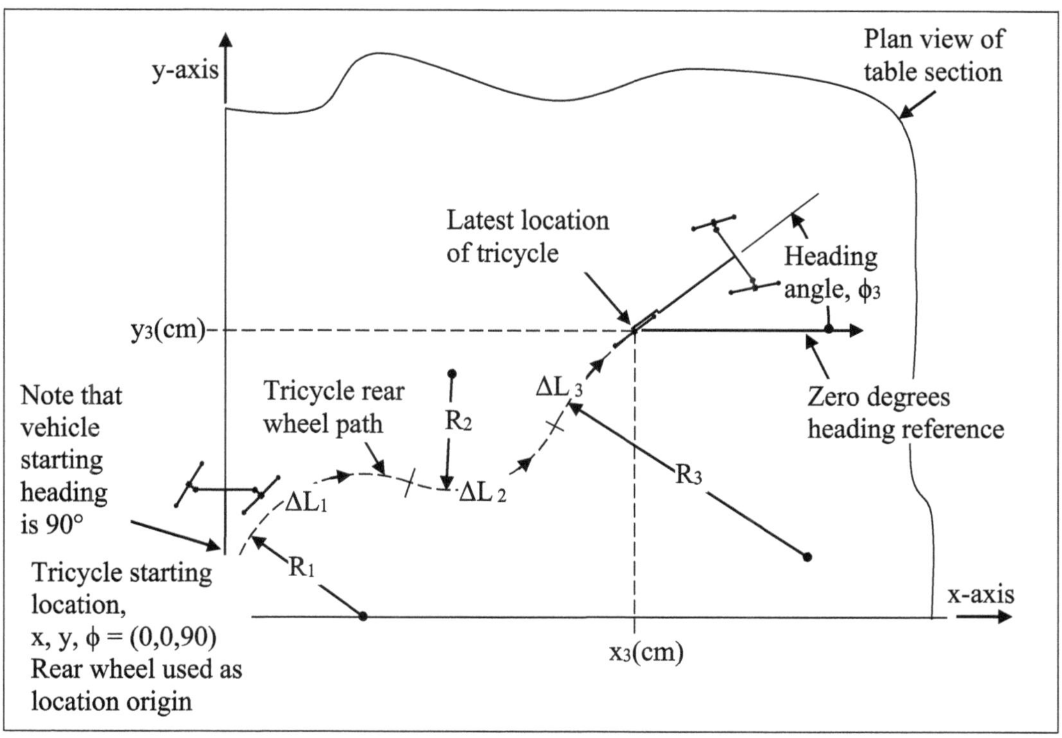

Figure 3.24 Plan view of tricycle located on table at coordinates, x, y, ϕ.

Note that the tricycle starting location is, x, y, ϕ = (0,0,90). This is an arbitrary starting position; it can be anywhere and at any angle on the table. We use cm units for x and y coordinates and degree units for the heading, ϕ. The heading angle is measured clockwise from the x-axis. The contact point of the rear drive wheel with the table is used as the x,y location coordinates. From the starting location, the vehicle undertakes a sequence of steering and displacement patterns annotated by, (R_1,d_1), (R_2,d_2), (R_3,d_3), before coming to a halt at x_3, y_3 coordinates. So the problem is; given the starting coordinates and the history of the steering and displacement patterns, what is the final location and heading of the vehicle?

We analyse the situation by using figure 3.25.

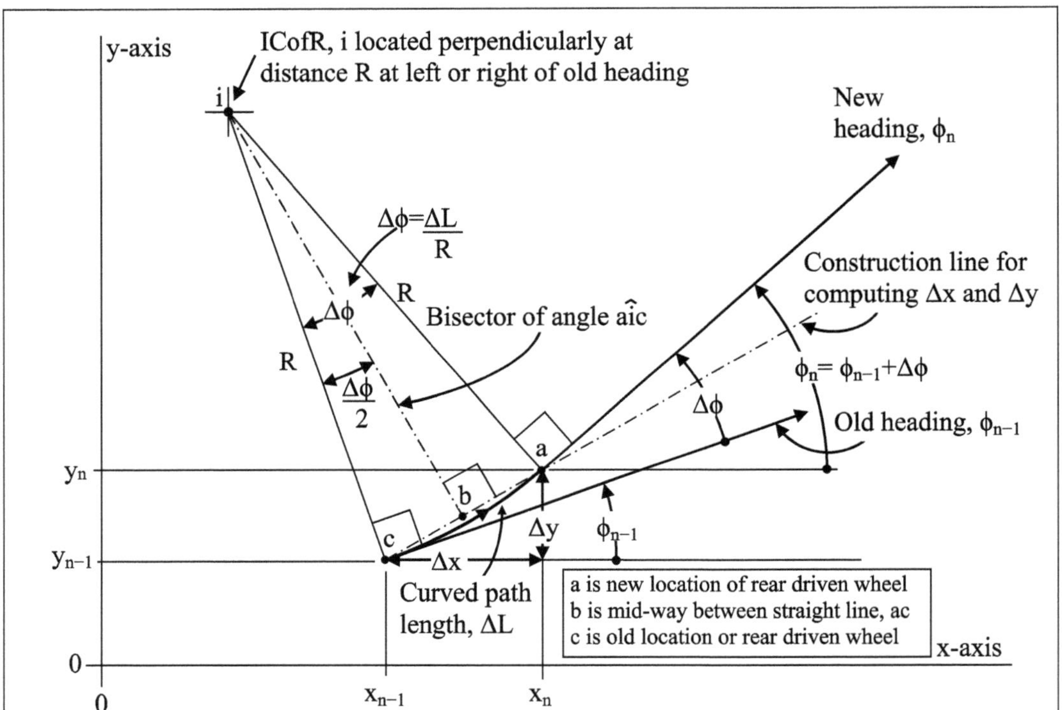

Definition of symbols

ϕ_{n-1} = initial heading of vehicle
ϕ_n = final heading of vehicle
Δx = change in x coordinate
x_{n-1} = initial x-axis coordinate
x_n = final x-axis coordinate = $x_{n-1} + \Delta x$
Δy = change in y coordinate
y_{n-1} = initial y-axis coordinate
y_n = final y-axis coordinate= $y_{n-1} + \Delta y$
R = radius about instantaneous centre of rotation, i
ΔL = length of path of rear drive wheel

Figure 3.25 Odometer computing of the tricycle coordinates, x, y and ϕ using distance travelled, L and radius, R, of ICofR. Vehicle starts at point c then travels small discrete distance to point a.

Here, in figure 3.25, we show a vehicle at initial 'old' coordinates, $(x_{n-1}, y_{n-1}, \phi_{n-1})$ travelling a short discrete distance to 'new' coordinates, (x_n, y_n, ϕ_n) with an ICofR, i located perpendicular to the old heading, ϕ_{n-1} and to the vehicle left side.

3.13 Odometer algorithm to compute x, y and ϕ

First let us list what quantities we know and what quantities we don't know and thus have to compute.

Four quantities that we know; $x_{n-1}, y_{n-1}, \phi_{n-1}$, R, ΔL

Three quantities that we don't know and must compute; $\Delta x, \Delta y, \Delta \phi$

To give; x_n, y_n, ϕ_n

Step 1	Compute $\Delta\phi$ based on known values, ΔL like this, $\Delta\phi = \dfrac{\Delta L}{R}$	**Note that,** we are doing discrete digital computations so $\Delta\phi$ should be small, say not larger than 0.05 radian, (about 3°). Smallest value of R is 20cm so ΔL should not exceed 1cm. Thus, odometrical values should be computed every 1cm of rear drive wheel travel.
Step 2	Compute length, ac = $2R\sin\left(\dfrac{\Delta\phi}{2}\right)$	
Step 3	Compute $\Delta x = ac.\cos\left(\phi_{n-1} \pm \dfrac{\Delta\phi}{2}\right)$	**Note:** +ve sign if ICofR is on the left of vehicle path and -ve if on the right
Step 4	Compute $\Delta y = ac.\sin\left(\phi_{n-1} \pm \dfrac{\Delta\phi}{2}\right)$	
Step 5	Compute new value of x-coordinate, $x_n = x_{n-1} + \Delta x$	
Step 6	Compute new value of y-coordinate, $y_n = y_{n-1} + \Delta y$	
Step 7	Compute new value of φ-coordinate, $\phi_n = \phi_{n-1} + \Delta\phi$	

The computation of this algorithm is left as a highly challenging problem for students to work out how it can be executed using the BS2px#3 microcomputer in figure 3.8.

We now conclude this chapter with an example of programming the tricycle to undertake a figure-of-eight shaped locomotion path in the ground plane.

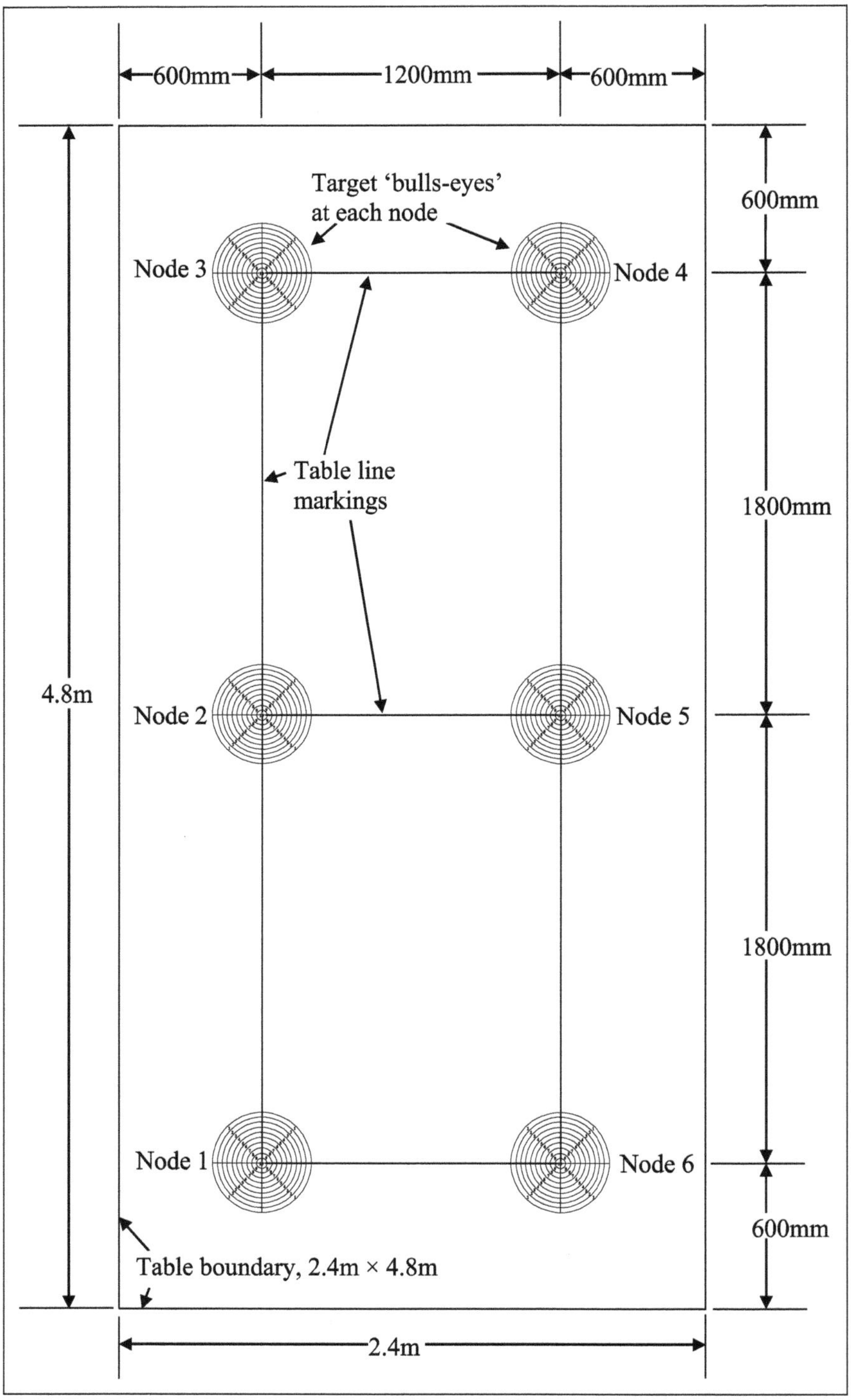

Figure 3.26 Diagram of the 'Vehicle Motion in a Ground Plane' table' showing nodes and target areas. This table is used to evaluate vehicle locomotion computer programmes.

3.14 Programming of the tricycle to undertake a figure-of-eight locomotion path

Figure 3.26 shows a ground plane table of dimensions 2.4m × 4.8m with six pieces of white paper bulls-eye target markings taped to the table at the locations shown in the figure. The idea is to programme the tricycle to undertake a series of locomotion patterns such that its rear wheel will pass through the centre of each of the six target bull's eyes as it carries out a figure-of-eight locomotion path. The six bull's-eyes act as nodes from node 1 to node 6 and the tricycle is to traverse through the centre of each bull's eye acquiring marks as it travels. The bull's eye is shown in detail in figure 3.27 where it can be seen that the marks acquired range from 10 marks for the rear drive wheel stopping at the bull's eye down to zero marks if outside the bull's eye radius of 20cm. The tricycle is to locomote itself through the figure-of-eight path shown in figure 3.28 that also shows various steering techniques for turning left and right. The tricycle is a non-holonomic vehicle, (cannot 'crab' sideways), so to turn left and right requires certain steering path techniques and a sample of these techniques is shown in figure 3.28 whereas the locomotion sequence algorithm is shown on the right side.

No line-following sensors are to be used to correct straying errors along the path and at the bull's eyes. The idea is for students to concern themselves with the calibration and programming of their tricycle to carry out precision locomotion patterns. Such locomotion path planning and enactment is called 'dead-reckoning' path following. It is a blind technique and is an excellent engineering discipline to develop before sensors are added to give added path following accuracy. The tricycle is to pass through 8 bull's eye target points during its travel where it can accumulate a maximum of 8 × 10 = 80 marks. You can see the author's attempt at programming the tricycle on YouTube here; https://youtu.be/jt5szC5ERes

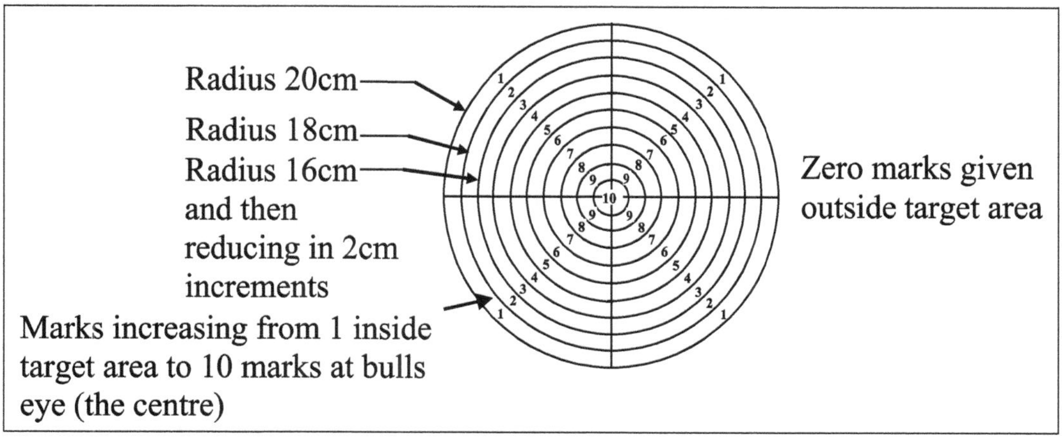

Figure 3.27 Detailed drawing of the 'Bulls Eye'.

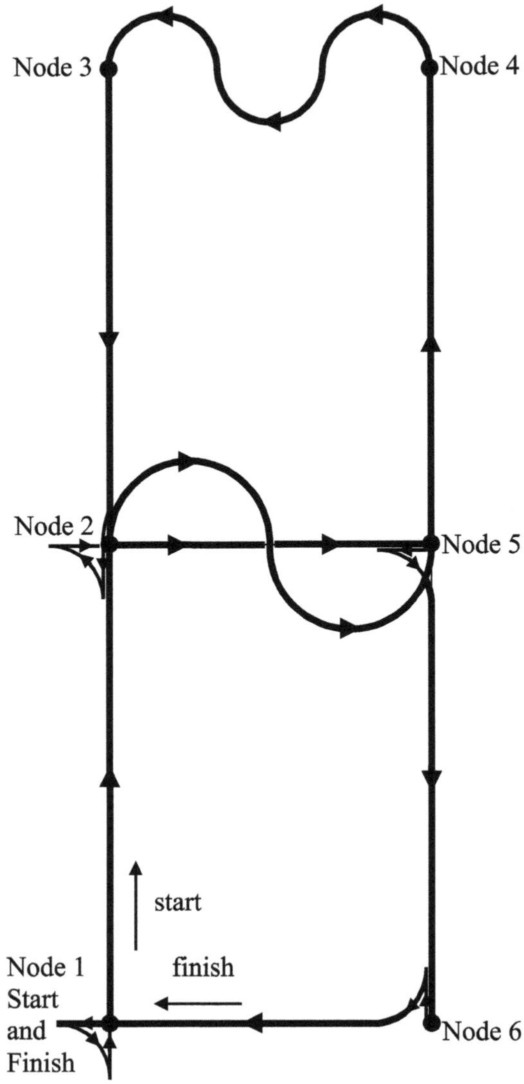

Figure-of-eight algorithm

Start
1. Node 1 to node 2
Go straight 1800mm
2. Stop for 5 seconds to enable position measurement at node 2 bulls eye
3. Node 2 to node5
(i) Turn right 180° with ICofR=30cm
(ii) Turn left 180° with ICofR=30cm
4. Stop for 5 seconds to enable position measurement at node 5 bulls eye
5. Node 5 to node 4
Go straight for 1800mm
6. Stop for 5 seconds to enable position measurement at node 4 bulls eye
7. Node 4 to node 3
(i) Turn left for 180° with ICofR=20cm
(ii) Turn right for 180° with ICofR=20cm
(iii) Turn left for 180° with ICofR=20cm
8. Stop for 5 seconds to enable position measurement at node 3 bulls eye
9. Node 3 to Node 2
Go straight for 1800mm
10. Stop for 5 seconds to enable position measurement at node 2 bulls eye
11. Node 2 to node 5
(i) Forward 2-point turn left by 90° with ICofR=20cm
(ii) Go straight for 1200mm
12. Stop for 5 seconds to enable position measurement at node 5 bulls eye
13. Node 5 to node 6
(i) Backward 1-point turn right by 90° with ICofR=20cm
(ii) Go forward by 1600mm
14. Stop for 5 seconds to enable position measurement at node 6 bulls eye
15. Node 6 to node 1
(i) Backward 1-point turn right by 90° with ICofR=20cm
(ii) Go forward 1200mm
9iii) Backward 2-point turn right by 90° with ICofR=20cm
16. Stop for 5 seconds to enable position measurement at node 1 bulls eye
Finish

Figure 3.28 Figure-of-eight path in the ground plane of table shown in figure 3.26. Computer programme gives locomotion to the tricycle using a 'dead reckoning' method that relies on the vehicle steering and displacement accuracy.

3.15 Some tips for odometer computing with the Basic Stamp

We can use the following two Taylor series for computing sinϕ and cosϕ to less than 1% error for angles of ϕ between 0° and 180° meaning 0centirads to 314centirads respectively.

$$\sin\phi = 0.988\phi + 0.000478\phi^2 - 0.0000231\phi^3 + 0.0000000367\phi^4$$

$$\cos\phi = 100 + 0.0617\phi - 0.00668\phi^2 + 0.0000142\phi^3$$

where, ϕ is in integer centiradian units meaning that ϕ ranges from 0 to 314 centiradians integer units and sinϕ ranges from 0 to 100 as in figure 3.29.

These equations can be computed using PBasic with the following code:

```
sinφ =  +(98*φ/100)                        + (8*φ/1000)
        +(47*φ/200*φ/500)                  + (8*φ/20*φ/50000)
        – (23*φ/200*φ/250*φ/20)            – (1*φ/20*φ/250*φ/2000)
        + (36*φ/200*φ/250*φ/500*φ/40)      – (7*φ/20*φ/250*φ/500*φ/4000)

cosφ =  100
        +(61*φ/1000)                       + (7*φ/10000)
        – (66*φ/200*φ/50)                  – (8*φ/20*φ/5000)
        + (14*φ/200*φ/250*φ/20)            + (2*φ/20*φ/250*φ/2000)
```

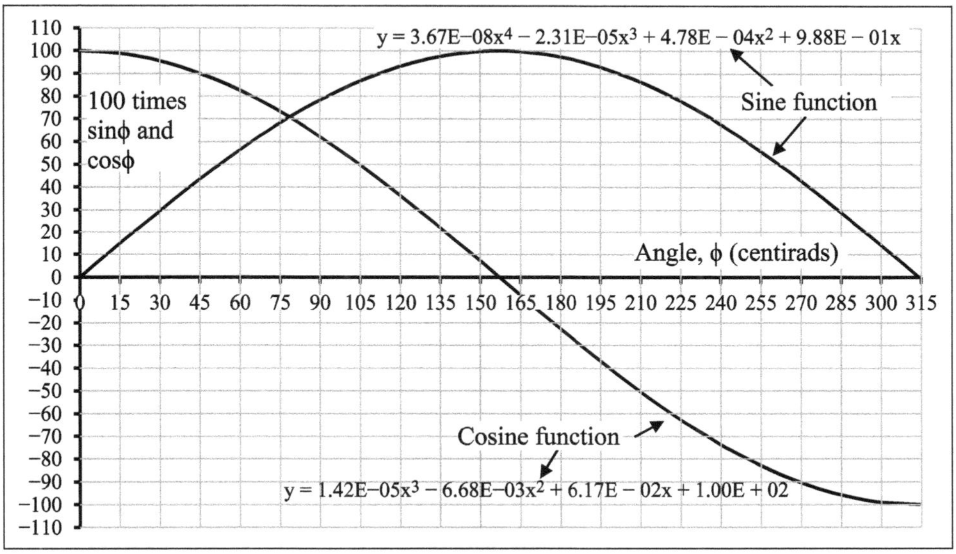

Figure 3.29 Sine and cosine of angle from 0 centirads (0°) to 314centirads (180°) and their Taylor series fit equations.

The complete code for implementing the algorithm in figure 3.28 is now given together with detailed comments and explanations.

3.16 Complete PBasic figure-of-eight programming code listing with comments and explanations

```
'Code for the Basic Stamp BS2px#1. This Stamp serves as a 20msec polled interrupt
'update rate clock. The variable,'intervalnumber' increments each cycle.
'{$STAMP BS2px}
intervalnumber VAR    Word
again: intervalnumber=intervalnumber+1
      SEROUT 8,16780,[intervalnumber.HIGHBYTE,intervalnumber.LOWBYTE]
      IF intervalnumber=0 THEN skip0
  PAUSE 17   'pause 17ms to produce overall 20ms update period
  GOTO again
```

'The following code is loaded into the BS2px#4 that controls the steering and wheel drive 'servos

```
' {$STAMP BS2px}
'note that Pbasic instructions are not case sensitive
'DEFINE PINS
'pin8=left servo drive
'pin9=right servo drive
'pin10=right steering wheel 1875=0deg calibration is 555=45deg 111/9 pu/deg
maximum angle is +/-55 deg
'pin11=left steering wheel

'DEFINE variables and constants
Steeringwheelpuangleold           VAR  Word  'pulsout angle units
Steeringwheelpuanglenew           VAR  Word  'pulsout angle units
ICofR                             VAR  Word  'minimum=20cm (maximum
                                             curvature)
                                             'maximum=10000cm (go
                                             straight ahead)
drivewheelangle                   VAR  Word  '0degrees to 359 degrees
delta                             VAR  Byte  '0degrees to 59degrees
section                           VAR  Nib   'one of six 60degree sections
drivewheelanglenew                VAR  Word
drivewheelanglenewrange           VAR  Word
drivewheelcumulativeangleold      VAR  Word
drivewheelcumulativeanglenew      VAR  Word
drivewheeldistancenewmm           VAR  Word
intervalnumber                    VAR  Bit
drivewheelspeed                   VAR  Nib   'maximum speed is 10

i                                 VAR  Byte  'counter for drive wheel, cannot be
                                             'shared
j                                 VAR  Byte  'counter for hold time, can be
                                             shared
k                                 VAR  Word  'counter for virtual steering wheel,
                                             'cannot be shared
steerdirection                    VAR  Bit   '0=left, 1=right
steeringrotatespeed               CON  20
holdtime                          CON  3     '3 second hold time, maximum is 5
                                             'seconds
```

'initialisation

```
drivewheelcumulativeangleold=0
drivewheelanglenew=0
```

```
    steerdirection=0
    steeringwheelpuangleold=0
    steeringwheelpuanglenew=0
    k=0

'stop at NODE 1 and set up steering wheel angles for straight ahead and set drive wheel
angle to zero degrees
    GOSUB holdposition              'drive wheel and steering wheels set at zero degrees
    PAUSE 5000                      'wait for 5 seconds to locate manually the tricycle
                                    at 'node 1
```

'START of FIGURE-of-EIGHT

'Node 1 to node 2

```
    drivewheeldistancenewmm=1800        'set travel distance to 1800mm
    GOSUB sectiontravel                 'execute travel motion
```

'**Stop for 5 seconds** to enable position measurement at node 2 bulls eye

```
    GOSUB holdposition
```

'Node 2 to node 5

'(i) Turn right 180° with ICofR=30cm (travel distance =π × 300mm =942mm)

```
    ICofR=30                        'set ICofR to 30cm
    steerdirection=1                'set steering to turn right
    GOSUB setsteeringangle          'execute steering direction
    drivewheeldistancenewmm=942     'set travel distance to 942mm
    GOSUB sectiontravel             'execute travel motion
```

'(ii) Turn left 180° with ICofR=30cm (travel distance =π × 30cm =942mm)

```
    ICofR=10000                     'set up steering go straight ahead, (very large
                                    'ICofR). To turn right must first go through
                                    'straight ahead before setting turn right
    GOSUB setsteeringangle          'execute steering direction
    ICofR=30                        'set ICof R back to 30cm
    steerdirection=0                'set steering turn right
    GOSUB setsteeringangle          'execute steering direction
    drivewheeldistancenewmm=942     'set travel distance to 942mm
    GOSUB sectiontravel             'execute travel motion
```

'End the manoeuvre by setting steering wheels straight

```
    ICofR=10000
    GOSUB setsteeringangle
```

'**Stop for 5 seconds** to enable position measurement at node 5 bulls eye

```
    GOSUB holdposition
```

'Node 5 to node 4

```
drivewheeldistancenewmm=1800     'set up travel distance of 1800mm
GOSUB sectiontravel              'execute travel motion
```

'Stop for 5 seconds to enable position measurement at node 4 bulls eye

```
GOSUB holdposition
```

'Node 4 to node 3

'(i) Turn left for 180° with ICofR=20cm (travel distance =π × 20cm =628mm)

```
ICofR=20                         'set ICofR to 20cm
steerdirection=0                 'set steering turn left
GOSUB setsteeringangle           'execute steering direction
drivewheeldistancenewmm=628      'set up travel distance of 628mm
GOSUB sectiontravel              'execute travel motion
ICofR=10000
GOSUB setsteeringangle
```

(ii) Turn right for 180° with ICofR=20cm (travel distance =π × 20cm =628mm)

```
ICofR=20                         'set ICofR to 20cm
steerdirection=1                 'set turn right
GOSUB setsteeringangle           'execute steering direction
drivewheeldistancenewmm=628      'set up travel distance of 628mm
GOSUB sectiontravel              'execute travel motion
ICofR=10000                      'set up ICofR representing straight ahead
GOSUB setsteeringangle           'execute steering straight ahead direction
```

(iii) Turn left for 180° with ICofR=20cm (travel distance =π × 20cm =628mm)

```
ICofR=20                         'set up 20cm ICofR
steerdirection=0                 'set up turn left
GOSUB setsteeringangle           'execute steering direction
drivewheeldistancenewmm=628      'set up travel distance of 628mm
GOSUB sectiontravel              'execute travel motion
ICofR=10000                      'set up ICofR to give straight ahead steering
GOSUB setsteeringangle           'execute steering straight ahead direction
```

'Stop for 5 seconds to enable position measurement at node 3 bulls eye

```
GOSUB holdposition
```

'Node 3 to Node 2

```
drivewheeldistancenewmm=1800     'set up travel distance of 1800mm
GOSUB sectiontravel              'execute travel motion
```

'Stop for 5 seconds to enable position measurement at node 2 bulls eye

```
GOSUB holdposition
```

'Node 2 to node 5
'(i) Forward 2-point turn left by 90° with ICofR=20cm
GOSUB twopointturnleft0 'see subroutine for instructions to do this

'(ii) Go straight for 1200mm
drivewheeldistancenewmm=1200 'set up travel distance of 1200mm
GOSUB sectiontravel 'execute travel motion

'Stop for 5 seconds to enable position measurement at node 5 bulls eye
GOSUB holdposition

'Node 5 to node 6
'(i) Backward 1-point turn right by 90° with ICofR=20cm followed by 1600mm straight 'ahead
GOSUB onepointturnright1800 'see subroutine for instructions to do this

'Stop for 5 seconds to enable position measurement at node 6 bulls eye
GOSUB holdposition

'Node 6 to node 1
'(i) Backward 1-point turn right by 90° with ICofR=20cm followed by 1200mm straight 'ahead
GOSUB onepointturnright1200 'see subroutine for instructions to do this

(ii) **Backward 2-point turn right by 90° with ICofR=20cm**
GOSUB twopointturnright0 'see subroutine for instructions to do this
stopvehicle:

'Stop for 5 seconds to enable position measurement at node 1 bulls eye
GOSUB holdposition
GOTO stopvehicle 'infinite loop to stop the tricycle moving

'FINISH of FIGURE-of-EIGHT

'THE SUBROUTINES ARE AS FOLLOWS

sectiontravel:
'this subroutine causes the vehicle to undertake a behaviour pattern defined as travelling a 'specified distance with a specified ICofR. The subroutine can also deal with a negative 'specified distance which means to travel backwards
IF drivewheeldistancenewmm>32767 THEN negativenewrange 'travel backwards
'if fail negative test above then compute drive wheel angle new range as follows
drivewheelanglenewrange=drivewheeldistancenewmm + (18*(drivewheeldistancenewmm/100)) 'convert mm range to drive wheel angle degrees, (drivewheelangle=drivewheeldistancenewmm*1.18)

GOTO skip102
negativenewrange:
drivewheelanglenewrange=–((0-drivewheeldistancenewmm) + ((0–
(18*(drivewheeldistancenewmm/100))'convert mm range to drivewheelangle degrees
skip102:
drivewheelcumulativeanglenew=drivewheelcumulativeangleold +
drivewheelanglenewrange 'update drive wheel cumulative angle
'instruction below checks for drive wheel angle being less than 180 mm either forwards or
'backwards in which case there is no acceleration or deceleration
IF drivewheeldistancenewmm<180 OR drivewheeldistancenewmm>65356 THEN
noaccel '

'The figures below show what the programme 'achieves in terms of drive wheel angle speed and acceleration

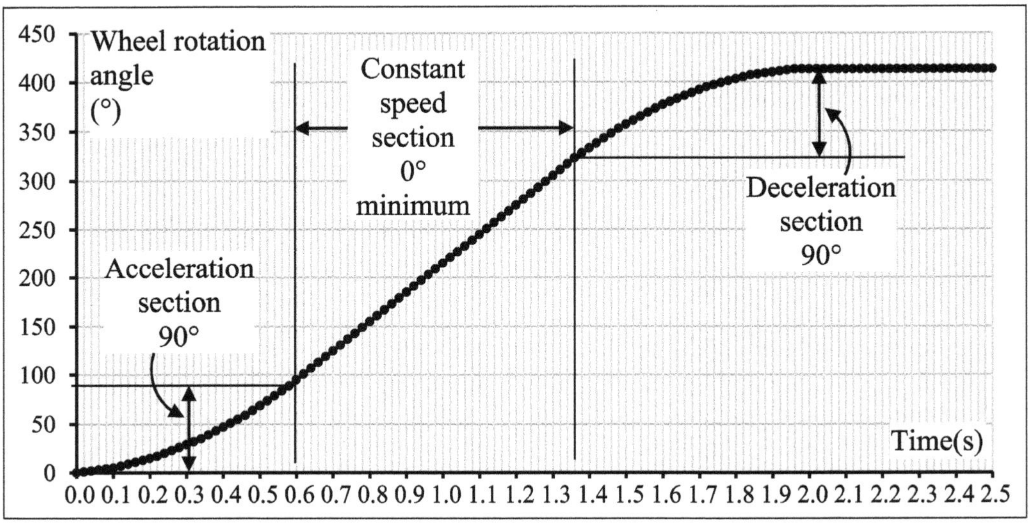

Figure 3.30 Drive wheel rotation angle against time in seconds.

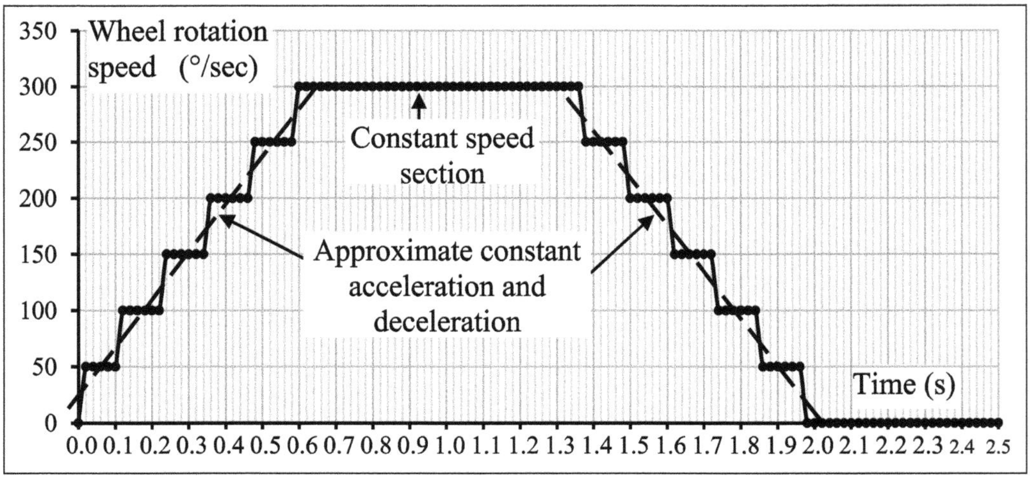

Figure 3.31 Drive wheel rotation speed against time in seconds.

```
'begin travel motion with acceleration.
'rotate drive wheel 6 times in 1 degree increments from 0 degrees to 5 degrees
'from starting position
'(note that vehicle already holding position drivewheelcumulativeangleold)
    drivewheelspeed=1                          'set speed low to begin travel motion
    FOR drivewheelanglenew=drivewheelcumulativeangleold + 1 TO
    drivewheelcumulativeangleold + 5 STEP drivewheelspeed
    SERIN 7,16780,[intervalnumber,intervalnumber]'wait for 20ms update rate signal
    GOSUB servodrivewheelangles                'execute wheel drive instruction
    GOSUB poutsteering                         'execute steering instruction (hold position)
    NEXT

'now rotate drive wheel 6 times in 2 degree increments from 5 degrees to 17 degrees
    drivewheelspeed=2
    FOR drivewheelanglenew=drivewheelcumulativeangleold+7 TO
    drivewheelcumulativeangleold+17 STEP drivewheelspeed
    SERIN 7,16780,[intervalnumber,intervalnumber]      'wait for 20ms update rate signal
    GOSUB servodrivewheelangles                'execute wheel drive instruction
    GOSUB poutsteering                         'execute steering instruction (hold position)
    NEXT

'now rotate drive wheel 6 times in 3 degree increments from 17 degrees to 35 degrees
    drivewheelspeed=3
    FOR drivewheelanglenew=drivewheelcumulativeangleold+20 TO
    drivewheelcumulativeangleold+35 STEP drivewheelspeed
    SERIN 7,16780,[intervalnumber,intervalnumber]'wait for 20ms update rate signal
    GOSUB servodrivewheelangles                'execute wheel drive instruction
    GOSUB poutsteering                         'execute steering instruction (hold position)
    NEXT

'now rotate drive wheel 6 times in 4 degree increments from 35 degrees to 59 degrees
    drivewheelspeed=4
    FOR drivewheelanglenew=drivewheelcumulativeangleold+39 TO
    drivewheelcumulativeangleold+59 STEP drivewheelspeed
    SERIN 7,16780,[intervalnumber,intervalnumber]'wait for 20ms update rate signal
    GOSUB servodrivewheelangles                'execute wheel drive instruction
    GOSUB poutsteering                         'execute steering instruction (hold position)
    NEXT

'now rotate drive wheel 6 times in 5 degree increments from 59 degrees to 89 degrees
    drivewheelspeed=5
    FOR drivewheelanglenew=drivewheelcumulativeangleold+64 TO drivewheelcumula-
    tiveangleold+89 STEP drivewheelspeed
    SERIN 7,16780,[intervalnumber,intervalnumber]'wait for 20ms update rate signal
    GOSUB servodrivewheelangles                'execute wheel drive instruction
    GOSUB poutsteering                         'execute steering instruction (hold position)
    NEXT
```

```
'now complete the last angle of the acceleration at drivewheelspeed=6
   SERIN 7,16780,[intervalnumber,intervalnumber] 'wait for 20ms update rate signal
   drivewheelanglenew=drivewheelcumulativeangleold+95
   GOSUB servodrivewheelangles      'execute wheel drive instruction
   GOSUB poutsteering               'execute steering instruction (hold position)

'that completes the acceleration which has taken 95 degrees of wheel drive rotation.
'Now rotate drive wheel at constant speed of 6 degrees per update (20ms) until
'90 degrees before completion of travel displacement (drivewheelanglenew–90)
'Note that 6 degree increments have only a 1-in-6 chance of falling at 90 degrees before
'completion of travel so there is a 5-in-6 chance of being a slight wheel speed disjoint at
'the start of deceleration
   drivewheelspeed=6           'maximum constant speed until start of deceleration
   FOR drivewheelanglenew=drivewheelcumulativeangleold+101 TO drivewheelcumula-
   tiveanglenew–90 STEP drivewheelspeed
   SERIN 7,16780,[intervalnumber,intervalnumber] 'wait for 20ms update rate signal
   GOSUB servodrivewheelangles      'execute wheel drive instruction
   GOSUB poutsteering               'execute steering instruction (hold position)
   NEXT

'End of maximum constant speed section now start deceleration
'Finish travel motion with deceleration which starts 90 degrees before end of travel
'rotate drive wheel 6 times in 5 degree increments from 90 degrees to 60 degrees
'before end of travel
   drivewheelspeed=5
   FOR drivewheelanglenew=drivewheelcumulativeanglenew–85 TO
   drivewheelcumulativeanglenew–60 STEP drivewheelspeed
   SERIN 7,16780,[intervalnumber,intervalnumber] 'wait for 20ms update rate signal
   GOSUB servodrivewheelangles
   GOSUB poutsteering
   NEXT

'rotate drive wheel 6 times in 4 degree increments from 60 degrees to 36 degrees
'before end of travel
   drivewheelspeed=4
   FOR drivewheelanglenew=drivewheelcumulativeanglenew–56 TO
   drivewheelcumulativeanglenew–36 STEP drivewheelspeed
   SERIN 7,16780,[intervalnumber,intervalnumber] 'wait for 20ms update rate signal
   GOSUB servodrivewheelangles
   GOSUB poutsteering
   NEXT

'rotate drive wheel 6 times in 3 degree increments from 36 degrees to 18 degrees
'before end of travel
   drivewheelspeed=3
   FOR drivewheelanglenew=drivewheelcumulativeanglenew–33 TO
   drivewheelcumulativeanglenew–18 STEP drivewheelspeed
```

```
    SERIN 7,16780,[intervalnumber,intervalnumber] 'wait for 20ms update rate signal
    GOSUB servodrivewheelangles
    GOSUB poutsteering
    NEXT

'rotate drive wheel 6 times in 2 degree increments from 18 degrees to 6 degrees
'before end of travel
    drivewheelspeed=2
    FOR drivewheelanglenew=drivewheelcumulativeanglenew–16 TO
    drivewheelcumulativeanglenew–6 STEP drivewheelspeed
    SERIN 7,16780,[intervalnumber,intervalnumber] 'wait for 20ms update rate signal
    GOSUB servodrivewheelangles
    GOSUB poutsteering
    NEXT

'rotate drive wheel 6 times in 1 degree increments from 6 degrees to 0 degrees
'before end of travel
    drivewheelspeed=1
    FOR drivewheelanglenew=drivewheelcumulativeanglenew–5 TO
     drivewheelcumulativeanglenew STEP drivewheelspeed
    SERIN 7,16780,[intervalnumber,intervalnumber] 'wait for 20ms update rate signal
    GOSUB servodrivewheelangles
    GOSUB poutsteering
    NEXT

'end of deceleration and end of behaviour pattern
    GOTO skip103
'if there is no acceleration or deceleration then the programme is much simplified as
'follows
    noaccel:            drivewheelspeed=2            'travel at low speed
    FOR drivewheelanglenew=drivewheelcumulativeangleold TO drivewheelcumulativean-
    glenew STEP drivewheelspeed
    SERIN 7,16780,[intervalnumber,intervalnumber]    'wait for 20ms update'rate signal
    GOSUB servodrivewheelangles
    GOSUB poutsteering
    NEXT

skip103:
'update the value of drivewhelangleold variable
    drivewheelcumulativeangleold=drivewheelcumulativeanglenew
    RETURN                     'end of sectiontravel subroutine

'Here is the subroutine that sets the steering wheels to the required steering angle and
'at the same time sets the angles of wheel drive servos
setsteeringangle:
    steeringwheelpuanglenew=((10000/ICofR/2*131/50)–
    (10000/ICofR/ICofR*572/100))
```

```
    FOR k=steeringwheelpuangleold TO steeringwheelpuanglenew STEP
     steeringrotatespeed
    SERIN 7,16780,[intervalnumber,intervalnumber]
    GOSUB poutsteering                   'actuate the left and right steering servos
    GOSUB servodrivewheelangles       'actuate the left and right drive wheel servos
    NEXT
    k=steeringwheelpuanglenew
    steeringwheelpuangleold=steeringwheelpuanglenew
    RETURN
'hold vehicle static for holdtime seconds
holdposition:   FOR j=1 TO holdtime*25
    SERIN 7,16780,[intervalnumber,intervalnumber]
    GOSUB servodrivewheelangles
    GOSUB poutsteering
    NEXT
    RETURN
Servodrivewheelangles:
    i=drivewheelanglenew/360                      'i variable will compute how many
                                                  drive wheel revolutions have been
                                                  completed
    drivewheelangle=drivewheelanglenew-(i*360)    'drivewheelangle variable is how
                                                  many'degrees into a new revolution
    section=drivewheelangle/60                    'compute which section of
                                                  360degrees
    delta=drivewheelangle-(section*60)            'compute how many degrees into
                                                  the'relevant section
    BRANCH section,[s0,s1,s2,s3,s4,s5]            'branch according to one of six
                                                  sections
'at each branch compute left and right wheel drive servo angles and output their values
to the servos for each of the six 60degree sections, s0, s1, s2, s3, s4, s5
s0:    PULSOUT 8,1833-(784*delta/100) - ( 0*delta/100*delta/10) +
        ( 33*delta/100*delta/50*delta/20)          'leftservodrive
       PULSOUT 9,2379+( 59*delta/100)- (144*delta/100*delta/10) +
       ( 75*delta/100*delta/50*delta/20)           'rightservodrive
       GOTO skip100                                'section 0-59degrees complete
s1:    PULSOUT 8,1435-(469*delta/100)+( 76*delta/100*delta/10)+
       ( 0*delta/100*delta/50*delta/20)            'leftservodrive
       PULSOUT 9,2058-(838*delta/100)- ( 12*delta/100*delta/10) +
       ( 31*delta/100*delta/50*delta/20)           'rightservodrive
       GOTO skip100                                'section 60-119degrees complete
```

```
s2:    PULSOUT 8,1425+(392*delta/100)+( 42*delta/100*delta/10)–
       ( 8*delta/100*delta/50*delta/20)            'leftservodrive
       PULSOUT 9,1579–(649*delta/100) + ( 30*delta/100*delta/10) +
       ( 25*delta/100*delta/50*delta/20)       'rightservodrive
       GOTO skip100                            'section 120–179degrees complete

s3:    PULSOUT 8,1793+(743*delta/100)+( 56*delta/100*delta/10)–
       ( 78*delta/100*delta/50*delta/20)       'leftservodrive
       PULSOUT 9,1350+( 1*delta/100)+( 76*delta/100*delta/10)–
       ( 17*delta/100*delta/50*delta/20)       'rightservodrive
       GOTO skip100                            'section 180–239degrees complete

s4:    PULSOUT 8,2271+(618*delta/100)–(111*delta/100*delta/ 10) +
       ( 17*delta/100*delta/50*delta/20)       'leftservodrive
       PULSOUT 9,1589+(691*delta/100)+( 42*delta/100*delta/10)–
       ( 42*delta/100*delta/50*delta/20)       'rightservodrive
       GOTO skip100                            'section 240–299degrees complete

s5:    PULSOUT 8,2278–(546*delta/100)–( 51*delta/100*delta/10) +
       ( 31*delta/100*delta/50*delta/20)       'leftservodrive
       PULSOUT 9,2066+(773*delta/100)–( 10*delta/100*delta/10)–
       ( 53*delta/100*delta/50*delta/20)       'rightservodrive
                                               'section 300–359degrees complete
skip100:        RETURN
```

'Subroutine to compute the left and right steering wheel servo angles

```
poutsteering:   BRANCH steerdirection,[poutturnleft,poutturnright]
poutturnleft:   PULSOUT 10,(1875–21)–((98*k/100)–
                (98*k/1000*k/387))'steeringwheelangleouter 'right steering wheel, (1875–
                21) is straight ahead servo calibration
                PULSOUT 11,(1875+37)–
                ((98*k/97)+(98*k/1000*k/282))'steeringwheelangleinner 'actuate the left
                steering wheel servo, (1875+37) is the straight ahead servo calibration
                GOTO skip101
poutturnright:  PULSOUT 10,(1875–21) +
                ((198*k/97)+(98*k/1000*k/282))'steeringwheelangleinner 'right steering
                wheel
                PULSOUT 11,(1875+37)+((98*k/100)–(98*k/1000*k/387))'steering-
                wheelangleouter 'left steering wheel
skip101:        RETURN
```

```
'turn left using 2–point turn
twopointturnleft0:
    drivewheeldistancenewmm=200
    GOSUB sectiontravel
    ICofR=20
    steerdirection=1
    GOSUB setsteeringangle
    drivewheeldistancenewmm=–314
    GOSUB sectiontravel
    ICofR=10000 MIN 20 MAX 10000
    GOSUB setsteeringangle
    drivewheeldistancenewmm=200
    GOSUB sectiontravel
    RETURN
'turn right using 2–point turn
twopointturnright0:
    drivewheeldistancenewmm=200
    GOSUB sectiontravel
    ICofR=20
    steerdirection=0
    GOSUB setsteeringangle
    drivewheeldistancenewmm=–300
    GOSUB sectiontravel
    ICofR=10000
    GOSUB setsteeringangle
    drivewheeldistancenewmm=200
    GOSUB sectiontravel
    RETURN
'turn right for 1800 mm using 1–point turn
onepointturnright1800:
    drivewheeldistancenewmm=–200
    GOSUB sectiontravel
    ICofR=20
    steerdirection=1
    GOSUB setsteeringangle
    drivewheeldistancenewmm=314
    GOSUB sectiontravel
    ICofR=10000
    GOSUB setsteeringangle
    drivewheeldistancenewmm=1600
    GOSUB sectiontravel
    RETURN
```

```
'turn right for 1200 mm using 1–point turn
onepointturnright1200:
    drivewheeldistancenewmm=–200
    GOSUB sectiontravel
    ICofR=20
    steerdirection=1
    GOSUB setsteeringangle
    drivewheeldistancenewmm=314
    GOSUB sectiontravel
    ICofR=10000 MIN 20 MAX 10000
    GOSUB setsteeringangle
    drivewheeldistancenewmm=1000
    GOSUB sectiontravel
    RETURN
```

4. Programming of walking legs, their leg tip motion and the leg tip digital locus

4.1 Introduction

Preceding chapters have laid the foundation for vehicle motion in a ground plane helped by a practical application of a wheeled robot in chapter 3. We will extend these ideas in successive chapters to a hybrid 2-wheeled/2-legged robot, figure 4.1, known as the Rickshaw Robot and a six legged robot, figure 4.2, the Hexapod Robot.

Figure 4.1 The Rickshaw Robot.

However, before this is done we need to explain how a leg articulates robot locomotion. This we do in this chapter by introducing a three degree-of-freedom, (3 dof) robot leg figure 4.3, shown mounted on a calibration stand and controlled by the ServoTeach module, described in chapter 2. The detailed inverse kinematics programming of this robot leg is explained, together with software coding developed in chapter 5. However, before the detailed programming of the leg is explained, this chapter outlines its motion and how it is used for robot locomotion.

Figure 4.2 The Hexapod Robot.

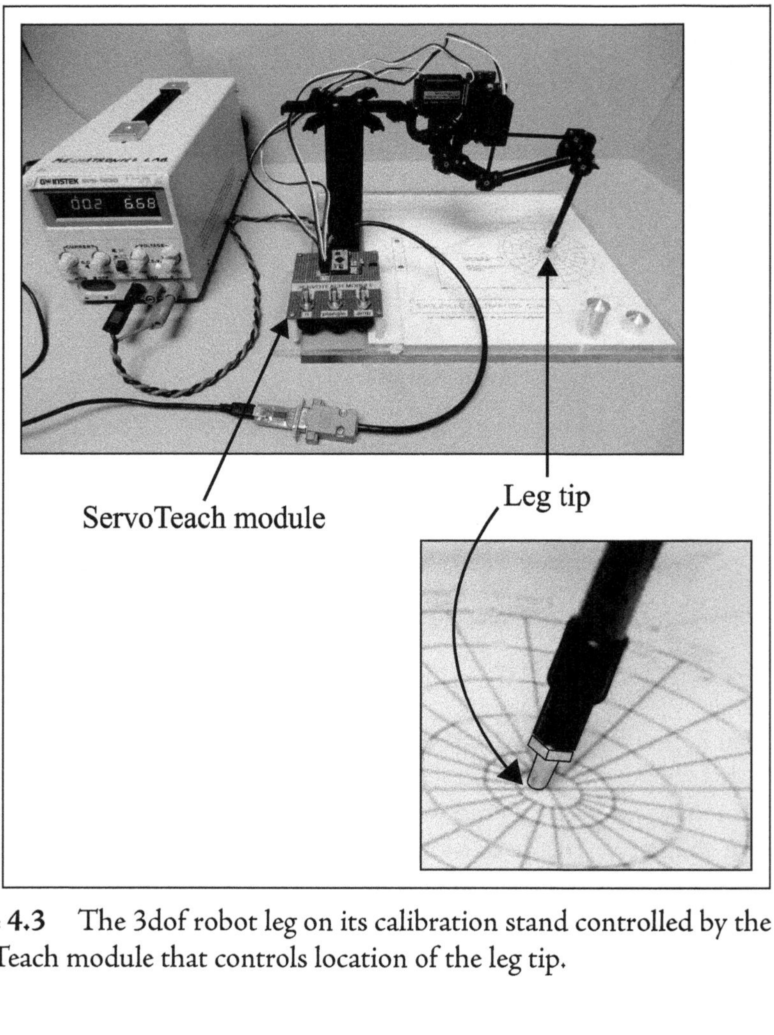

Figure 4.3 The 3dof robot leg on its calibration stand controlled by the ServoTeach module that controls location of the leg tip.

The leg has a 'leg tip, i.e. the 'foot' that supports the robot and provides traction. The leg tip is capable of accessing three axis coordinate space and is programmed to travel through a defined locus that produces robot locomotion. Each one of the leg's three degrees of freedom is actuated by a servo. We start by turning our attention to omni-directional motion of a six-legged robot and how the six legs are articulated to provide locomotion. We will not, just yet in this chapter, talk about the locomotion of the Rickshaw Robot because the hexapod locomotion techniques are easily applied to the Rickshaw Robot.

We show in this chapter that legs can be programmed to produce the same locomotion effect as steerable wheels. Hexapods, i.e. six-legged creatures, manifested as insects in the animal kingdom, together with eight-legged spiders, are highly successful prolific living organisms. Man's attempt to mimic insects as hexapod robots serves as providing interesting machine platforms to programme.

Such hexapod robots provide much fun and mathematical challenge for students. Their six legs mean that hexapod robots are inherently stable and students do not have to worry about their stability as they would do, for example, with bipedal humanoid robots. As a final introductory note, the author is frequently asked by parents during university Open Days; "What are these robots for? What is their purpose and application?" The answer given is,

"The practical application of such robots is for the training of students to solve difficult programming problems in engineering such as, to name just a few; computer controlled manufacturing machines, (also known as CNC machines), 3-d printers, assembly robots, and driverless autonomous vehicles."

Now we move on to introducing the three degree-of-freedom leg and its leg tip motion that provides locomotion for the Rickshaw and Hexapod robots.

4.2 The 3 degree-of-freedom leg, (3dof leg)

Figure 4.4 shows a schematic view of the 3dof leg which provides locomotion for the Rickshaw and Hexapod robots. The leg having three degrees of freedom, means that the hexapod robot can 'crab', i.e. walk sideways. More precisely, it means that the robot is omnidirectional so it can translate and rotate simultaneously, like the 'Viennese waltz' action described in chapter 1. Scientifically speaking, a robot with 3dof legs means that the ICofR can be placed anywhere in its locomotion plane. The leg tip can access any point inside its working volume which is purposely designed to be a cylindrical coordinate frame because such coordinates are convenient for omnidirectional walking. The frame could be a Cartesian but it turns out that such a frame is not convenient for omnidirectional walking.

The leg tip can be programmed to move to a point inside the working volume by programming the angles of the S1, S2 and S3 servos. The mathematical methodology of calculating the values of the three servo angles, in order to place the leg tip at a certain point in the working volume, is known as 'Inverse Kinematics'. The inverse kinematics method and its computation will be explained in the next chapter.

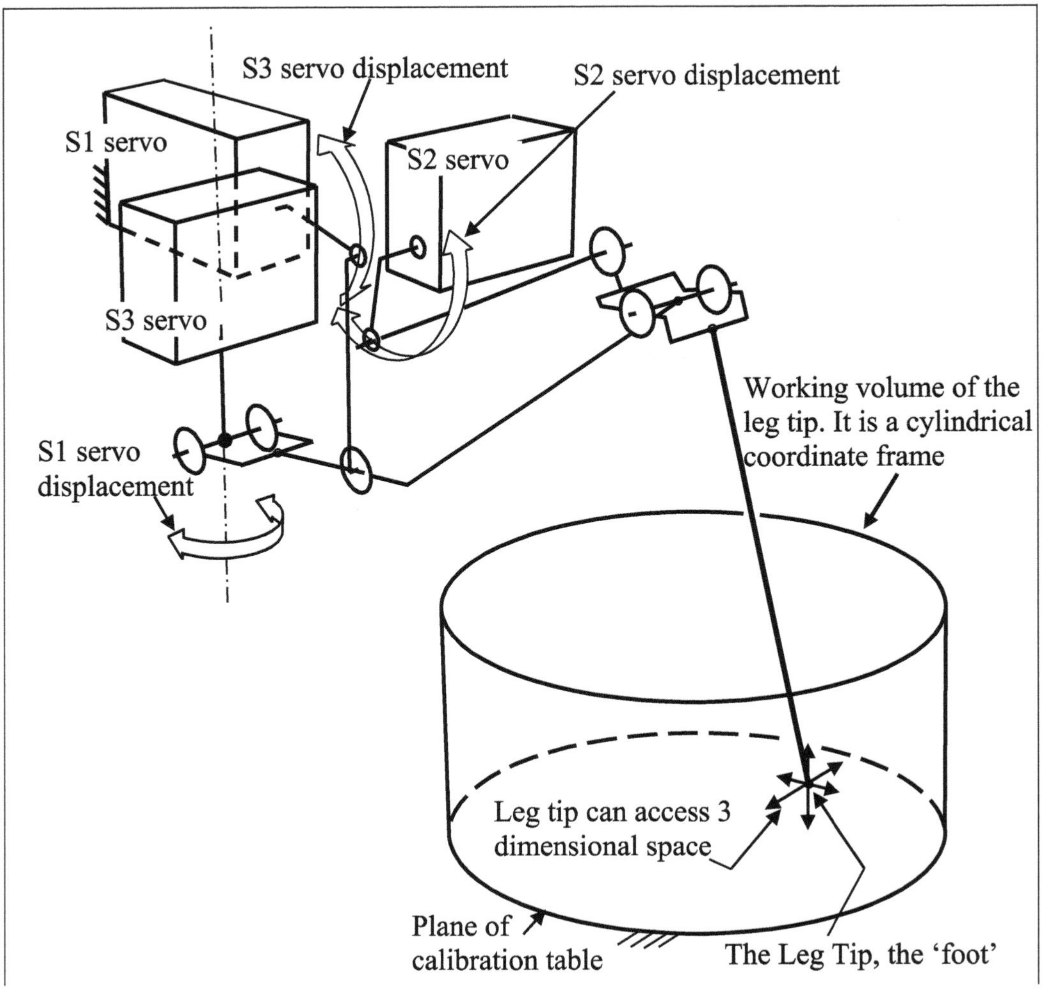

Figure 4.4 Schematic view of 3dof leg system showing the leg tip accessing 3 dimensional space inside a cylindrical coordinate frame. The leg is actuated by three servos, S1, S2 and S3 with corresponding displacements as shown.

The Rickshaw robot cannot carry out omnidirectional motion because it is very similar to the Tricycle robot in the previous chapter in that wheels do not permit 'crabbing' sideways motion. Alternatively speaking, the ICofR cannot be anywhere in the locomotion plane; it can only lie along the wheel axes. However, the Rickshaw robot, having two 3dof legs means that these legs can replicate the two steering wheels of the Tricycle. Thus the vehicle motion of the Tricycle and the Rickshaw are very similar. We now continue by describing hexapod robot legged locomotion in a ground plane from which Rickshaw locomotion can also be extrapolated.

4.3 Leg layout of omni-directional robot

Figure 4.5 shows a schematic layout of the six legs of the omni-directional robot. This diagram will be progressively analysed so as to give understanding of the leg tip walking motion.

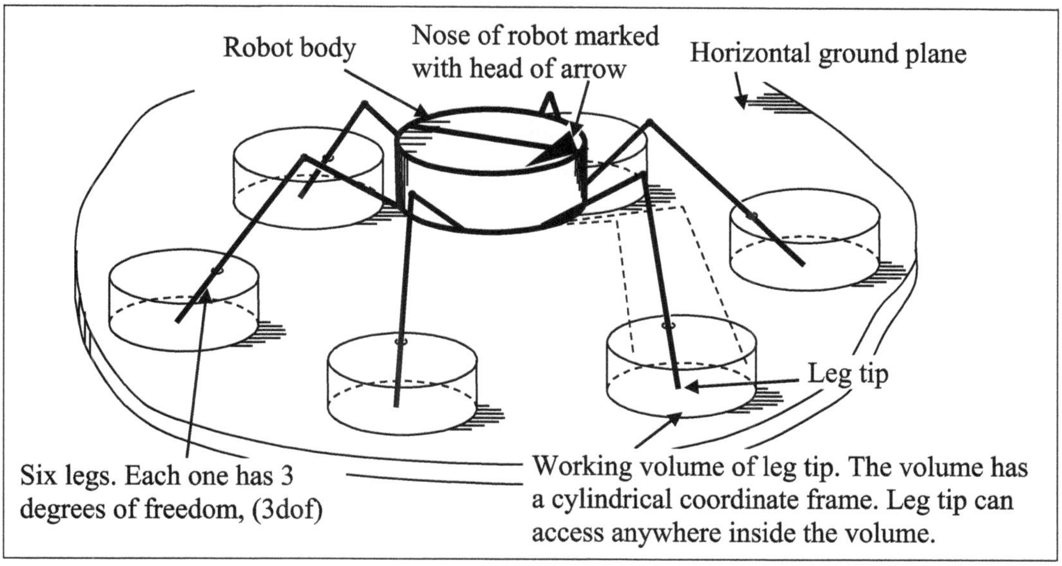

Figure 4.5 Schematic layout of six legs of the omni-directional hexapod robot. Each leg is programmed to place its leg tip in a cylindrical coordinate frame.

We will now take a closer look at how the leg tip is given motion to enable robot locomotion.

4.4 The leg tip locus

The leg tip articulation system shown in figure 4.5 above is used to obtain robot walking motion. For example, figure 4.6 below shows the robot walking forward in a straight line due to the leg tips following the orange coloured loci.

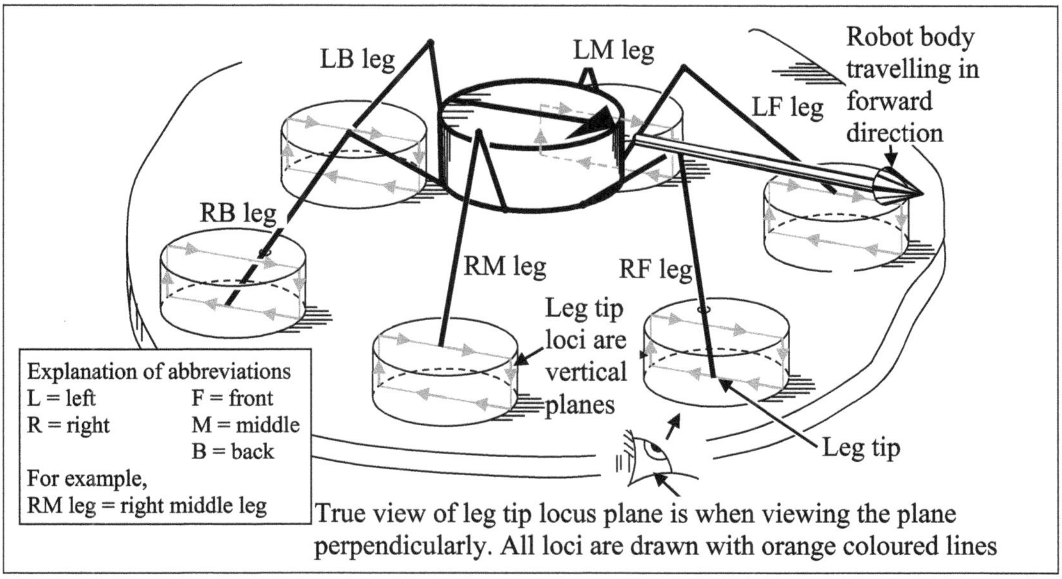

Figure 4.6 Diagram showing leg tip loci for omni-directional robot walking in straight line using double-tripod gait.

The robot is walking on two tripods, known as the left and right tripods. The walking style is known as double tripod gait. The left tripod is composed of the LF, LB and RM legs and the right tripod is composed of the RF, RB and LM legs. It can be seen that the left tripod is currently in the air on its return stroke and the right tripod is currently on the ground supporting the robot. Later, as the robot body moves forward, the tripod situation will reverse. Note that the leg tip loci are vertical planes and we will be discussing the true view of these planes later.

The two tripods making up the double tripod gait are shown in simplified plan view form in figure 4.7.

The figure shows that alternate legs make up each tripod. In the same way that human beings walk, which is to walk by placing one foot on the ground as the other foot is lifted off the ground, a six legged robot places one tripod on the ground as the other tripod is lifted off the ground. In other words each tripod is in opposite phase to the other as shown in figures 4.8a and 4.8b, which show just one each of the legs of the left and right tripods.

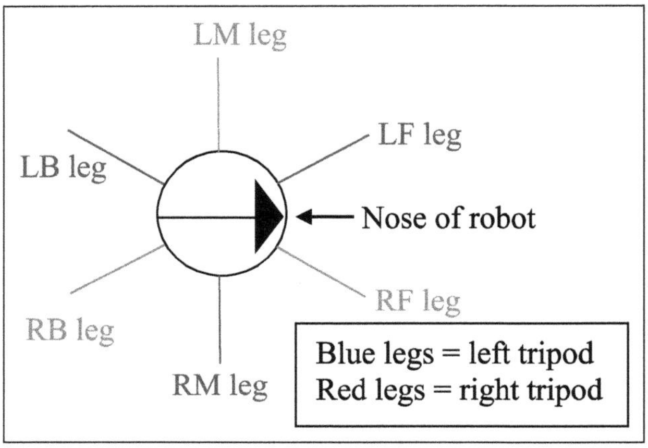

Figure 4.7 Simplified plan view of robot showing the two tripods used for double-tripod gait.

The leg tip locus in figure below is divided into six sections, which are, ab, bc, cd, de, ef, and fa. The points, a, b, c, d, e and f, which separate the sections are called, nodes. The two diagrams show the leg tips, firstly with respect to the robot body and, secondly with respect to the ground. This is important because the leg tips are programmed to move with respect to the robot body but the effect is seen as robot motion with respect to the ground.

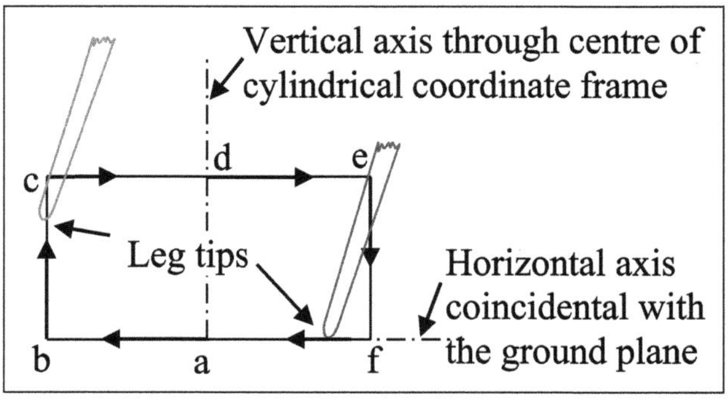

Figure 4.8a True view of leg tip locus with respect to robot body. This is the programmed leg tip motion and this is how we see the leg tip moving when sitting on the robot body.

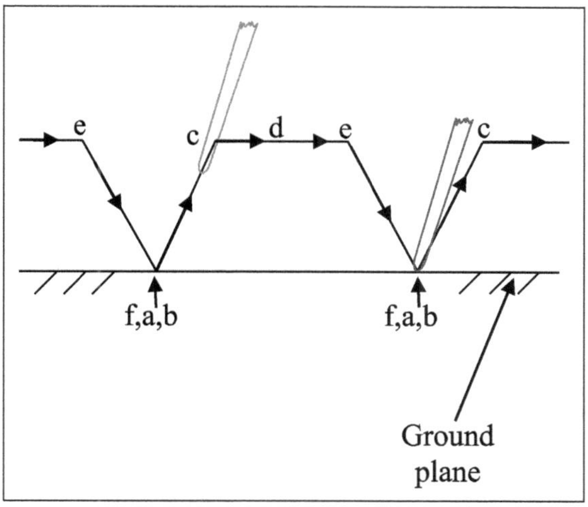

Figure 4.8b True view of leg tip locus with respect to the ground. This how we see the leg tip moving when we are standing on the ground.

4.5 Translation of the robot body

We can programme the robot to translate, i.e. to walk without rotating to the left or right, in different directions by turning all six planes of the leg tip locus to coincide with the required direction of motion. Thus all planes are parallel. This can be seen in figure 4.9 where the robot is "crabbing" sideways to the right. Let the angle between the forward direction of the robot body and the leg tip locus plane be equal to θ. For example, $\theta = 90^0$, in figure 4.9. Thus by setting the leg tip locus planes to any angle, θ between 0° and 360°, the robot can be made 'to crab', i.e. to translate in any direction in the ground plane.

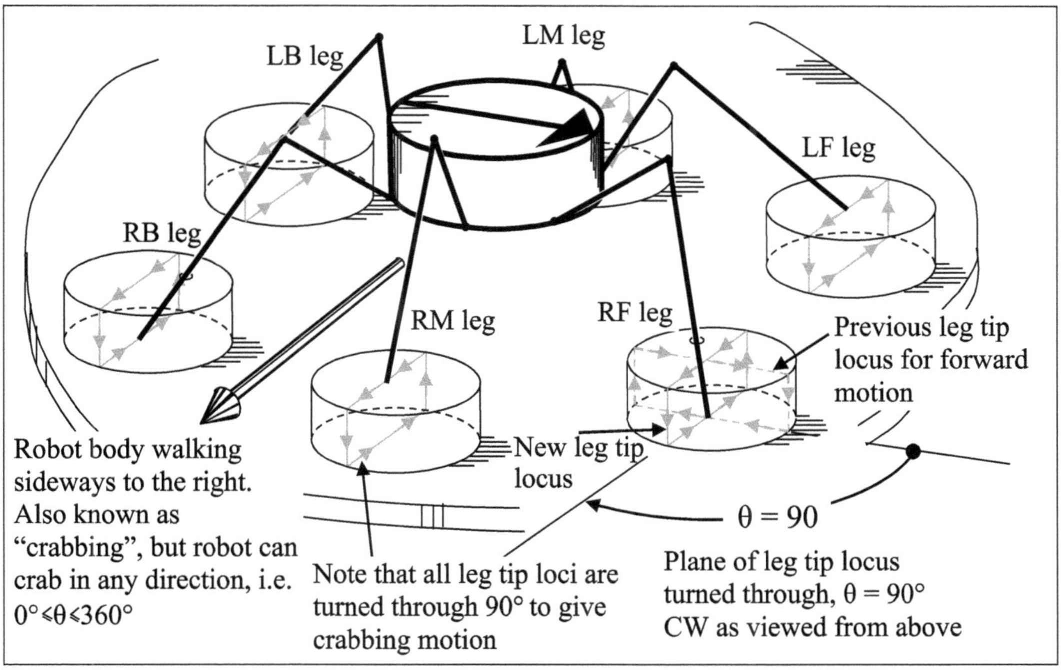

Figure 4.9 The omni-directional robot crabbing sideways to the right.

The following diagram, figure 4.10, showing a plan view of the robot as it translates, makes the idea clearer. As an example, the 'crabbing' or translating angle is set to 200°.

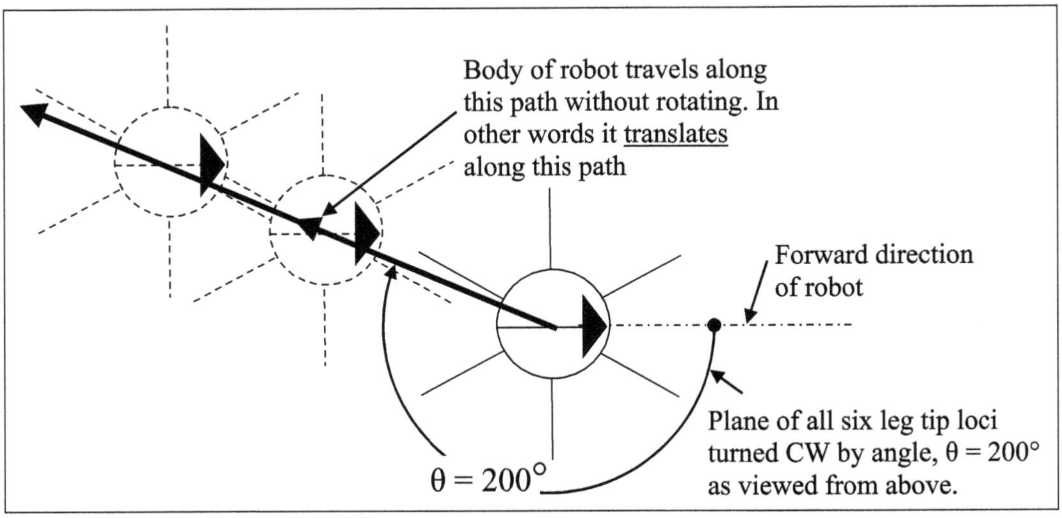

Figure 4.10 Plan view showing translation motion of robot with translation direction, θ set to 200°.

Robot translation is a very useful behaviour pattern but gives limited performance since rotation is missing. Adding the ability of robot rotation to behaviour patterns very much improves robot performance and even more so is for the robot to be capable of combined translation and rotation as already described in chapter 1. In a later chapter we will show how to do this. However, the next step is to explain details of the leg tip motion in order eventually to achieve these behaviour patterns.

4.6 Details of the leg tip locus

We now come to an important concept which is the time that a leg tip spends on the ground supporting the robot, figure 4.8a, nodes f, a, b, compared to the time that it spends in the air going back to its starting point on the ground, figure 4.8a, nodes b, c, d, e, f. The double tripod gait alternately supports the robot on the left tripod then the right tripod, figure 4.7. On the assumption that the walking effort is equally shared between each tripod in the double tripod gait, we can conclude that each tripod spends 50% of its time on the ground and 50% of its time in the air. Alternatively, we can say, for the double tripod gait, that the time any one leg tip spends on the ground is equal to the time it spends in the air.

If we look at the leg tip locus with respect to the robot body, figure 4.8a, we notice that the distance the leg tip travels in the air, b, c, d, e, f, is greater than the distance travelled, f, a, b. However, the time spent in the air, i.e. off the ground, is equal to the time spent in contact with the ground which means that the leg tip speed in the air must be greater than the leg tip speed when in contact with the ground. As an aside note, the leg tip locus does not have to be rectangular in shape. For example, it could be triangular or like a running track shape, figures 4.11a and 4.11b.

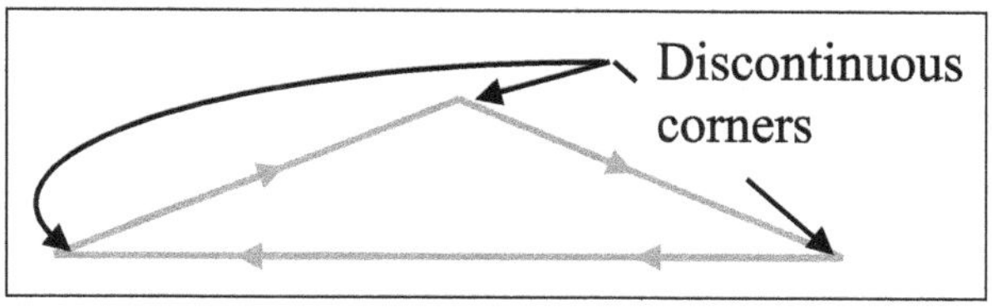

Figure 4.11a Triangular shape.

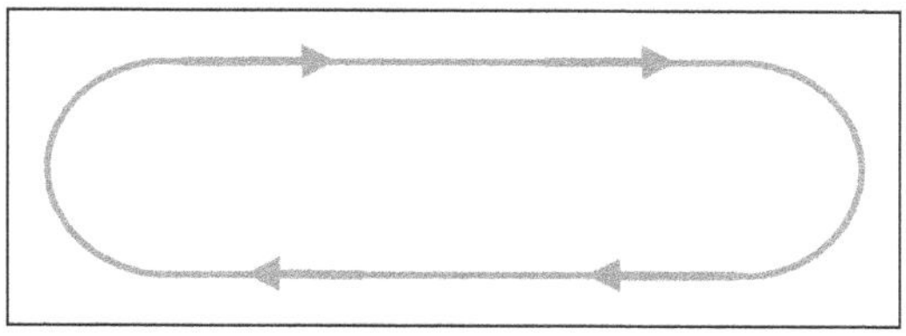

Figure 4.11b Running track shape.

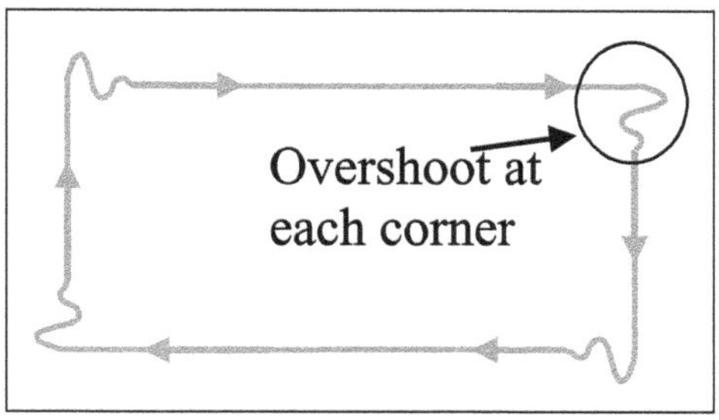

Figure 4.11c Corrupted locus at its discontinuous corners.

Since the leg tip is driven at a finite speed along the leg tip locus, there will be infinite accelerations at any discontinuous corners, e.g figure 4.11a. This is not possible in reality so the shapes will become corrupted at corners, e.g. with overshoots for the rectangular locus, figure 4.11c.

On the other hand the curved running track shape, figure 4.11b, has the advantage of more accurate control of the leg tip position along the locus due to the less discontinuous corners. However, its disadvantages are, (i) the leg is neither quickly detached from the ground nor quickly placed onto the ground and (ii) the added computing complexity to achieve its shape. As a consequence the curved shape is left as a research project for students. Hence for the time being we will use a rectangular profile and live with the inaccuracies at the corners. A triangular profile could also be used but that also can be left as a further research project for students.

4.7 Height modification of the rectangular locus

We programme rectangular locus shape to vary its height, i.e. the path length, bc, which equals ef, figure 4.12.

Why should the locus height be varied? Well, the locus height is similar to you picking your feet up when the walking terrain is uneven. If there are no trip hazards then the locus height can be shallow but the opposite is true if the terrain is highly irregular or when walking over obstacles. In that case the locus height should be increased.

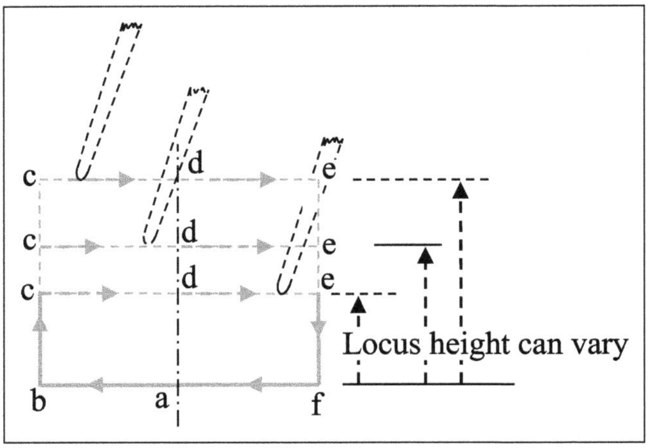

Figure 4.12 Varying the height of the locus.

4.8 Leg tip speed along the locus

In order to create a design rule, it is decided to maintain a constant leg tip speed whilst the leg tip is in the air. In other words, the speed of the leg tip is constant for the sections, bc, cd, de, and ef. Creating this design rule decreases the computing complexity. Similarly, the leg tip speed is kept constant whilst the leg tip is in contact with the ground because this is what determines the robot body locomotion speed. The two speeds of the leg tip whilst in the air and on the ground are constant, (but of different values), for a given cycle through the locus but the speeds can be changed if necessary. For example, let us assume the following maximum dimensions for the locus, ab = bc = cd = de = ef = fa = 40mm, figure 4.13.

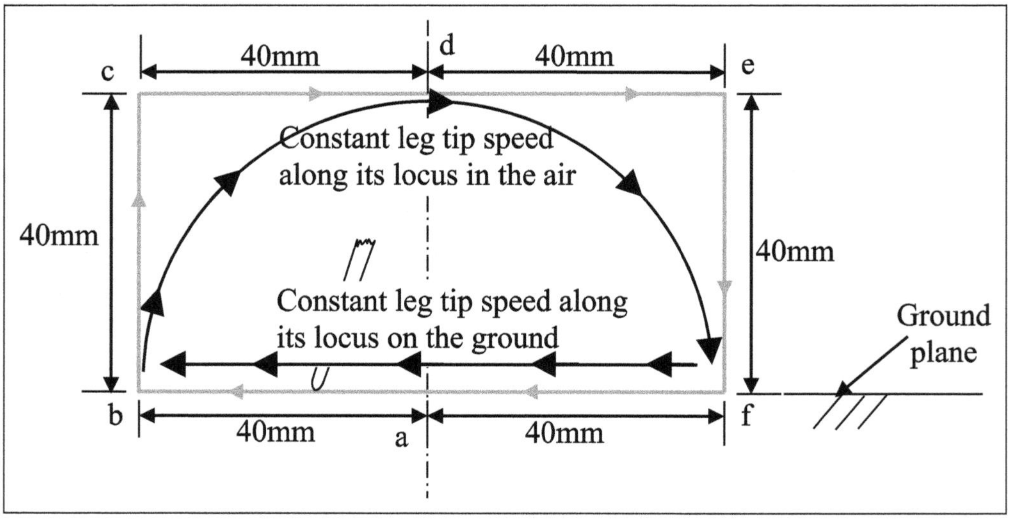

Figure 4.13 Leg tip speed along its locus. Note that the speed is higher in the air and lower on the ground and that the locus size is at its maximum.

Assume that we want a 1second cycle time for the locus. In other words the time taken for the leg tip to traverse the locus through points a, b, c, d, e, f, and back to, a, is 1 second. Assume that the robot is using the double tripod gait, which means that the leg tip time on the ground is equal to the leg tip time in the air.

Thus the leg tip spends 0.5 second on the ground and 0.5 second in the air.
From the diagram above we can see that the leg tip distance travelled in one cycle is:

1. 80mm when the leg tip is in contact with the ground, (fa+ab = 40mm+40mm)
2. 160mm when the leg tip is in the air, (bc+cd+de+ef =40mm+40mm+40mm+40mm)

..and as already stated:

1. The speed of the leg tip whilst in contact with the ground remains constant.
2. and the speed of the leg tip when in the air also remains constant (but different to the ground speed).

Hence the leg tip speeds are:

1. 80mm/0.5sec = 160mm/sec when the leg tip is in contact with the ground.
2. 160mm/0.5sec = 320mm/sec when the leg tip is in the air.

So for this case, figure 4.13, the leg tip speed doubles when leaving contact with the ground. We now formalise the leg tip speed as a function of the shape of the locus by giving variable names to its height and the amplitude of its step size, where the latter is given by the length of ab which is equal to fa. These values are known as the step height and step amplitude.

4.9 Computing leg tip speed as a function of step height and step amplitude

We make the locus height and amplitude to be programmable and these quantities are given variable names, 'zag' and 'amp' respectively, figure 4.14. A vertical axis related to the leg tip locus is denoted the Z-axis that runs through the centre of the locus. The locus height, which determines the step height, is the variable called 'zag', which is a synonym for 'z air-to-ground'. The step amplitude, denoted as the variable, amp, represents half of the step size since the leg tip will remain on the ground and travel all the way from f through a to b thus completing a distance of 2 × amp.

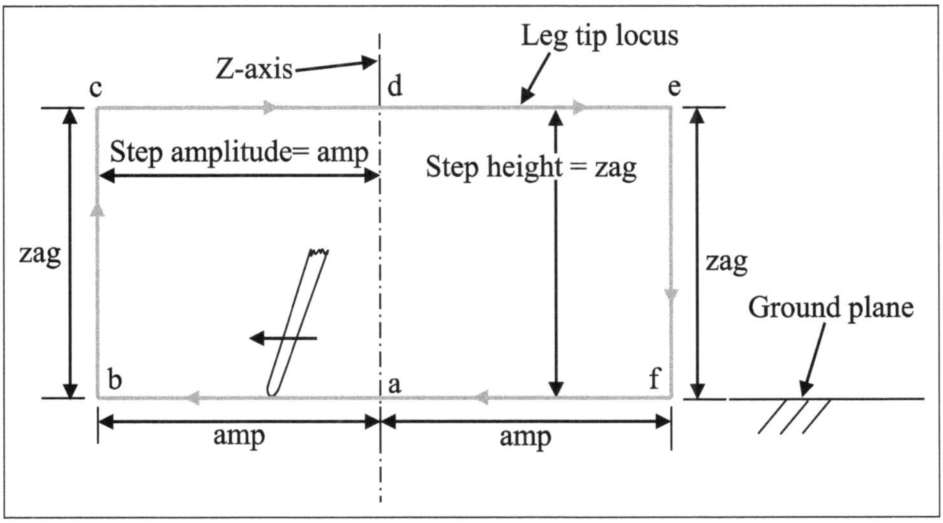

Figure 4.14 Leg tip locus showing the zag and amp variables that represent step height and step amplitude respectively.

Now let us compute leg tip speed as a function of step height, zag, and step amplitude, amp.

Assume that zag and amp are given variables.

Figure 4.14 shows the z-axis, which is vertical, at the centre of the locus. The distances, ab, and, af, represent the amplitude of the walking step. We set,

ab = af = amp

The height of the locus, zag, ("z air-to-ground"), represents the walking step height. We set,

bc = fe = zag

Let the cycle time of the locus = T

Let leg tip speed when in contact with the ground = V_G

Let leg tip speed when in the air = V_A

Hence,

$$V_G = \frac{2 \times \text{amp}}{T/2} \text{ hence, } V_G = \frac{4 \times \text{amp}}{T} \qquad \text{Eq. 4.1}$$

$$V_A = \frac{2 \times (\text{amp}+\text{zag})}{T/2} \text{ hence, } V_A = \frac{4 \times (\text{amp}+\text{zag})}{T} \qquad \text{Eq. 4.2}$$

The idea is to program a leg tip locus based on given values for, amp, zag, and, T. The equations above, eq 4.1 and 4.2 will tell us the speed we need to drive the leg tip but we need to know when to change directions at nodes, b, c, e, and, f. For example, assume we programme the locus to look like the following sequence in figure 4.15 starting from node a.

We now design a leg tip locus specification that will enable our hexapod to carry out many different and flexible behaviour patterns.

4.10 Leg tip locus design specification

1. Leg tip is to trace CW or CCW around a rectangular locus with variable values of, amp, zag, and, T.
2. The values of, amp, zag, and T, can be changed at any point on the locus.
3. The leg tip can be stopped at any point on its locus
4. The direction of the leg tip can be reversed at any point on the locus.
5. The speed of the leg tip can be increased or decreased at any point on the locus
6. The entire locus can be biased up or down by a programmable amount using a variable named, zbias, as shown in figure 4.16. This ability allows the hexapod robot, for example, to walk along a sloped pavement and to keep its body level.

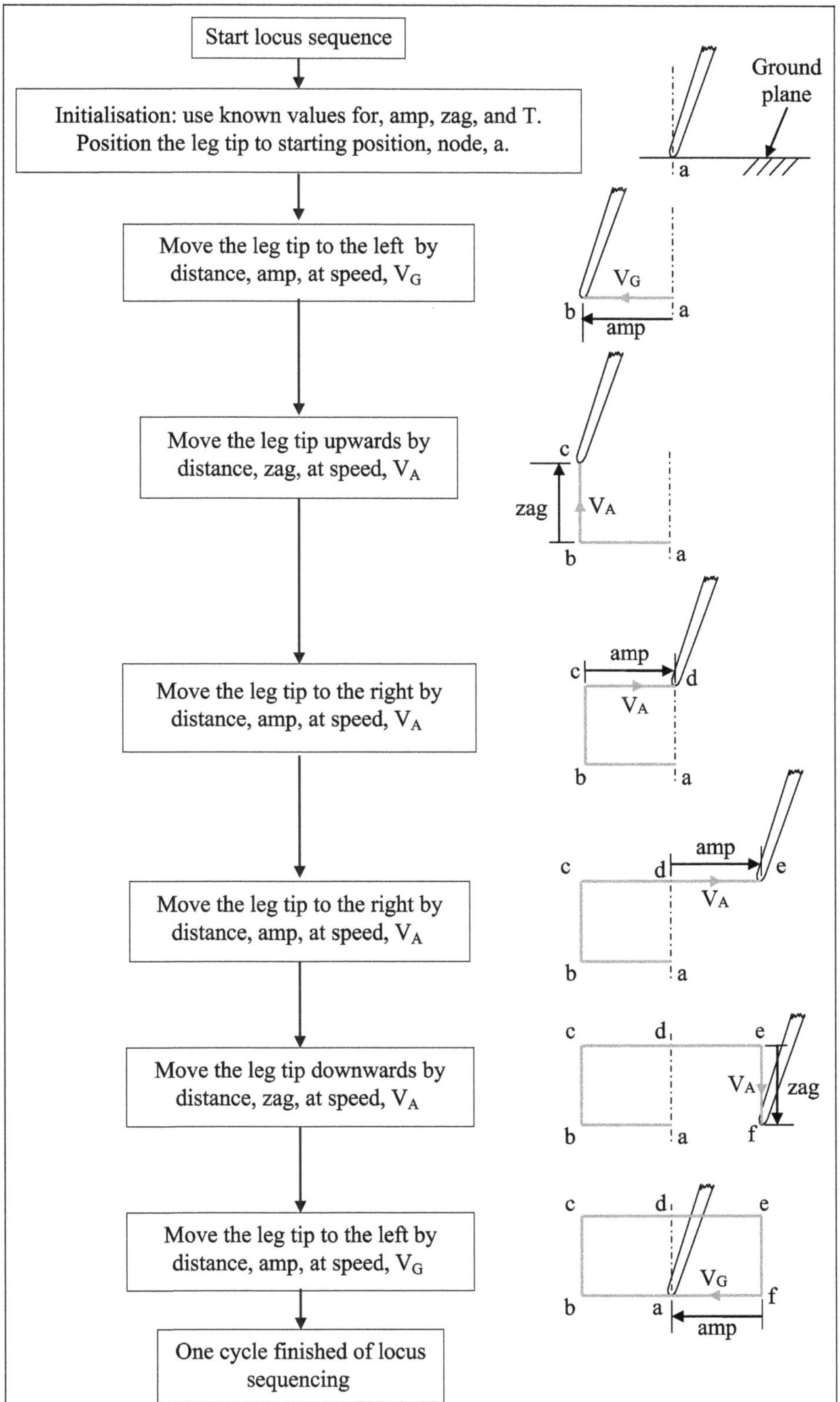

Figure 4.15 Sequence for driving leg tip along its walking locus.

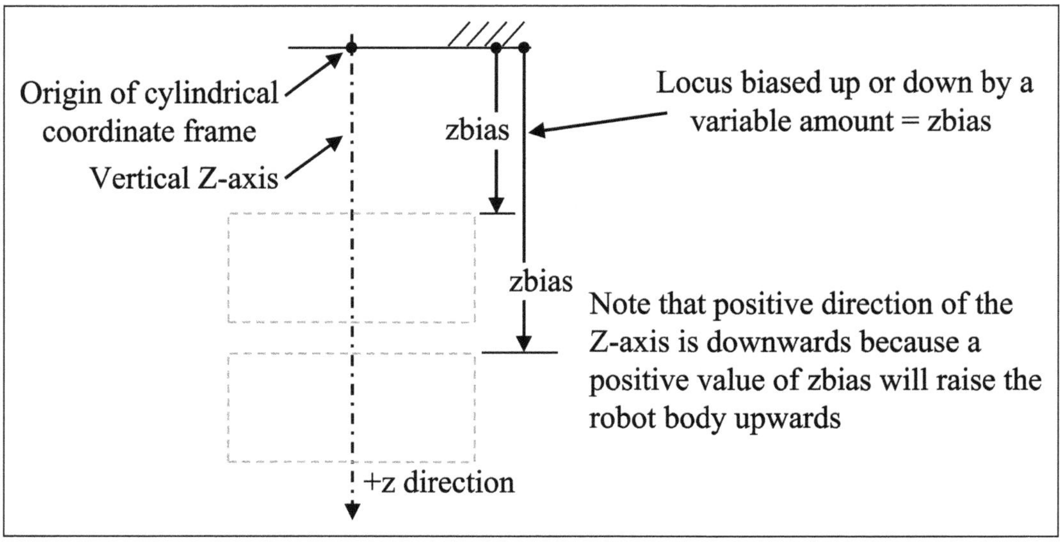

Figure 4.16 Vertical biasing of the leg tip locus.

7. The plane of the locus is to be turned about the Z-axis by an angle θ, named 'plangle', as shown in figure 4.9.

Now that the locus design is specified we now set about planning how to achieve it.

4.11 Achieving the leg tip locus specification via a Digital Locus

The plan is to divide up the locus into 256 displacement increments as in figure 4.17. Thus the leg tip locus becomes a discretised version of the continuous locus as currently described. Such a locus is named the digital locus. Each increment of displacement through the locus is given an integer number value represented by the variable, n, and is an integer that ranges from 0 to 255; which is a total of 256 values. Each value of n represents a discretised position along the leg tip locus. The increments are leg tip displacement values that are not necessarily equal since they are functions of the variables, zag, and amp that cause the locus to heighten and lengthen respectively. The value of, n, represents the non-dimensional position of the leg tip through one cycle of the locus. For example, we could have said that, n, ranged from 0 to 100, in which case, n would have represented the percentage of the leg tip through one locus cycle and n=100 means that the leg tip is back at its starting point, i.e. n=0. Instead of using 100 percentage units we use 256 incremental units. Figure 4.17 shows each of the 256 incremental units as little dots. Note that there are two points along the digital locus called 'breakpoint 1, (bp1), and breakpoint 2, (bp2). These points occur at certain pre-computed values of n and represent a change of direction of the leg tip from travelling vertical to horizontal. The computation of these breakpoint values will be explained shortly.

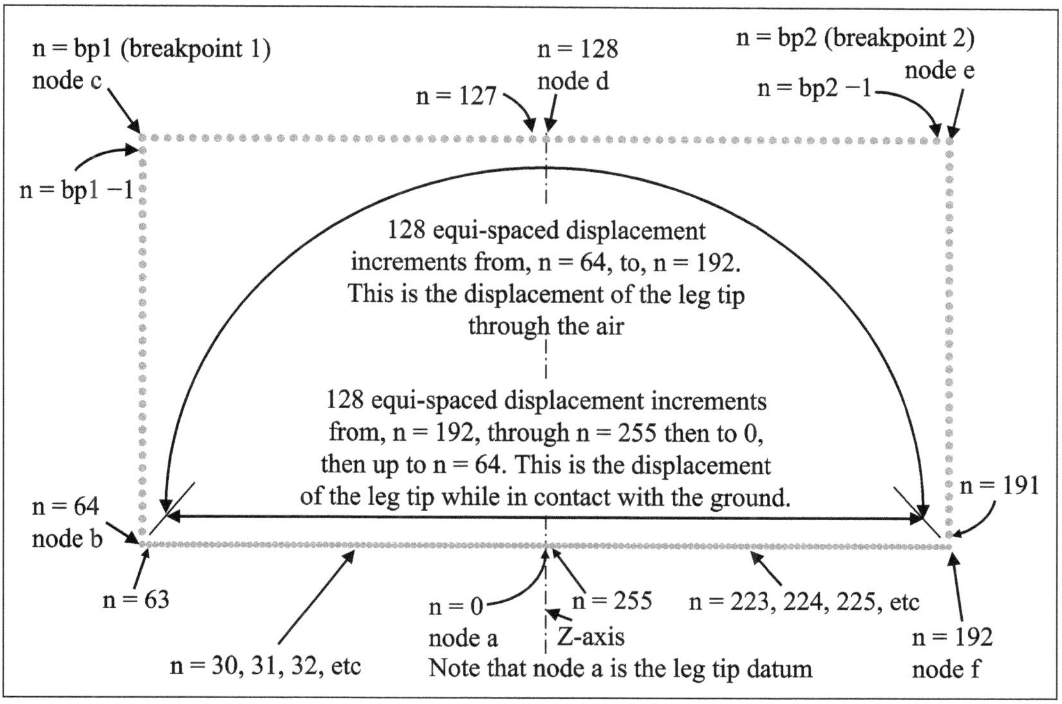

Figure 4.17 The digital locus showing how it is divided into 256 increments.

Figure 4.17 shows that 128 discretised increment units are used for the leg tip in contact with the ground and the remaining 128 units used for the leg tip in the air. As with all digital systems there is a digitisation error and the digital locus is no exception. Here the digitisation error means that the leg tip positioning accuracy is limited to +/− half the distance between the dots. If greater accuracy is required then the locus should be divided up into a greater number of increments, e.g., 2560 incremental units. However it is decided that 256 values of, n, on the digital locus are adequate and serve as a practical engineering solution. In fact 256 values for the whole locus will give a leg positioning accuracy on the ground is 0.4% relative error and 0.3mm absolute error, (½ × 1/128 × 100% = 0.4% and ½ × 80mm/128 = 0.31mm).

An important feature of the normalised locus is that 50% of the digital locus n-values locate the leg tip in contact with the ground and the remaining 50% locate the leg tip in the air. This means that 128, n, values, i.e. 192 to 255 and 0 to 63 represent the leg tip in contact with the ground and, 128, n, values, i.e. 64 to 191 represent the leg tip in the air. Figure 4.18 is helpful to see these numbers in brief.

Note that, due to discretisation there are anomalies in the digital locus, figure 4.18, such as when n=64, the leg tip should be travelling horizontal and be part of the ground section of the locus. Instead, the leg tip is part of the air section of the locus and is travelling vertically. A similar effect occurs at n=192 which is part of the ground section but should be included as part of the air section. The effect of the anomaly is that for a small duration of time two leg tips are in contact with the ground; one leg tip changing from ground to air and the other changing from air to ground. The effect of this changeover will produce a slight jerking walking action and can be overcome by developing an improved, more complicated walking algorithm. Developing such a piece of code is a good piece of research for students.

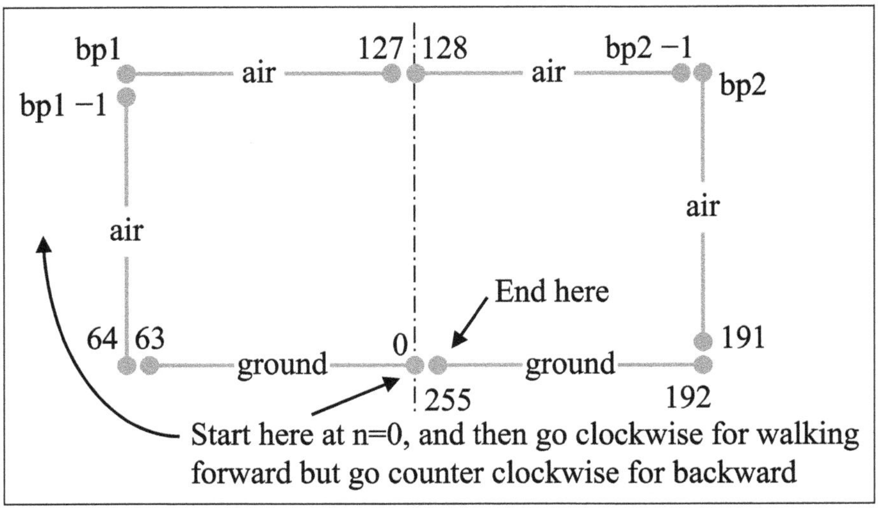

Figure 4.18 n-values along the digital locus. (bp1 and bp2 explained later).

Since, n, is an integer, and because the 3dof uses position feedback controlled servos, (namely, Futaba servos, chapter 2), then as we displace the leg tip through its locus we will see incremental step displacements in the motion of the leg tip. If we choose sufficiently small incremental steps then the leg tip motion satisfies a specified accuracy, looks smooth and is not a problem. The next stage is to describe how we use the n-variable to enable walking.

4.12 Using the n-variable to increment the leg tip through a walking locus

The n-variable is a powerful concept that enables hexapod walking in many modes that will be described later. A unique n-variable is assigned to each of the six legs. We choose an 8-bit integer variable known as a 'byte' for the n-variable. A byte variable is an integer that varies from 0 to 255 and if 1 is added to 255 the variable returns to zero. Another example is if the current value of n is 200 and we add 255 then the new value of n becomes 199. A further example is if the current value of n is 0 and we decrement, (subtract) by 1 then n becomes 255. The smart idea behind the n-variable is that we choose values of, amp and zag, and a starting value for, n, which is usually, but not always, the datum position, n = 0. Then all we need to do is to increment, n to walk forward; decrement, n to walk backwards or keep n constant to stay still. The n-variable is a high level control variable and relies on lower level computation algorithms to compute values of the breakpoints, bp1 and bp2 and ultimately the inverse kinematics of the three servos that control leg tip position.

For example, suppose we are driving our six-legged robot in double-tripod gait with separate, n values, n_{left} and n_{right} for the left and right tripods respectively, (see the red and blue legs in figure 4.7). We drive the left tripod as, n_{left} = n+0, and the right tripod as, n_{right}=n+128, which puts the two tripods in anti-phase with the left tripod in ground contact and the right tripod in the air, figure 4.19. If we increment n by 1 and keep doing this we notice that by the time n_{left} gets to 64, n_{right} gets to 192, which means that the left tripod is about to leave contact with the ground and the right tripod has just made contact with the ground. The graph, figure 4.19 below illustrates the sequencing.

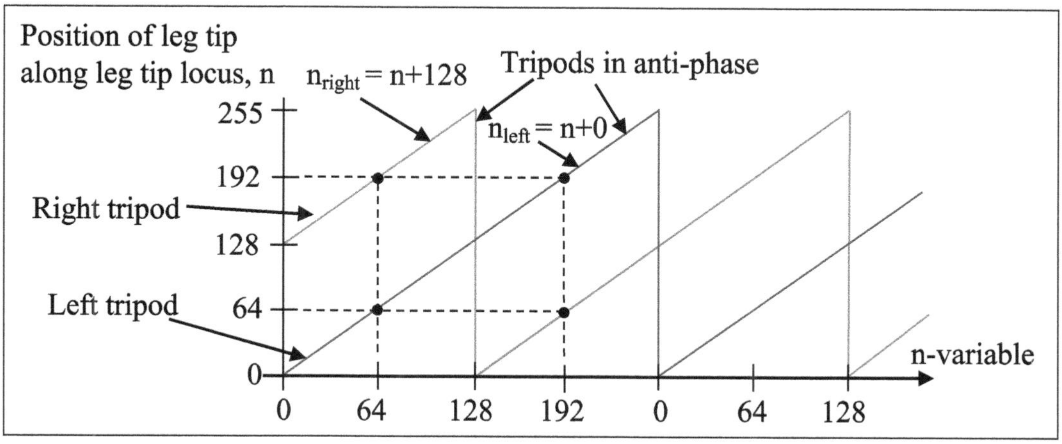

Figure 4.19 Two anti-phase n-value waveforms for the double tripod gait.

By the time we have incremented n_{left} by a further 128 then, $n_{left} = 64 + 128 = 192$, and $n_{right} = 192 + 128 = 320 = 256 + 64 = 0 + 64$ (0 because n = 256 means that the locus has returned to its starting point, i.e. n = 0). At this point, $n_{left} = 192$ and $n_{right} = 64$, and the tripods are about to exchange duties either from ground to air or from air to ground. Just by incrementing or decrementing the n-variable applied to each leg tip we can drive each leg independently along its digital locus.

Thus the concept of the n-value and the digital locus simplifies control of a walking leg robot by automatically placing the leg tips in the right place at the right time. However before we can sit back and use n-values to control the leg positions, we need to explain and calculate the breakpoints, bp1, and, bp2.

4.13 The breakpoints of the digital locus, bp1 and bp2

There are two points on the digital locus that are called, bp1, and bp2, which are abbreviations for, "breakpoint 1", and "breakpoint 2", respectively. The breakpoints are values of, n, on the digital locus that correspond with nodes c and e on the leg tip locus, figure 4.17. These nodes represent a 90° change of direction of the leg tip but no change in speed. The n-values for the other nodes are as follows:

1. Node a, corresponds to, n = 0, also known as the leg tip datum
2. Node b, corresponds to, n = 64, the point at which the leg tip is about to lift off the ground going CW or just put down on the ground going CCW
3. Node d, corresponds to, n = 128, which is exactly half way through the leg tip locus.
4. Node f, corresponds to, n = 192, the point at which the leg tip is just about to put down on the ground going CW or lift off the ground going CCW.

It is important to realise that the 128 distance increments of the locus from n = 64 to n = 192, inclusive, are of equal length. The idea is to get the leg tip to travel at a constant speed in the air when the n-values are incremented or decremented by a constant frequency interrupt timer which will be set to 50Hz, i.e. 20ms clock period. However, the leg tip air distance is different to the ground distance as shown in figure 4.20 below. Now let's do the calculations for the n-variable at the breakpoints which determine the change of locus directions. Referring to figure 4.20, the distance in the air, going CW, from, n = 64, to, n = 192, is equal to: 2 × (zag + amp) (mm), which corresponds to 128 n-values.

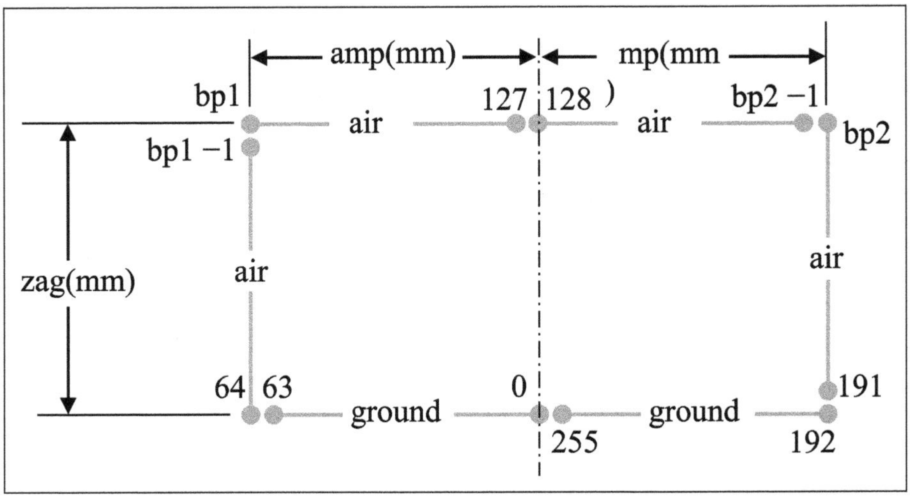

Figure 4.20 Separating the digital locus using breakpoints, bp1 and bp2.

Hence, each distance increment is equal to:

$(2 \times (zag + amp)/128)$ mm/n-value

Thus, $bp1 = 64 + zag/(2 \times (zag + amp)/128)$

Simplifying,

$$bp1 = 64 \times 1 + \left(\frac{zag}{zag + amp)}\right)$$

Or, more suitable for computation,

$$bp1 = 64 + 1 + \left(\frac{64 \times zag}{zag + amp)}\right) \quad \text{Eq. 4.3}$$

Don't forget the values for, bp1, and, bp2, are to be worked out to the nearest integer because the n-values on the digital locus are all integers.

A similar analysis for bp2gives,

$$bp2 = 64 \times \left(\frac{1 + zag + (2 \times amp)}{zag + amp)}\right)$$

Or, more suitable for computation,

$$bp2 = 192 - 64 \times \left(\frac{zag}{zag + amp}\right) \quad \text{Eq. 4.4}$$

Here are a few examples:

Assume the following values;

amp = 40mm

zag = 40mm

Using equations 4.3 and 4.4 above, we obtain; bp1 = 96, and, bp2 = 160

So the digital locus n-values are like this below, figure 4.21.

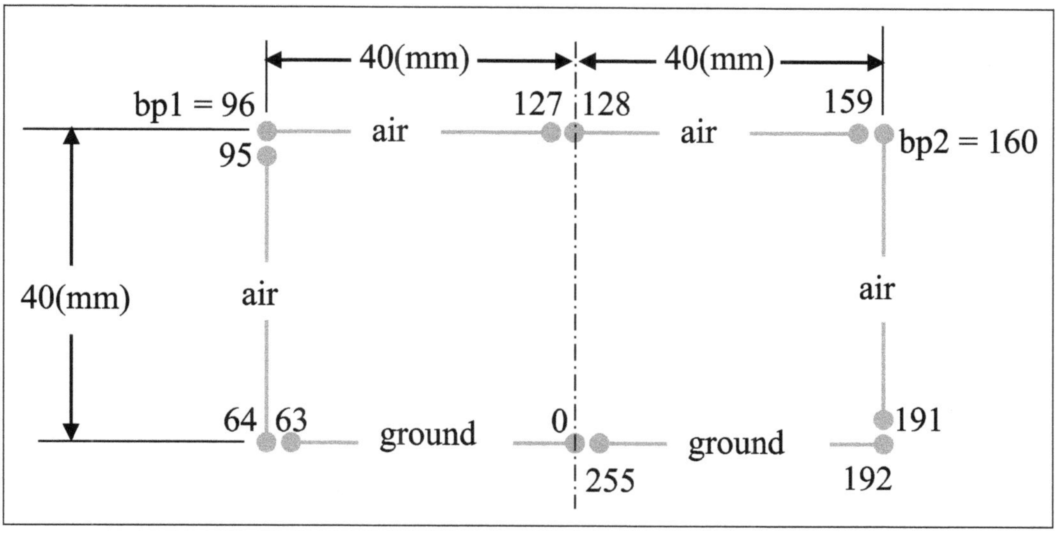

Figure 4.21 Digital locus for: amp=40mm zag=40mm.

Now try;
amp = 20mm
zag = 10mm
Eqs 4.3 and 4.4; bp1 = 85.3 = 85 (nearest integer) and, bp2 = 170.7 = 171 (nearest integer)
So the digital locus n-values are like this.

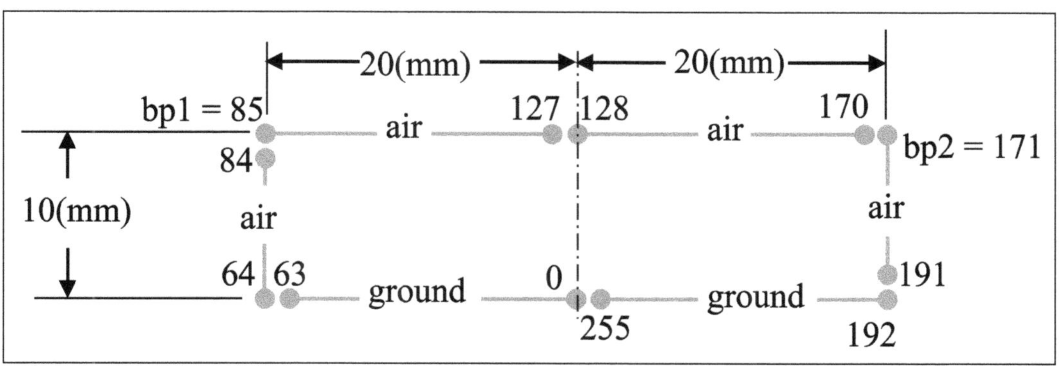

Figure 4.22 Digital locus for: amp=20mm zag=10mm

Now try a special case with zero amplitude;
amp = 0mm
zag = 17mm
Eqs 4.3 and 4.4; bp1 = 128 and, bp2 = 128
So the digital locus n-values are like this below, figure 4.23.

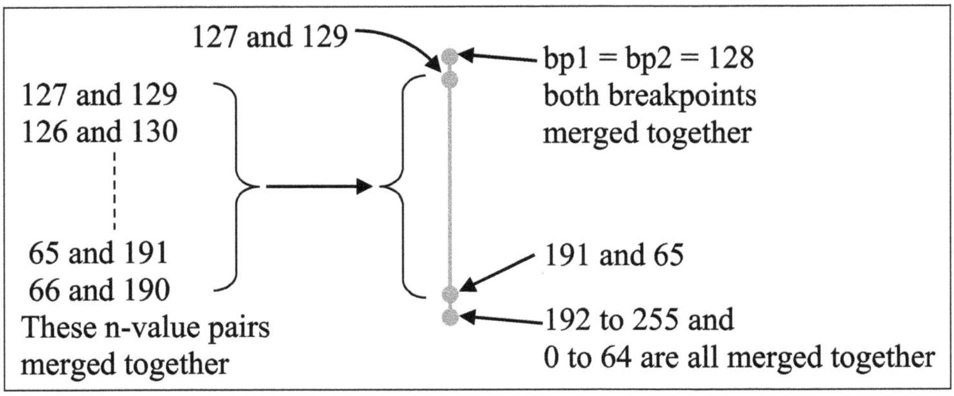

Figure 4.23 Digital locus for: amp=0mm zag=17mm.

Now try another special case with zero zag;
amp = 31mm
zag = 0mm
Using equations 4.3 and 4.4 above, we obtain; bp1 = 64 and, bp2 = 192
So the digital locus n-values are like this below, figure 4.24

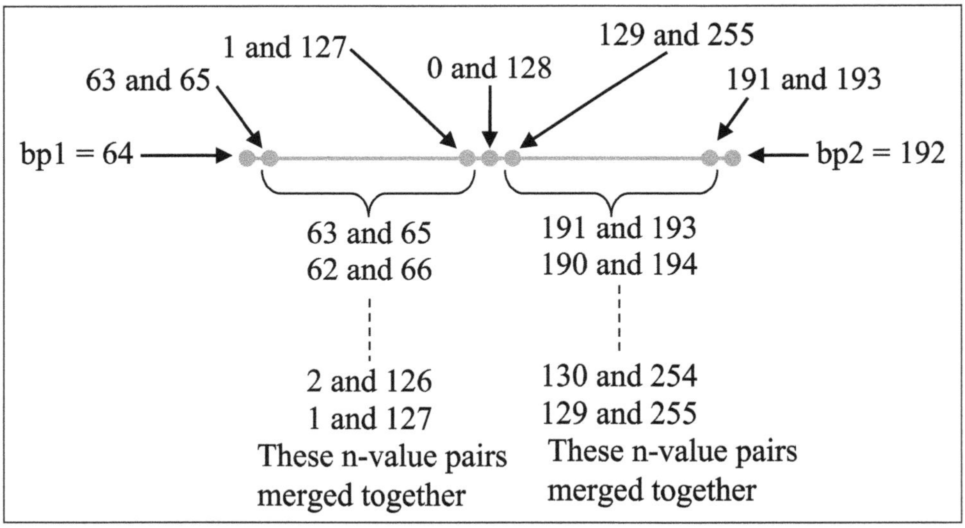

Figure 4.24 Digital locus for: amp=31mm zag=0mm.

4.14 Distance travelled by leg tip along the digital locus

Looking back at figure 4.9 showing the six-legged robot with each leg tip in a cylindrical coordinate frame, we recall that the leg tip locus is a plane that is coincident with the Z-axis of the cylindrical coordinate frame. This plane can be seen more clearly with an analogy. Imagine cutting a cake in two equal halves through the centre at an angle θ clockwise from an angle datum and exposing an inside plane as in figure 4.25. The exposed plane is the plane of the leg tip locus and points in this plane are described with a Cartesian coordinate frame using R, Z coordinates.

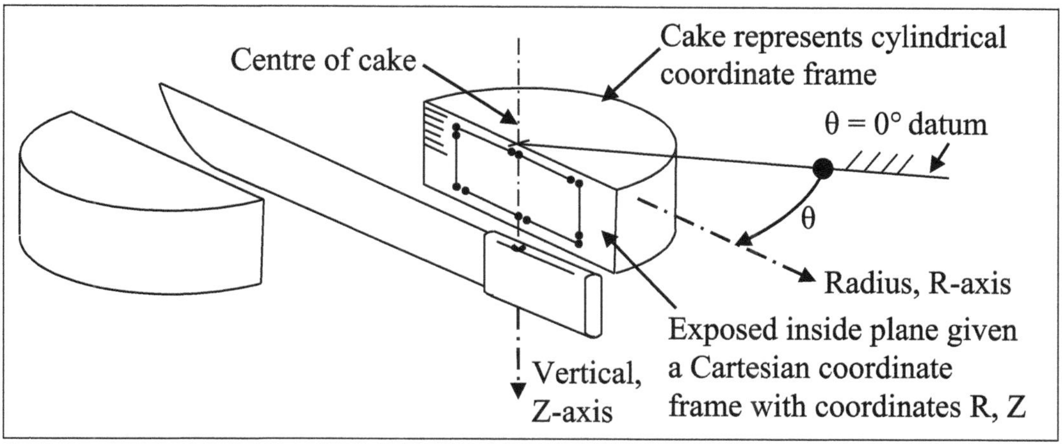

Figure 4.25 Leg tip locus plane is analogous to a plane exposed when cutting a cake into two equal halves through the Z-axis at an angle, θ clockwise from an angle datum.

The following diagram, figure 4.26, is a scientific representation of the cut cake. The diagram shows the digital locus and its four geometrical variables, i.e. θ, zag, amp and zbias. We now split digital locus into two halves. The first half is in a plane at angle θ and the second half, which is a coplanar continuation of the first half, is in the plane θ + 180°. We split the digital locus like this for computational convenience.

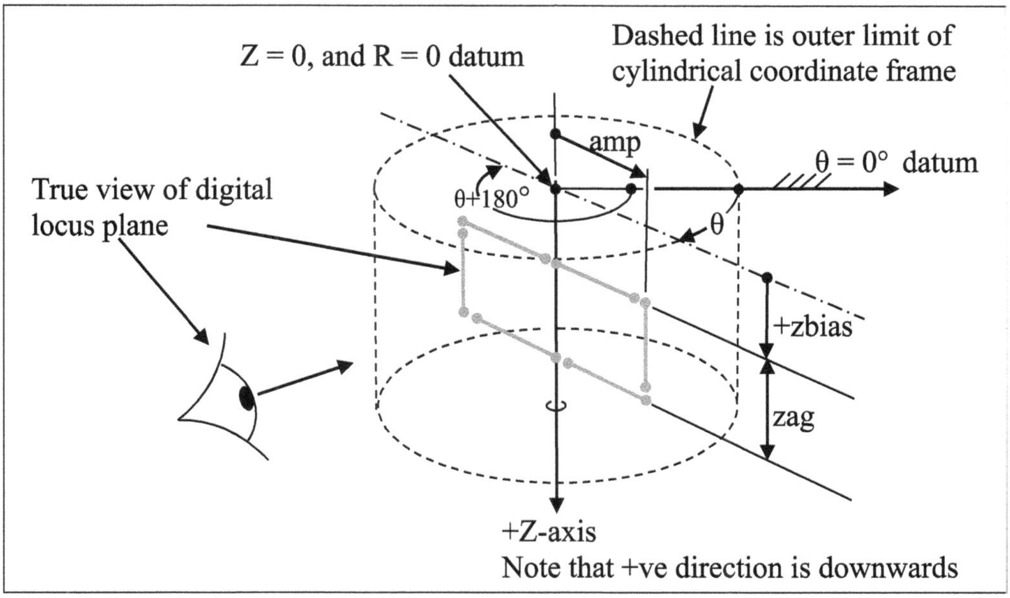

Figure 4.26 Scientific representation of the 'cut cake' indication the true view of the digital locus and its geometrical variables, θ, zag, amp and zbias.

The first half of the locus at angle θ is named as being located in the right plane of the complete digital locus and the second half of the locus at angle θ +180° is named as being in the left plane of the digital locus, figure 4.27 below.

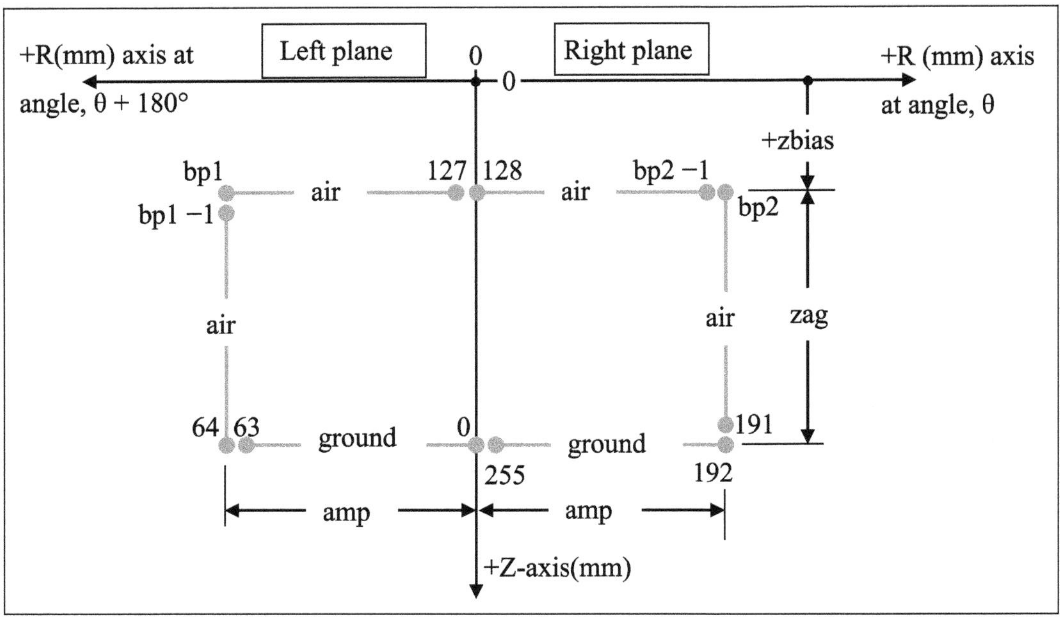

Figure 4.27 Left and right planes of the digital locus at angle θ and angle θ+180° in the leg tip cylindrical coordinate frame.

The digital locus shown in a Cartesian coordinate frame, figure 4.27 above, has two coordinates for R, where R represents the radius of the leg tip position, and Z, represents the vertical position of the leg tip. Note however, that there are no negative values of R. Instead, negative values of R are qualified as lying in the $\theta + 180^0$, side of the cylindrical coordinate frame. Similarly, positive values of R lie in the θ side. A much simpler way of expressing whether R is positive or negative is to specify the right plane or the left plane respectively as shown in figure 4.27. Note that the positive value of Z is in the downward direction because driving the leg tip downwards will raise the robot which is intuitively associated with the positive Z direction. It is decided for convenience that there should be no negative values of Z since Z=0 will be considered a lower limit for the Z direction.

Referring to the table in figure 4.28 below, we arrange the highest part of the leg tip locus, i.e. n-values from, bp1, through 128, to bp2, to be equal to the value of zbias. Hence if zbias is zero then the highest part of the locus will coincide with the R axis, i.e. at Z = 0.

> We are now in an important position which is to compute the R, Z, position of the leg tip in the Cartesian coordinate frame in actual units of millimetres as a function of four given variables, which are, n, amp, zag, and zbias

4.15 Computation of, R, and, Z, values of the leg tip on the digital locus

Note that values of R and Z are computed in units of millimetres. The digital locus is divided up into 6 segments which are, figure 4.27a;

n= 0 to 63	segment 1
n= 64 to bp1-1	segment 2
n=bp1 to 127	segment 3
n=128 to bp2-1	segment 4
n=bp2 to 191	segment 5
n=192 to 255	segment 6

Figure 4.28 Table of digital locus segments.

Values of R and Z are computed separately for each segment as now shown.

Segment 1, n = 0 to n = 63 ← Note that segment 1 just stops short of reaching, n = 64

Segment 1, figure 4.29, is located in the left plane and on a line in the left plane where, Z = zbias + zag. For convenience we will call the R values in this segment, $R_{Leftplane}$. When, n = 64, the leg tip position, R = amp(mm). Don't forget, though, that the point at n = 64 does not belong to segment 1. Instead, it belongs to segment 2.

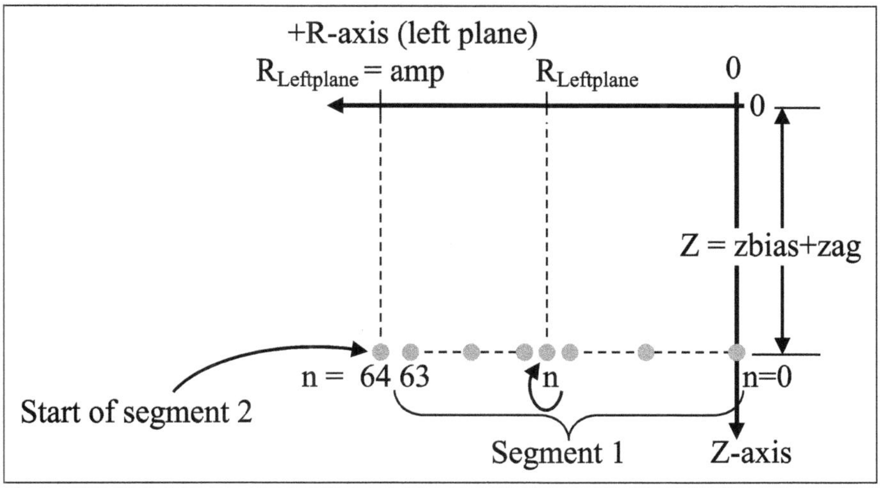

Figure 4.29 Detail of the digital locus at segment 1.

The digital locus divides up segment 1, which is of length amp(mm), into 64 equal distance increments. Hence the position of the leg tip in the radius, $R_{Leftplane}$, direction is proportional to the value of, n, figure 4.30.

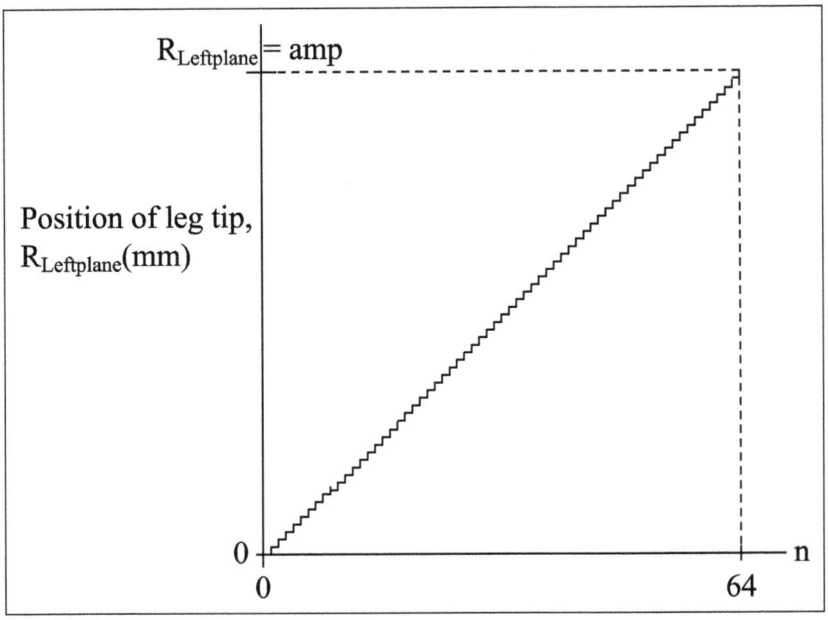

Figure 4.30 Detail of segment 1 in the left plane showing leg tip radius, $R_{Leftplane}$ is proportional to variable n.

The coordinates of the leg tip in segment 1 of the digital locus are given as follows.

$R_{Leftplane}\,(mm) = \dfrac{amp(mm) \times n}{64}$	$Z\,(mm) = zbias(mm) + zag(mm)$

Eqs. 4.5

Check equation, $n = 0,\ R_{Leftplane} = 0(mm)$ ✓ $n = 64,\ R_{Leftplane} = amp(mm)$ ✓	Z stays constant in segment 1

Referring to the above figure 4.27 we can work out equations for the remaining segments as follows.

Segment 2, n = 64 to n = bp1-1 ← Note that segment 2 just stops short of reaching n = bp1

$R_{Leftplane}\,(mm) = amp(mm)$	$Z\,(mm) = zbias(mm) + zag(mm) - \left(\dfrac{zag(mm)}{(bp1-64)} \times (n-64)\right)$

Eqs. 4.6

RLeftplane stays constant in segment 2	Check equation, $n = 64,\ Z = zbias(mm) + zag(mm)$ ✓ $n = bp1,\ Z = zbias(mm)$ ✓

Segment 3, n = bp1 to n = 127

$R_{Leftplane}(mm) = amp(mm) - \left(\frac{amp(mm)}{128-bp1}\times(n-bp1)\right)$	$Z(mm) = zbias(mm)$	Eqs. 4.7

Check equation, n = bp1, $R_{Leftplane}$ = amp(mm) ✓ n = 128, $R_{Leftplane}$ = 0(mm) ✓	Z stays constant in segment 2

Segment 4, n = 128 to n = bp2−1 This segment moves into the right plane.

$R_{Rightplane}(mm) = \frac{amp(mm)}{(bp2-128)} \times (n-128)$	$Z(mm) = zbias(mm)$	Eqs. 4.8

Check equation, n = bp1, $R_{Leftplane}$ = amp(mm) ✓ n = 128, $R_{Leftplane}$ = 0(mm) ✓	Z stays constant in segment 2

Segment 5, n =bp2, to, n = 191 ← Note that segment 5 just stops short of reaching n = 192

$R_{Rightplane}(mm) = amp(mm)$	$Z(mm)=zbias(mm)+\left(\frac{zag(mm)}{(192-bp2)}\times(n-bp2)\right)$	Eqs. 4.9

$R_{Righttplane}$ stays constant in segment 5	Check equation, n = 64, Z = zbias(mm) + zag(mm) ✓ n = bp1, Z = zbias(mm) ✓

Segment 6, n =192 to n = 255

$R_{Rightplane}(mm)=amp(mm)-\left(\frac{amp(mm)}{64}\times(n-192)\right)$	$Z(mm)=zbias(mm)+zag(mm)$	Eqs. 4.10

Check equation, $n = bp1, R_{Leftplane} = amp(mm)$ ✓ $n = 128, R_{Leftplane} = 0(mm)$ ✓	Z stays constant in segment 6

Ok so that's the digital locus sorted out with six equations, eqs. 4.5 to eqs.4.10 inclusive, in terms of positioning the leg tip in terms of R and Z coordinates in the left or right planes given the variables of n, amp, zag, and zbias. However, now we need to introduce a clock that will be used to shift incrementally the leg tip through the digital locus by integer changes in variable n so we can control leg tip velocity. Let's spend some time explaining the clock.

4.16 The clock; adding time to the control of leg tip position

The clock, in digital systems, doesn't give you the time of the day. (It can do if need be, though). Instead, it is a regular series of impulses that is used to regulate the actions of the robot leg to give the correct leg tip velocity or to put the leg tip in the right place at the right time. In our system the clock is representative of most digital systems and produces short 8μs long TTL pulses at 20msec intervals (= 50Hz pulse repetition frequency), figure 4.31.

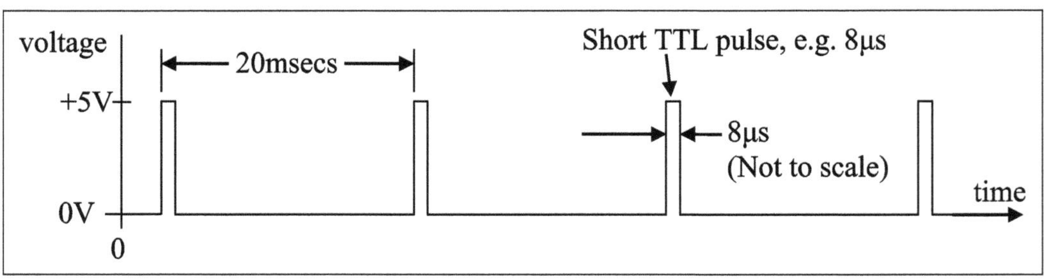

Figure 4.31 The TTL clock pulses used to increment the leg tip through the digital locus.

(Note that TTL means 'transistor-transistor-logic where 0V is a logic 0 and +5Vis a logic 1). A 50Hz clock is used because the Futaba servomechanisms, (see chapter 2), that will be used to actuate the leg mechanism require a 50Hz signal update rate. A 50Hz clock frequency is quite popular. It has its roots in the maximum speed that the human eye and brain can process images, so television screens run at this speed (a bit faster in the USA where televisions run at 60Hz) to give the effect of apparently smooth motion even though the screen is giving a flicking action.

Returning to our clock pulse; we use the clock pulse to instigate one of the following actions:

1. At each clock pulse, move the leg tip by one or more integer displacement increments in the CW direction or the CCW direction, e.g. n = n + 1 will move the leg tip CW around the digital locus, or, n = n – 2, will move the leg tip CCW at twice the speed.
2. Jump the leg tip to a new position on the locus, e.g. suppose the present value of, n, is 36. At the next clock pulse, n = 112.
3. Don't move the leg tip. Leave it where it is, e.g. suppose the present value of, n, is 73. At the next clock pulse, n = 73. Note that even if we want no change in the position of the leg tip, we still must send the same value to the servomechanism. If you don't do this then the servomechanism will become limp and drift away under load. Some new Futaba servomechanisms are not like this; they do not have to be regularly updated.

4.17 Using microcomputers to step the digital locus through a sequence

Let's give an example of how we can obtain the two anti-phase sawtooth n-value waveforms as shown in figure 4.19 to produce a hexapod double tripod gait. The horizontal axis uses units of n so if clock pulses replace these numbers then the horizontal axis becomes a time axis. First we show how to create a 50Hz, (20ms period) clock then we show how to create the anti-phase waveforms.

4.18 The 50Hz, 20ms clock

Figure 4.32 shows two Basic Stamp Microcontrollers, BS2SX#1 and BS2SX#2, connected together to form an elementary system that will be used to create the two anti-phase sawtooth n-value waveforms. The BS2SX#1 is responsible for creating the clock pulses whereas BS2SX#2 creates the anti-phase n-values.

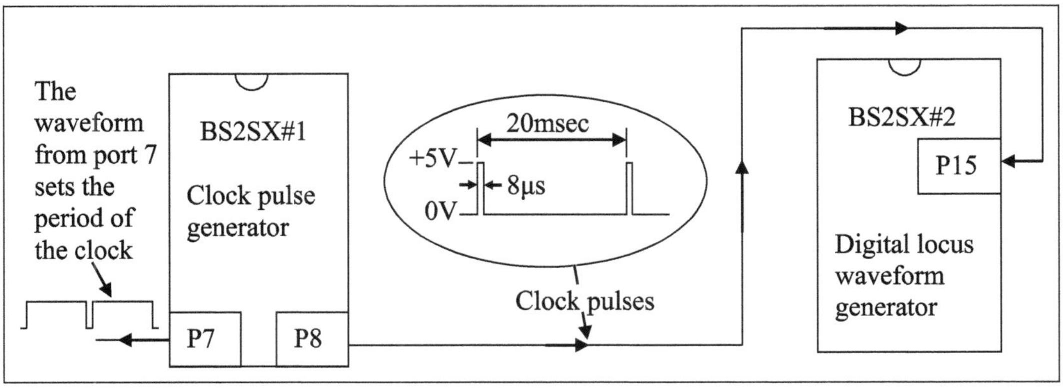

Figure 4.32 A BS2SX microcomputer programmed as a clock that steps a second BS2SX through the digital locus.

The BS2SX#1 is used to produce the repeating clock pulses which are outputted from port 8 and received by port 15 of BS2SX#2 which acts as the digital locus generator. These clock pulses act as a polled interrupt for the BS2SX#2. Figure 4.33 is a PBasic program for the BS2SX#1 microcomputer that will produce the 8μs repeating clock pulses of 20msec period.

```
'BS2SX#1, clock pulse generator program
rtcomp     con    2510                  'realtime delay compensation
again:     pulsout 8,10                 'send out one 8μs pulse from port 8
           pulsout 7,25000– rtcomp      'delay time to give 20msec period
           goto again
```

Figure 4.33 Clockpulse generator programme.

The BS2SX microcomputer utilises its own internal clock that runs at 1.25MHz, which thus has a period of 0.8μsec. Be careful with this internal clock because it has a significant tolerance band which should be calibrated with an oscilloscope if accurate timing is required. The internal clock is used to create the "pulsout" instruction. The program line, "pulsout 8,10", produces a pulse of length, 10 × 0.8μsec = 8μs that is output from port 8. You can see one of these pulses in the port 8 output waveform, figure 4.34. Note that the "goto again" instruction produces a continuous stream of pulses.

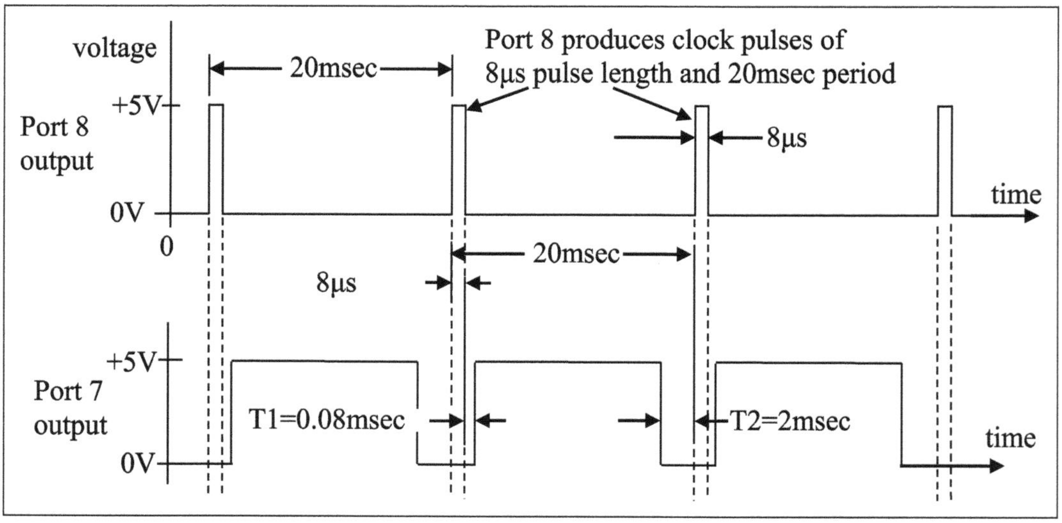

Figure 4.34 Waveforms of the BS2SX microcomputer that is programmed as a real-time clock.

But what is the purpose of port 7 that is connected to nothing and also outputs a pulse? The answer is that port 7 acts as a dummy delay signal that sets the 20ms period of the port 8 pulses. It should be realised when designing real-time digital systems that there is always a time delay in sequential programming line-to-line time delays, figure 4.35. If these values are not published by the manufacturer then an oscilloscope can be used to measure them. Parallax, the manufacturers of the Basic Stamp microcomputer, does not publish these delay times so an oscilloscope was used to measure time delays. The oscilloscope showed two delays, T1 and T2, equal to 0.08ms and 2.00ms respectively, as shown in figure 4.35.

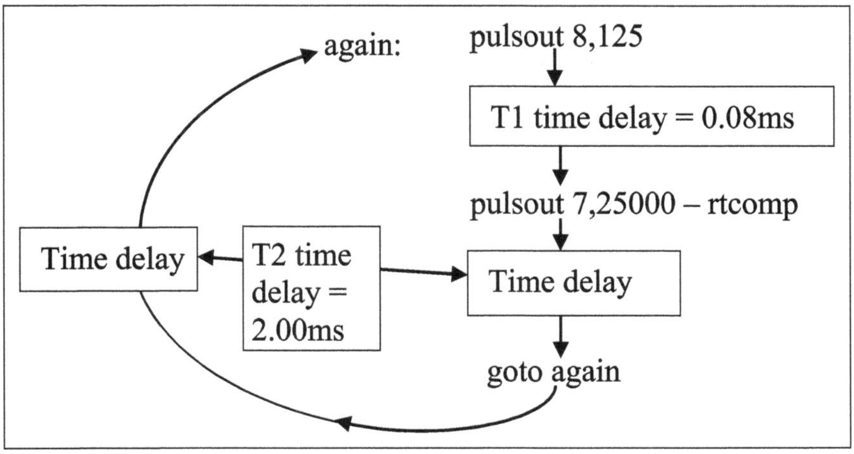

Figure 4.35 Sequential programme line-to-line time delays.

The time, T1, is the delay between the end of program line, "pulsout 8,10" and the start of, "pulsout 7,25000– rtcomp".

The time, T2, is the delay between the end of program line, "pulsout 7,25000 – 125 – rtcomp", and the start of, "goto again", plus the delay between, "goto again" and, "pulsout 8,125".

Since a periodic time of 20msec is required by the clock then the delay between clock pulses to be produced by port 7should be,

Period – (pulse length + T1 + T2) = period – rtcomp

Where, rtcomp = real-time compensation for the delay.

In this case, rtcomp = (0.008 + 0.08 + 2.00)ms = 2.008ms.

The BS2SX microcomputer produces a time delay of 2.008msec by utilising its internal clock of 0.8μsec.

Hence we set, rtcomp = 2.008msec/0.8 μsec = 2510.

A 20msec period utilising the internal clock is produced by 20msec/0.8 μsec = 25000 internal clock counts.

Hence the required instruction to give a 20msec clock period is,

pulsout 7,25000 – 2510

Alternatively, a more convenient method is to write the delay like this using a realtime compensation constant, 'rtcomp'

```
rtcomp          con    2510              'realtime delay compensation
                pulsout 7,25000 - rtcomp
```

Ok, that explains the BS2SX#1 clock pulse generator so let's move on to the BS2SX#2 which is the digital locus waveform generator.

4.19 The two anti-phase sawtooth waveforms to produce the double tripod gait

For the time being we will just write a program that will step the leg tip through the digital locus such that, n = 0, 1, 2, 3.....253, 254, 255, 0, 1, 2, 3...and keep going, figure 4.36. This will create a single saw-tooth waveform for the value of, n, and will drive the leg tip continuously around the digital locus circuit.

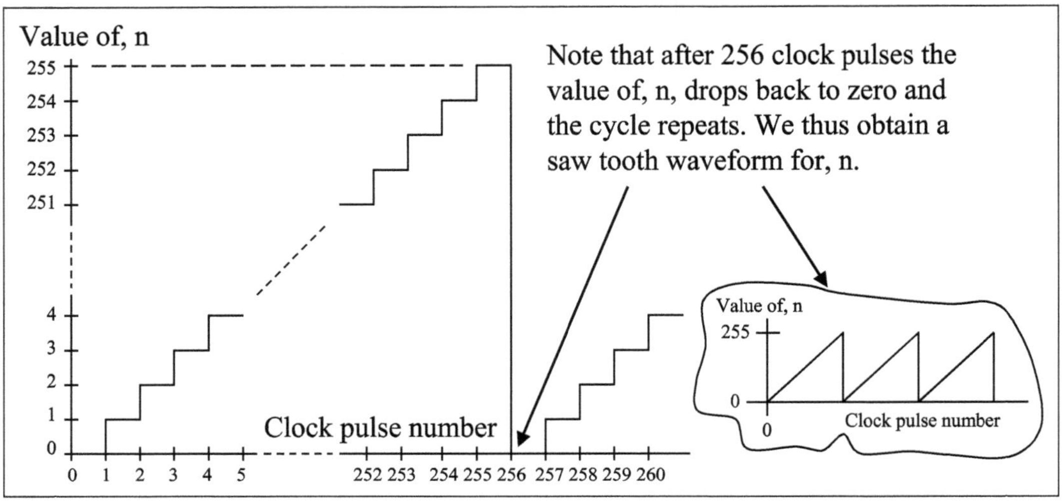

Figure 4.36 Clocking a 256-bit n-variable byte through a cycle.

The following program, figure 4.37, will do this.

```
'BS2SX#2, program to create saw-tooth waveform for n
n       var    byte
j       var    nib
speed   con    1
        n = 0          'set starting value of, n, equal to zero
again:  pulsin 15,1, j
        n = n +speed
        goto again
```

Figure 4.37 Sawtooth waveform for the n-variable.

The "pulsin 15,1, j" instruction means "wait for a positive-going pulse on port 15, (the 8μs pulse clock signal from BS2SX#1) and start measuring the pulse length"; the "j" means store the measured length of the pulse in variable j. However, the value of, j, is not required in this application). As soon as the pulse measurement has finished the programme moves to the next line which is 'n=n+speed', i.e. increment, n, by "speed", which equals 1, so, n = 0+1=1. The program then goes to "again", waits until the next clock pulse appears, which will be 20msec later, then adds another 1 to, n, so, n = 2. You can think of the program as a pulse counter with the count value being stored in the byte variable, n. Only trouble is that it can only count up to 255, which is a deliberate ploy to match the digital locus. Note that when, n, gets to 255, the next value of, n, is 0, not 256, because, n, has been specified as a byte, which has a maximum value of 255. If you add 1 to 255 it will reset back to zero. You can see this more clearly by seeing what happens to an 8-bit binary number at a stage when it is equal to 254 as follows in figure 4.38.

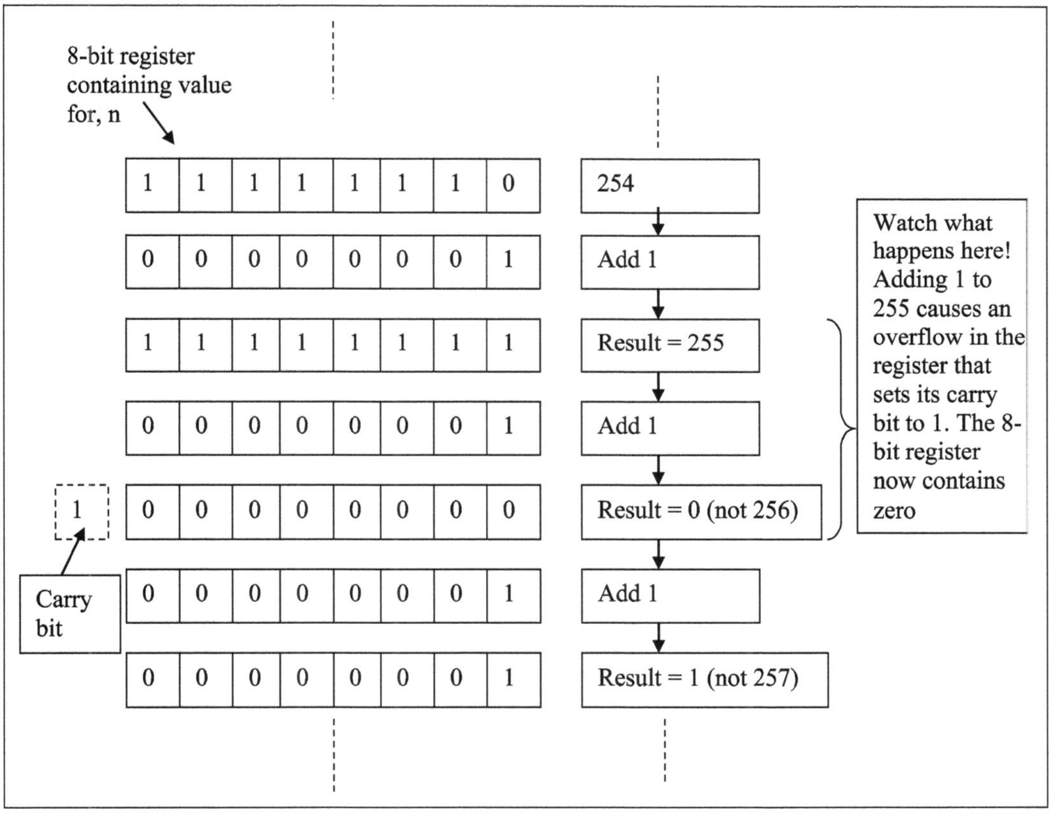

Figure 4.38 Overflow in the n-counter.

Some examples to make things clearer

Assumptions:

amp = 40mm
zag = 40mm
speed = 1
clock period = 20msec

Using previously derived equations for bp1 and bp2 (eq.4.3 and eq. 4.4), we obtain the following values.

$$bp1 = 64 + \left(64 \times \frac{zag}{zag + amp}\right) \qquad = 64 + \left(64 \times \frac{40}{40+40}\right)$$

Hence, $\underline{bp1 = 96}$

$$bp2 = 192 - 64 \times \left(\frac{zag}{zag + amp}\right) \qquad = 192 - 64 \times \left(\frac{40}{40+40}\right)$$

Hence, $\underline{bp2 = 160}$

Thus the digital locus looks like this, figure 4.39, below.

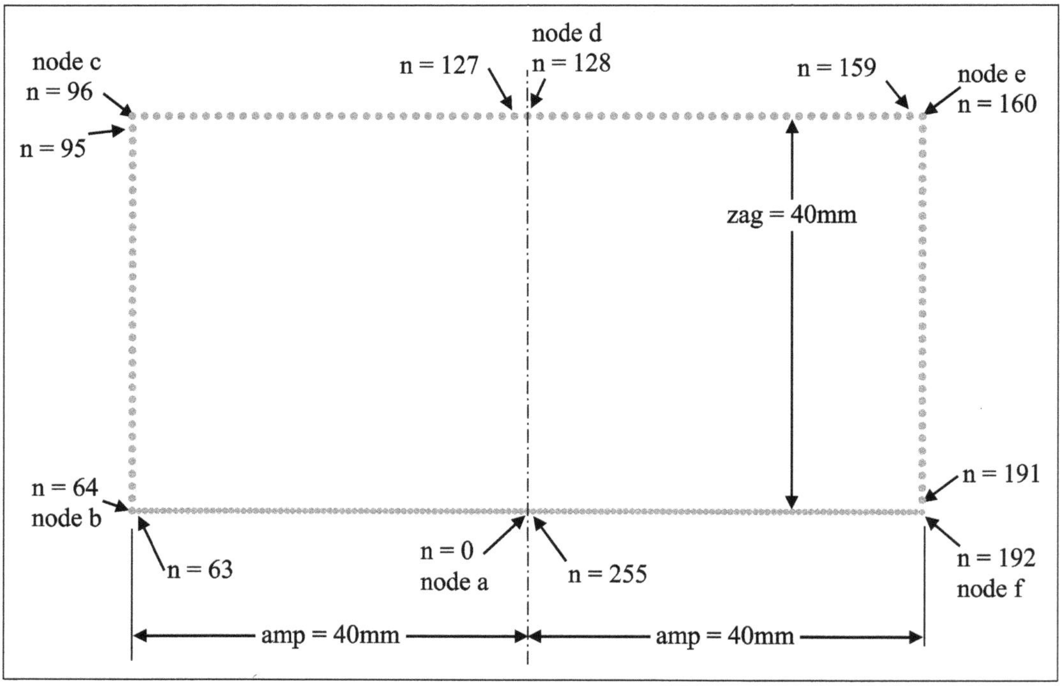

Figure 4.39 The digital locus showing the relationship between the n-value and location of its nodes.

As the clock pulses are received by BS2SX#2, the leg tip will be sequentially positioned on each dot of the above diagram starting at node a with n=0. Later, during its circuit, at n=192, the leg tip will move from node f, through node a, to node b in 128 clock pulses. During this time the leg tip will be in contact with the ground and supporting the robot body and will produce a robot body displacement along the ground of 2 × 40mm = 80mm. The period between each of the clock pulses is 20msec, hence the leg tip motion will produce a robot body locomotion speed, V_{loco}, of,

$$V_{loco} = \frac{2 \times amp}{128 \times \text{clock period}} = \frac{2 \times 40mm}{128 \times 20ms} = 31.3mm/s$$

In order to increase the speed of the robot body the value of "speed" in the BS2SX#2 program, figure 4.35, is increased to 2 with the program line,

```
speed          con          2
```

At each 20ms period clock pulse the value of n will now be sequenced, 0, 2, 4, 6,.......252, 254,0,2,4..... Hence 128/2 = 64 clock pulses, instead of 128 pulses, will cause the leg tip to move from node f , through node a, to node b.

As a consequence the locomotion speed is now doubled to,

$$V_{loco} = \frac{80mm}{64 \times 20ms} = 62.5mm/s$$

With speed=1, the leg tip will follow the, n, value by placing itself at the corresponding position on the locus and will complete a cycle in 256 × 20ms = 5.12secs.

If now, we double the speed so that speed = 2, i.e. n = n + 2, then, n, will have the values, 0, 2, 4, 6,...., and will complete a cycle in (256/2) × 20ms = 2.56secs

The following graph, figure 4.40 makes things clearer by showing the horizontal leg tip position that is dependent on the value of n showing speed 1 and speed 2 for travelling through the digital locus. Such a graph is normalised giving leg tip position as a function only of the n-variable meaning that there is no mention of time.

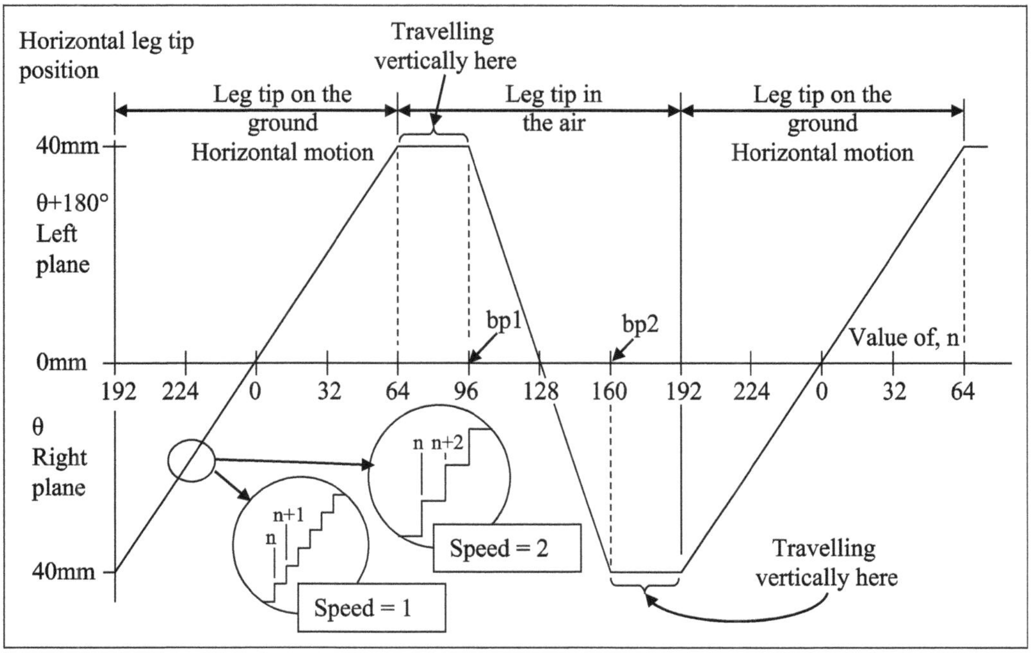

Figure 4.40 Relationship between horizontal leg tip position and value of n showing speed 1 and speed 2 for travelling through the digital locus.

The significant feature of the method is that the horizontal component of position of the leg tip is shown versus the value of, n, as n, clocks the leg tip along the digital locus. It doesn't matter if the leg tip is on the ground or off the ground; the horizontal component of the leg tip is still shown. We ignore any vertical component on the basis that, assuming the value of zag is large enough to lift the leg tip off the ground, the vertical component of leg tip position when in the air is not important for locomotion on a flat plane

It is important to realise that the above graph is true for both the preceding examples, i.e. speed = 1 and speed = 2. The only difference is in the resolution of the leg tip position with respect to the n value, as shown in the circular inset windows of figure 4.38 above..

However, if we display the horizontal position of the leg tip versus time, instead of the n value, then the following two graphs, figure 4.41 and figure 4.42, show the result for, speed = 1 and speed = 2, respectively.

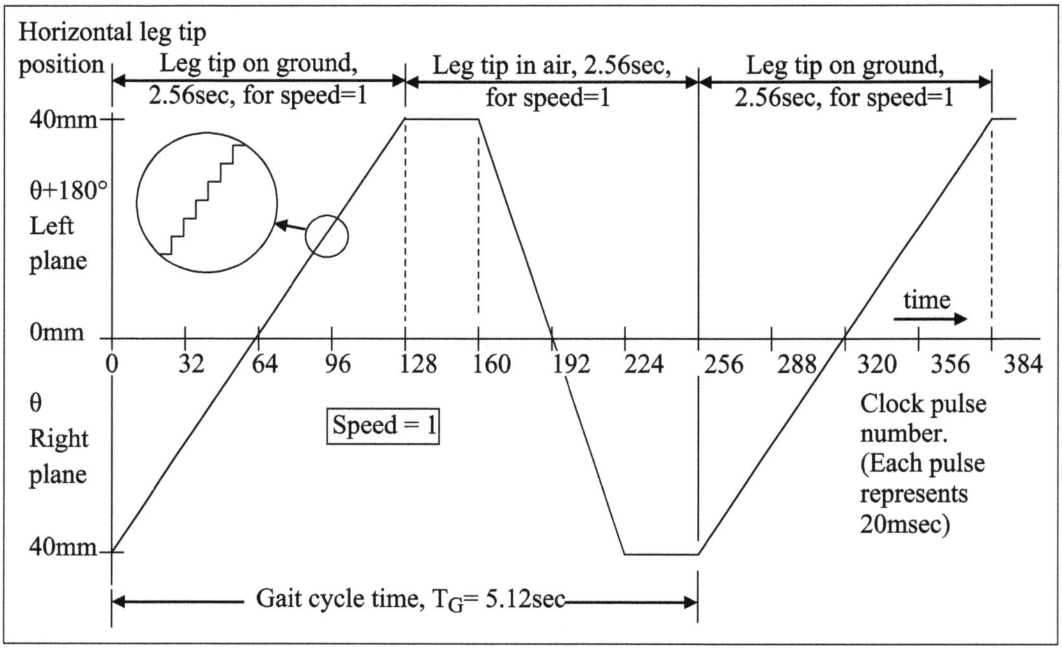

Figure 4.41 Relationship between horizontal leg tip position and clock pulse for speed = 1.

Hence by increasing "speed", which is the number added to, n, we can increase the locomotion speed of the walking robot. Remember that "speed" can only be an integer, so we cannot obtain speeds associated with fractional values of "speed". For example, with amp=40mm, setting, speed = 1, produces a locomotion speed of 31.25mm/s and speed = 2, gives a locomotion speed of 62.5mm/s. Suppose we want a locomotion speed of 53.6mm/s. How do we get that?

The answer is to choose the value of "speed" that will exceed 53.6mm/sec and then reduce the value of "amp" to get as close as possible to the required value. However, "amp" must also be an integer. Here is how we do the calculation:-

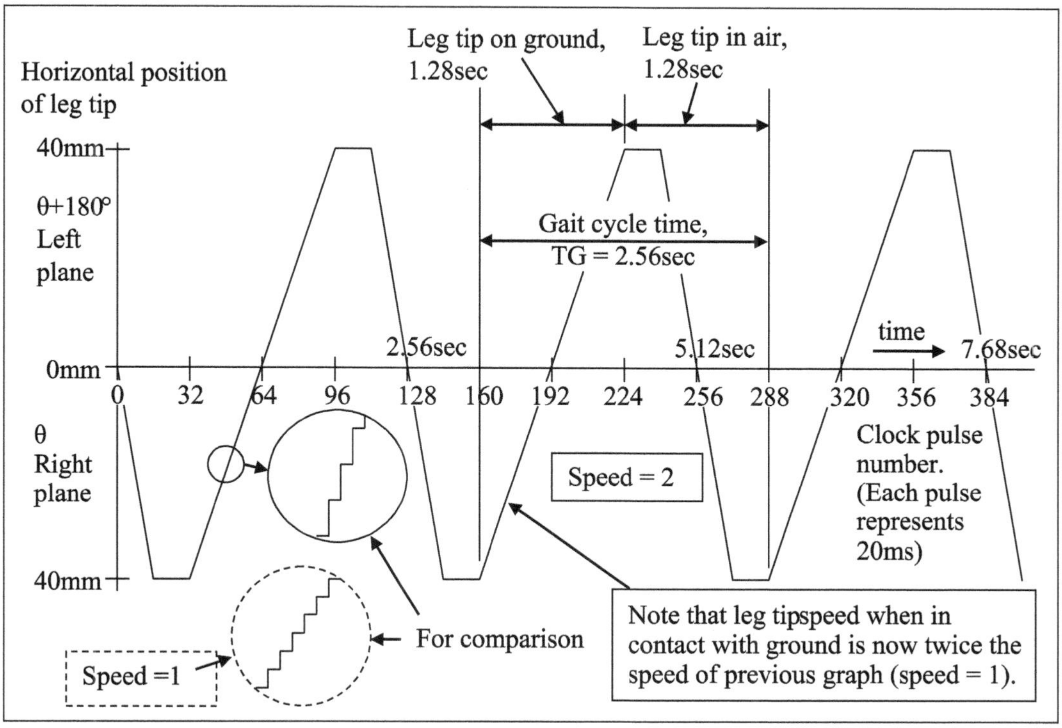

Figure 4.42 Relationship between horizontal leg tip position and clock pulse for speed = 2.

Setting, speed = 2, will give a locomotion speed of 62.5mm/s that exceeds the 53.6mm/s requirement

$$\text{Locomotion speed, } V_{loco} = 53.6\text{mm/sec} \quad = \frac{2 \times \text{amp} \times \text{speed}}{128 \times \text{clock period}} \quad = \frac{2 \times \text{amp} \times 2}{128 \times 20\text{ms}}$$

$$\text{Hence, amp} = \frac{53.6\text{mm/s} \times 128 \times 0.020\text{sec} = 34.3\text{mm}}{4}$$

But amp must be an integer, so nearest value is, amp = 34mm

It is natural to think why not adjust the clock periodic time to obtain the required locomotion speed. However, this cannot be done because the clock period should remain constant in real-time computing systems. The reason is due to the dynamic response performance of the system imposing the restriction that the periodic time should not increase and neither should it decrease because this will not allow the computer to complete all necessary calculations.

By this time the reader will be getting hold of the discipline of working with integers. Living without the decimal point pushes the engineering student into making decisions about the required resolution of a system. Furthermore, students gain an understanding of engineering limits.

4.20 Dynamic response of the leg servos

A couple of points should be made here about the response of the servomechanisms when they are asked to follow the digital locus. The digital locus requires the leg tip to make displacement steps as in the following figure 4.43 which is concerned with the horizontal component of the digital locus.

However, figure 4.43 shows that the leg tip response does not closely follow the digital locus. If the speed is low, e.g. speed = 1, then the leg tip has time to settle to the correct displacement after a slight overshoot.

If the speed is increased, e.g. speed = 6, then the leg tip deviates significantly from the digital locus. This is due to the servomechanisms that actuate the leg tip having more difficulty in following a faster changing digital locus. All controlled systems have this problem to a greater or lesser degree and there are ways to reduce the error between the digital locus and the response. Alternatively, the speed can be kept low enough such that the error is acceptable. At higher speeds the digital locus can be advanced in time in order to "fool" the leg tip to get to the right place at the right time. For the time being we will live with the errors but take care to stay within the servo slewing speed limits.

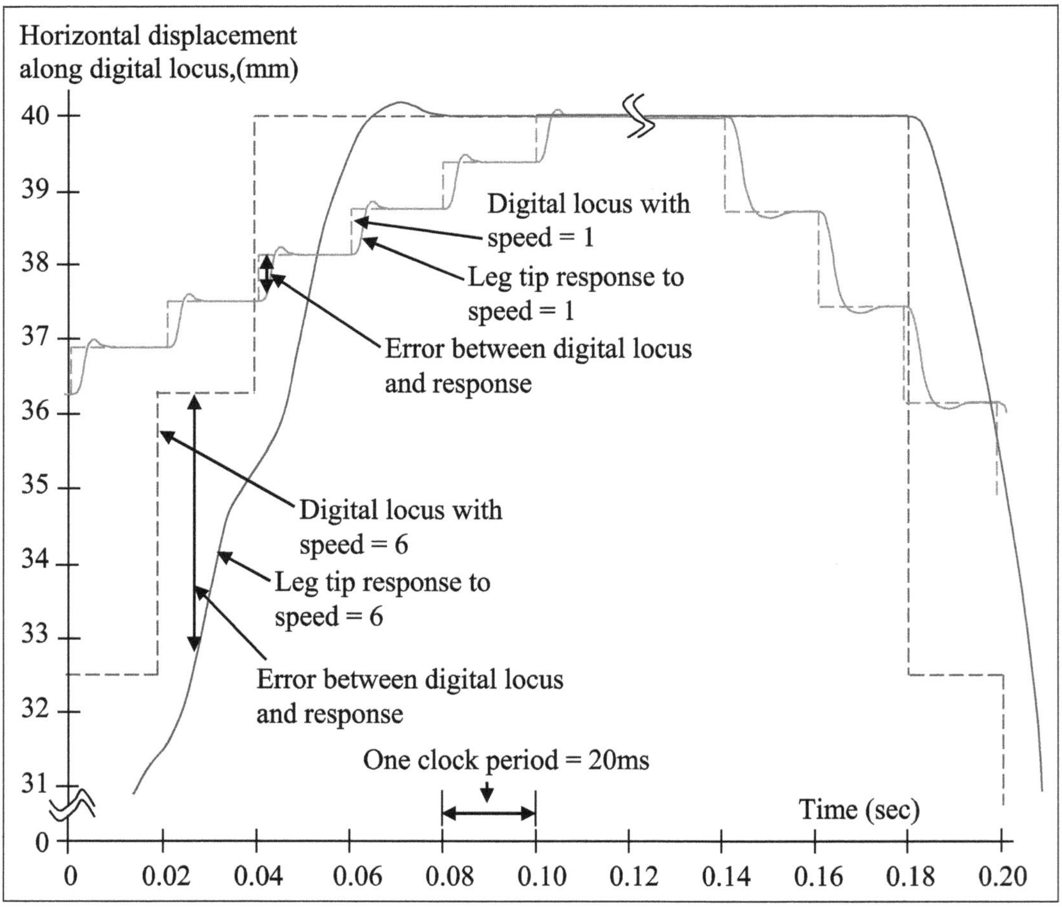

Figure 4.43 Real-time response of leg tip position.

Let us now devise a BS2SX computer program that will work out values of R and Z of the digital locus based on input values of, n, amp, zag, and zbias.

4.21 Computing, R, and, Z, values of the digital locus using a BS2SX computer program

Note that we denote the digital locus plan angle, θ, figure 4.26 as variable 'plangle'.
'Program to compute R and Z based on input values of, n, amp, zag and zbias.
'Assume polled interrupt clock timing pulses coming in at port 15

```
'Initialisation
n        var   byte   'digital locus position variable
R        var   byte   'realtime radius of the leg tip in cyl. coord. frame
Z        var   byte   'realtime vertical position of leg tip in cyl.coord frame
amp      var   byte   'amplitude of walking step
zag      var   byte   'height of walking step
zbias    var   byte   'vertical bias of digital locus
plangle  var   word   'digital locus walking step direction
j        var   nib    ' variable used to accept polled timing interrupt
speed    con   1      ' set walking speed to speed 1
'End of initialisation

again: n=n+speed

'compute values of bp1 and bp2
    bp1=zag*64/(zag+amp)+ 64          'equation 4.3
    bp2=192–zag*64/(zag+amp)          'equation 4.4

'compute in which section is n
    if n<64 then sect1     \
    if n<bp1 then sect2    |
    if n<128 then sect3    > See figure 4.28
    if n<bp2 then sect4    |
    if n<192 then sect5    /
    goto sect6
```

```
sect1: R=amp*n/64
       Z=zag+zbias
       plangle=180
       goto skip0

state2:R=amp
       Z=zbias+zag-(zag*(n-64)/(bp1-64))
       plangle=180
       goto skip0

state3:R=amp-(amp*(n-bp1)/(128-bp1))
       Z=zbias
       plangle=180
       goto skip0

state4:R=amp*(n-128)/(bp2-128)
       Z=zbias
       plangle=0
       goto skip0

state5:R=amp
       Z=zbias+(zag*(n-bp2)/(192-bp2))
       plangle=0
       goto skip0

state6:R=amp-(amp*(n-192)/64)
       Z=zag+zbias
       plangle=0
skip0:
'R, Z , and plangle all worked out
```

Equations 4.5 to 4.10 inclusive

Note that plangle=0 means the digital locus right plane and plangle=180 means the left plane

Computer code to compute R,Z and plangle based on variables, n, amp, zag, zbias and plangle

The above computer program has just computed output values, R, Z, and plangle based on input values of, n, amp, zag, and zbias, as illustrated in the following figure 4.44.

Stage 2, figure 4.44 is also called the inverse kinematics stage and will be described in the immediate following chapters 5, 6 and 7. After this we will talk, in chapter 8, about the generalised motion of the legs which includes walking with different gaits.

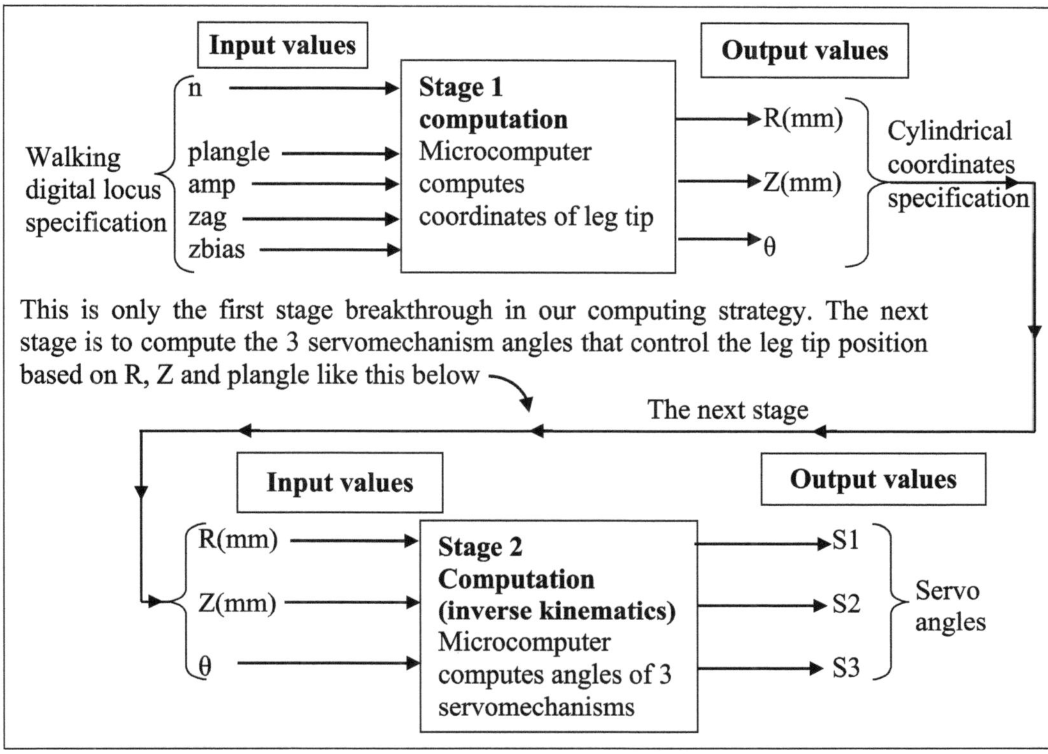

Figure 4.44 Illustration of the computing structure to compute the three servo angles of a 3dof leg to enable walking. This chapter and the code in figure 4.42 cover stage 1. Stage 2 is to be formulated in chapter 6.

5. Inverse kinematics analysis

Servo angles, S1, S2, S3 as a function of Leg Tip cylindrical coordinates, R, Z, θ

5.1 Introduction: The story so far

Let us recap the story so far as regards programming a leg tip to follow a path, i.e. a digital locus trajectory of a leg tip that makes it possible for a robot to walk. Figure 5.1 shows the overall generalised design of the single leg computational algorithm that drives just one leg through its digital locus trajectory using given variables of n, amp, zag, plangle, and zbias from which the algorithm computes servo angles, S1, S2 and S3. These three servos actuate a 3dof mechanical system which controls the position of a leg tip. This is the design we are aiming to achieve for each of the hexapod's six legs in preparation for the six legs to be mutually choreographed to achieve behaviour patterns. Figure 5.2 breaks down the single leg computational algorithm into a block diagram showing more detail of how it works. There are two stages of the algorithm. The first stage, Stage 1, computes the cylindrical coordinates, R, Z, θ from the digital locus variables, n, amp, zag, plangle, and zbias. Stage 1 has just been described in detail in the previous chapter, chapter 4. Stage 2 concerns computing the servo angles, S1, S2, S3 from R, Z, θ. Stage 2 is known as the inverse kinematics stage and will be described in this chapter.

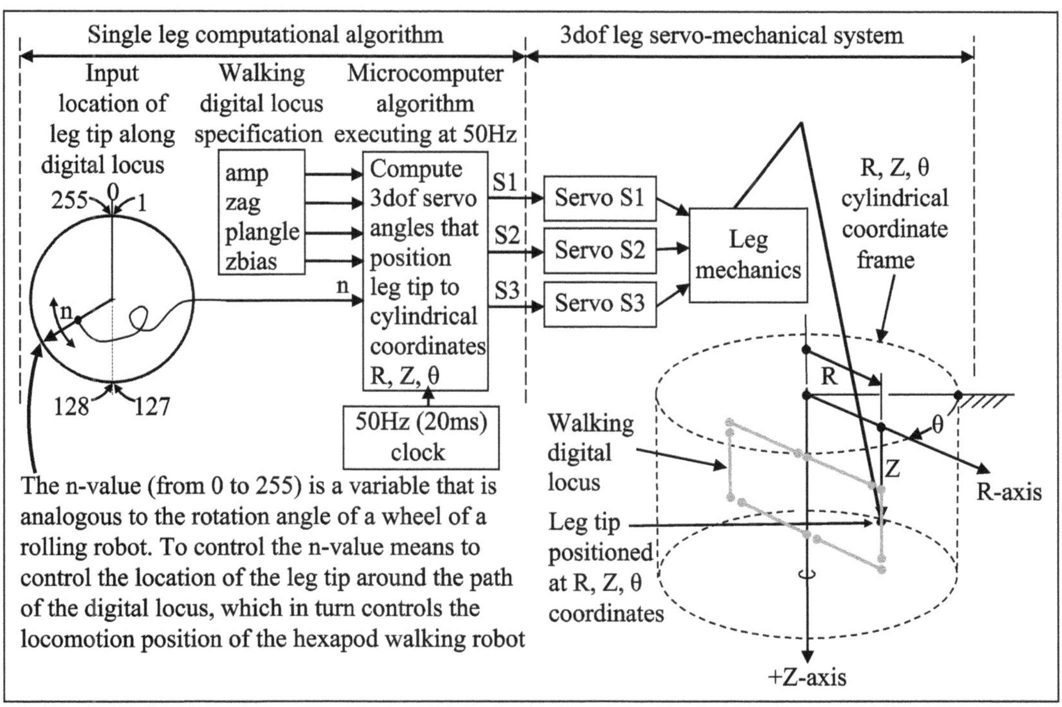

Figure 5.1 Overall generalised design of the computational algorithm to drive one leg through a walking digital locus.

Note, importantly, in figure 5.2 that there are two microcomputers that compute the single leg algorithm in figure 5.1. The two microcomputers work independently of each other but have different and complementary computation roles that produce a dual-level hierarchical computation architecture. The first microcomputer, the Local Pattern Generator, (the LPG), is analogous to a dancing choreographer who designs higher level behaviour pattern instructions and the second is analogous to the dancer who implements the details of these instructions, i.e. the lower level. In our application, the first microcomputer is solely responsible for the higher level behaviour patterns by deriving digital locus variable values, n, amp, zag, plangle, and zbias. These variables are sent to the second microcomputer which is solely responsible for the lower level inverse kinematics computation.

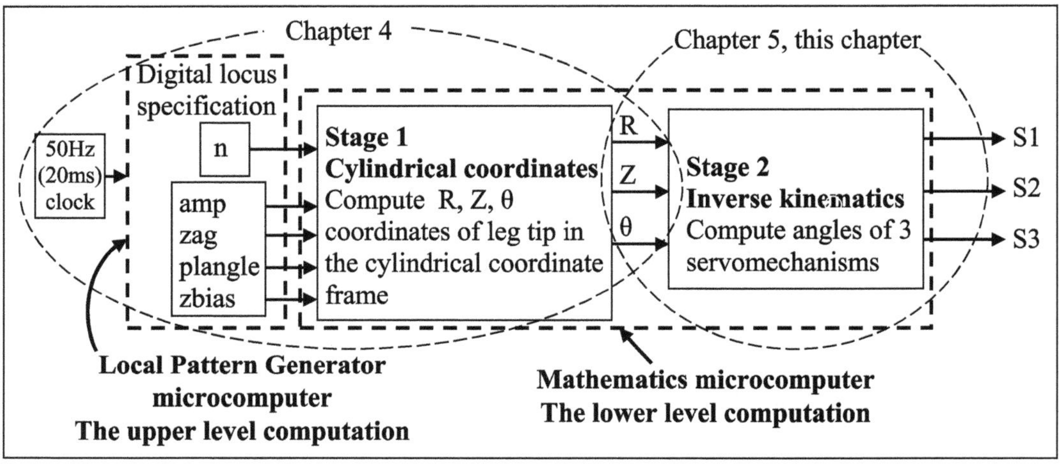

Figure 5.2 Block diagram of the single leg computational algorithm. Note the two microcomputers that produce dual-level hierarchical computation.

5.2 The meaning of 'inverse kinematics'

Inverse kinematics means that, "given the position of the leg tip in R, Z, θ coordinates, what are the S1, S2, S3 servo angles to give these leg tip coordinates?" On the other hand, forward kinematics means, "given the servo angles, S1, S2, S3, what is the leg tip position in R, Z, θ coordinates?" Forward kinematics is of little use in robotic systems since almost always we want to know the servo angles to position an end effector, (leg tip), at the required coordinates and that means inverse kinematics is the necessary computational requirement. However, inverse kinematics equations are inherently more difficult to solve because they require inverse trigonometrical solutions which give non-unique results. For example, arcsin(0.5) is equal to 30° and 150°; so which is the correct solution? The answer has to be found with human intervention so automating the computation of solutions is not easy. Furthermore, inverse kinematic solutions invariably comprise complicated mathematical formulations. In this book we will show a method that simplifies the formulations and produces unique solutions. We will use truncated Taylor polynomial series equations that produce unique solutions and are much easier and faster to compute whilst still maintaining accuracy.

Before we can do any inverse kinematics computation, it is necessary to describe in more detail the leg 3dof servo-mechanical system that has been illustrated in figure 5.1.

5.3 Description of the three degree-of-freedom (3dof) leg system

Figure 5.3 shows a 3dof leg system which is mounted on a calibration stand together with a DC power supply and a servoteach module. The system is used by students in a tutorial that is concerned with inverse kinematics and the real-time control of the leg tip.

The other figures, 5.4 and 5.5 show how the leg is used in two walking robot applications. The first one is a 2-wheeled-2-legged Rickshaw robot and the second is a Hexapod robot; both to be described in this book.

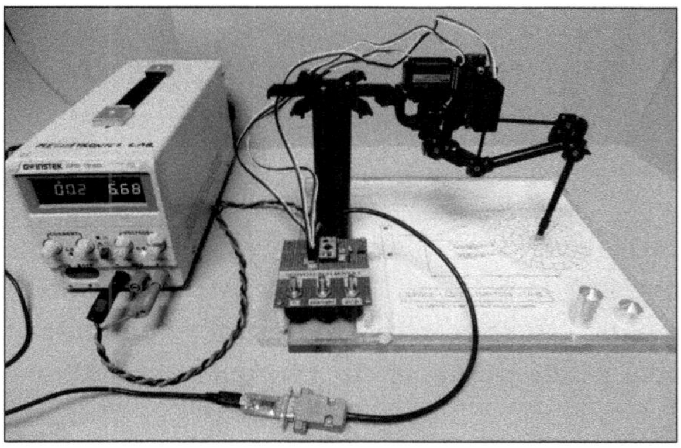

Figure 5.3 View showing a 3dof leg mounted on a calibration stand together with DC power supply and ServoTeach module.

Figure 5.4 Two of the legs fixed to a body that ambulates a Rickshaw robot.

Figure 5.5 Six of the legs fixed to a body that ambulates a Hexapod robot.

We now have the question; how do we synchronise and control a multitude of legs? The answer lies in figure 5.2 which shows a single leg being controlled by a behaviour pattern microcomputer. The role of this microcomputer can be extended to control a multitude of legs such as the 2 legs of the Rickshaw robot or the six legs of the Hexapod robot as now described in block diagram form.

5.4 Microcomputer architecture to control a multi-legged robot

Figure 5.6 shows a block diagram of the microcomputer architecture that controls the ambulation of a walking hexapod robot. The architecture uses seven single-core Basic Stamp microcomputers to form a parallel processing multicore computer system with a dual-level hierarchical design based on the elemental format in figure 5.2.

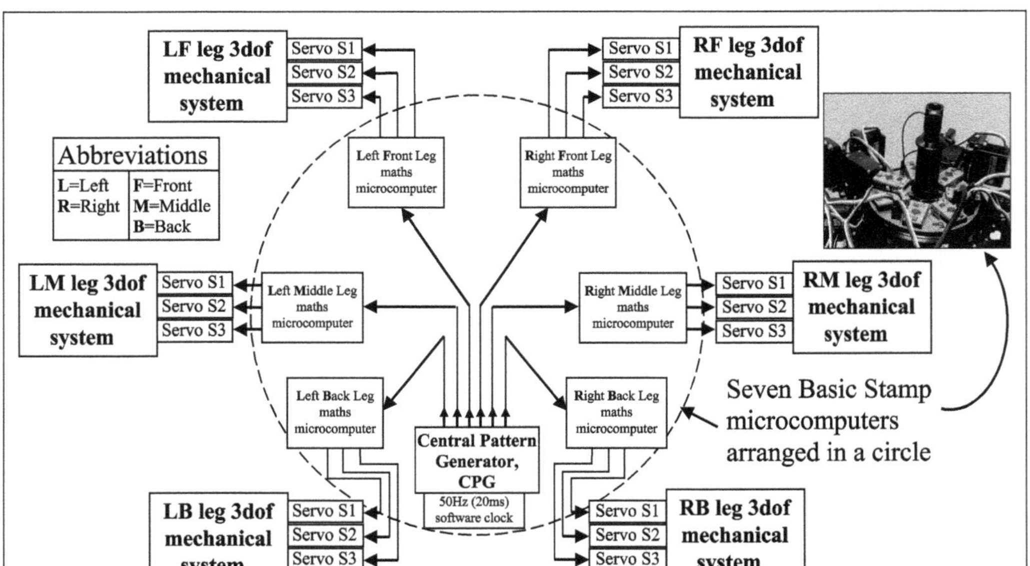

Figure 5.6 Block diagram of the 7-microcomputer parallel processing dual-level hierarchical architecture that controls the ambulation of a Hexapod robot. The upper level CPG microcomputer sends instructions to the six lower level Leg maths microcomputers.

5.5 The central pattern generator, the CPG

Of the seven microcomputers in figure 5.6, one is used as the higher level behaviour pattern choreographer; it is called the Central Pattern Generator, the CPG. The CPG sends six independent sets of instructions at 50 times per second to the remaining six lower level microcomputers that are used to control the motion of the six legs.

The separation of computing responsibilities, figure 5.6, carried out separately, independently and synchronously, coupled together with the behaviour pattern microcomputer that derives the five variables, n, amp, plangle, zag and zbias uniquely for each leg, is such a format that produces a computing system that enables a wide range of behaviour patterns to be easily programmed.

We now move on to explain the mechanical system geometry of the 3 dof leg.

5.6 Mechanical system geometry of the 3dof leg

We have already shown a schematic view of the 3dof leg in the previous chapter, figure 4.4. It is shown again in more detail, figure 5.7, in order to carry out geometrical analysis. This time we specify the leg tip in R, Z, θ coordinates in its cylindrical coordinate frame. Our task is to do inverse kinematics computation so that given, R, Z, θ leg tip coordinates, the values of S1, S2, S3 servo angles are computed to place the leg tip at these given coordinates.

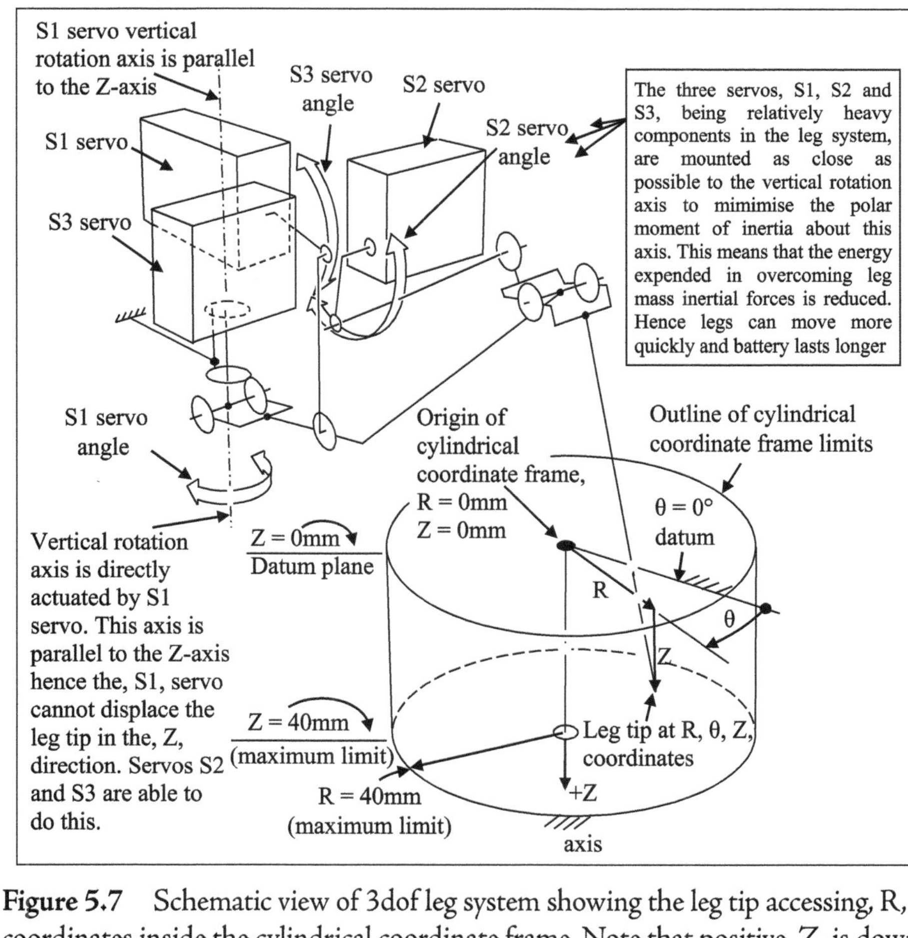

Figure 5.7 Schematic view of 3dof leg system showing the leg tip accessing, R, θ, Z, coordinates inside the cylindrical coordinate frame. Note that positive, Z, is downwards.

A cylindrical coordinate frame for the leg tip positioning is chosen, rather than, for example, a Cartesian coordinate frame, because it is the basis of a convenient method for producing complex walking gaits and body motions of the robot. Note that (i) the cylindrical coordinate frame has a radius limit, R and cylinder length, Z of 40mm for both axes and (ii) the positive Z-axis direction is downwards. The latter reason is because as the leg tip is positioned more positive in the Z direction then the robot body is elevated more upwards off the ground. Lifting the robot body upwards as the leg tip Z value is increased is considered intuitively more correct so this is the reason for the downwards Z-axis positive direction.

5.7 Kinematics analysis of the 3dof leg system

The next step in analysing the 3dof leg system is to show the geometrical details that will enable a kinematics analysis. Figure 5.8 below shows a true view side elevation of the leg system relative to a middle section through the cylindrical coordinate frame. Note that the origin of the cylindrical coordinate frame is 40mm above the ground

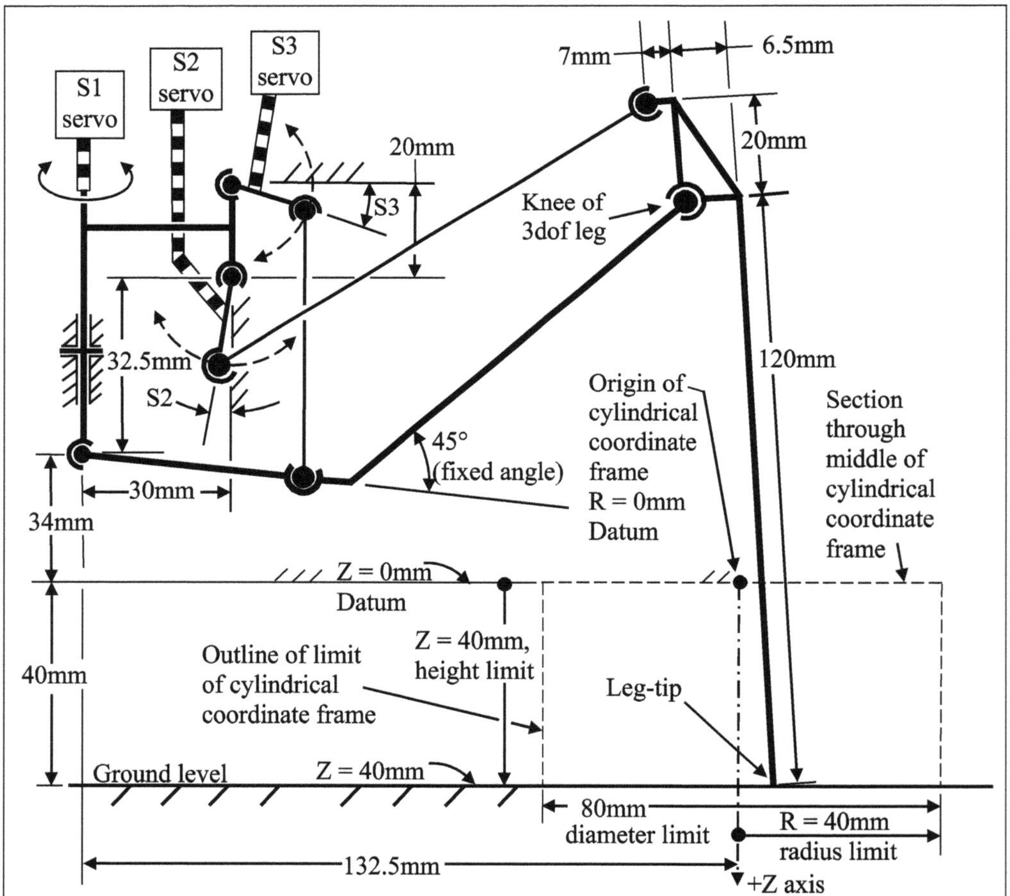

Figure 5.8 Side elevation, (not to scale), of 3dof leg system showing the three servos, S1, S2, and S3, actuating the mechanical system which place the leg tip at coordinates, R, θ, Z. Also shown is a middle section through the cylindrical coordinate frame which indicates the, R = 40mm, radius limit and the, Z = 40mm, height limit.

To avoid cluttering figure 5.8 above with too many geometrical details, another side elevation is shown below, figure 5.9. This figure, together with figure 5.8, provide sufficient geometrical information to allow an inverse kinematics analysis relating the leg tip position to the three servo angles, S1, S2, and S3.

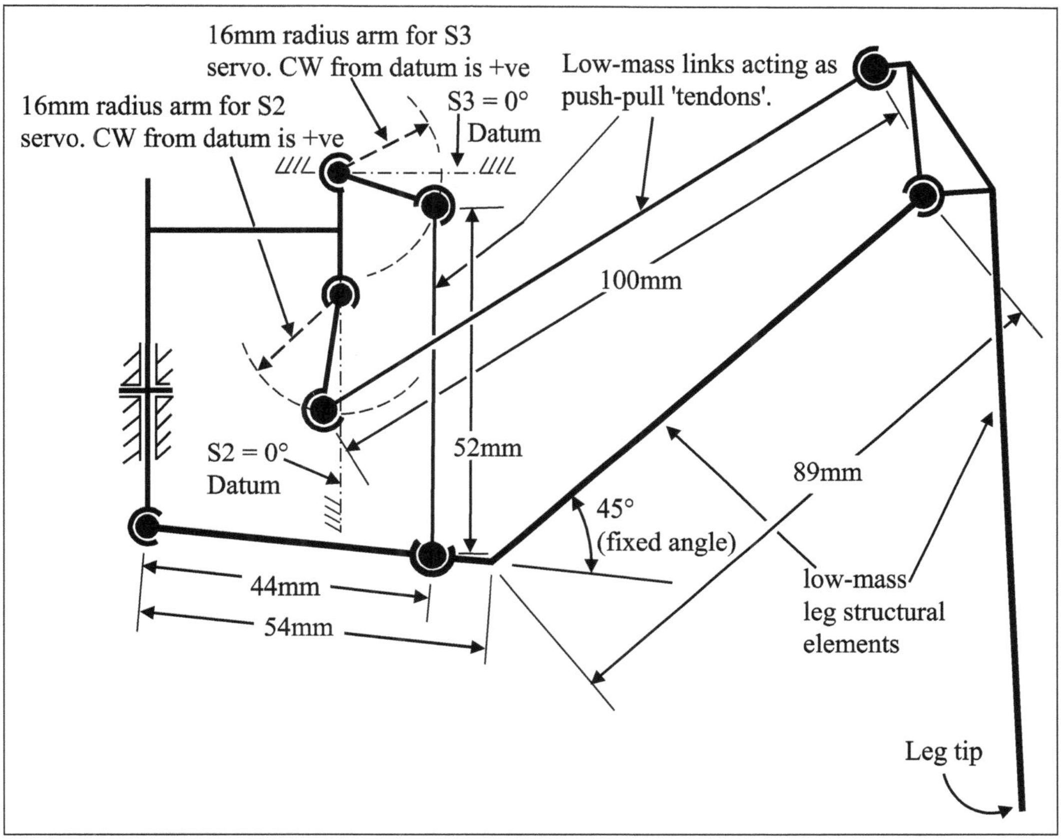

Figure 5.9 Another side elevation of the 3dof leg showing some more dimensions in order to provide enough information to obtain the leg tip position as a function of the three servo angles, S1, S2, and S3. Note the datum positions of the, S2, and, S3, servos, i.e., S2 = 0^0, and S3 = 0^0.

5.8 Minimising the polar second moment of inertia of the leg system

There is one characteristic of the mechanical system that is worth mentioning. It concerns the polar moment of inertia of the leg system that has been kept to a minimum by mounting the mass of the S2 and S3 servos as close as possible to the vertical rotation axis, (see figure 5.7). This means that servo, S1, can position the leg tip more quickly in, R, and, θ, coordinates and less energy is consumed in overcoming inertial forces. Remember the parallel axes theorem which indicates that the polar moment of inertia will increase with the square of the distance of the centre of gravity of a mass from an axis of rotation. Hence there will be a significant reduction in polar moment of inertia if design effort is taken to bring the leg mass as close as possible to the rotation axis.

If the S2 and S3 servos are brought close to the vertical axis then the leg tips must be actuated via low-mass links connected to servos S2 and S3 figure 5.9. This means that we follow the design of a horse leg which places its muscles, i.e. the servos, close to its body at the hip. That's why a horse leg is thin from the knee down and the foot is highly elongated. These powerful muscles, because they are close to the hip axis, oscillate with low amplitude during galloping so there is little power lost in overcoming inertial forces. Instead, they actuate low-mass tendons, (= the low-mass links in figure 5.9), and low-mass leg bones, (= the low-mass leg structural elements in figure 5.9), that are a long distance away from the muscles. These tendons and bones actuate the horse's hoof, i.e. the leg tip, back and forth through a large displacement at high velocity with minimum power consumed in overcoming the inertial forces produced by the muscle mass. In fact Nature sets up a natural frequency spring-mass system whereby even less energy is required to oscillate the reciprocating masses. The wing beats of insects, and humming birds are a prime example where there will be springs alongside the muscles to give a natural frequency of oscillation where energy is reversibly exchanged between the kinetic energy of the mass and strain energy of the spring.

5.9 Coupled displacement of the leg tip

The three servo angles, S1, S2, and, S3, do not directly displace the leg tip in the principal axes, R, θ, Z, respectively, of the cylindrical coordinate system. Instead, combined and coordinated displacement of two or three servos is required to produce displacement of the leg tip along a principal axis. This means that the servos are 'coupled'. We have to 'uncouple' the servos using software so as to obtain displacement of the leg tip along the principal axes. You can see why the servos are coupled by studying the diagrams in figures 5.10 to 5.13 inclusive. For example, figure 5.10 shows that displacement of, S1, servo alone, will displace the leg tip in both, R, and, θ coordinates. Also note that the S1 servo cannot displace the leg tip along the Z axis, i.e. S1 servo displacement is independent of, Z. This is because the, S1, rotation axis is parallel to the, Z-axis, see figure 5.7.

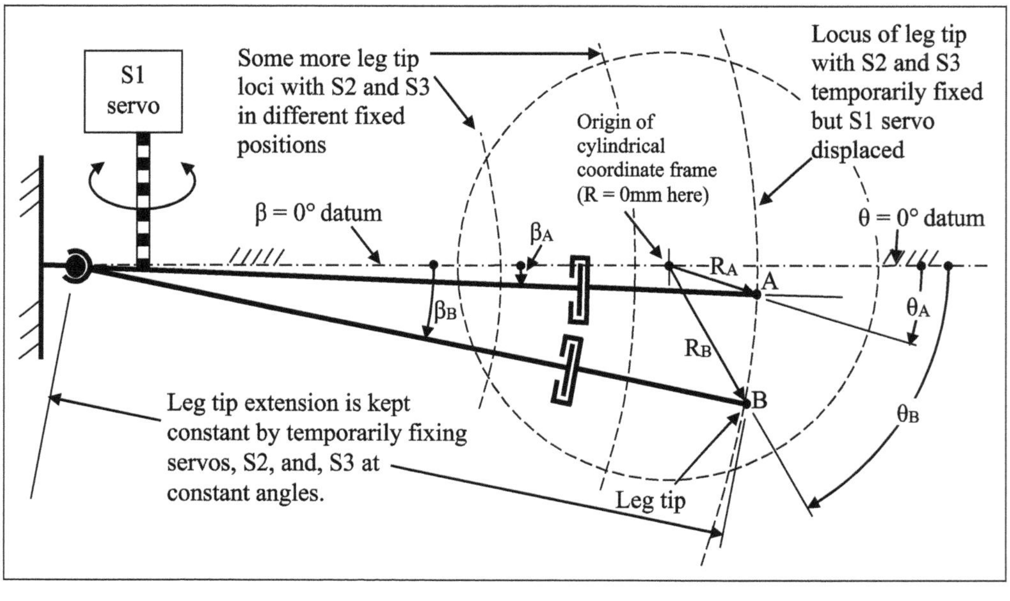

Figure 5.10 Plan view of 3dof leg system showing **only the S1 servo actuating the mechanical system.**

The angles of the other two servos, S2 and S3 are fixed temporarily. Actuating the S1 servo causes the leg tip to be simultaneously displaced in, R and θ coordinates. For example, the leg tip is positioned to point, A, by positioning, S1, to angle, β_A. This produces leg tip coordinates of, R_A, and, θ_A. Now, S1, is driven to angle, β_B, with servo angles, S2, and S3, remaining the same. The leg tip is now positioned at point, B, which produces leg tip coordinates, R_B, and, θ_B. The action produces significant changes in, R, and, θ, and in this case, no change along the, Z-axis, which lies perpendicular to the paper. Thus it can be seen that the S1 servo does not displace the leg tip in only one of the principal coordinates, either R, θ, or Z, to the exclusion of the other two principal coordinates. This is of no use to us. In contrast, we want a system that *does* drive the leg tip along any one, and only one, of the three principal axes to the exclusion of the other two. This is achieved using real-time software that coordinates all three servos to drive the leg tip along each principal axis. The action is called software 'decoupling' of the servos.

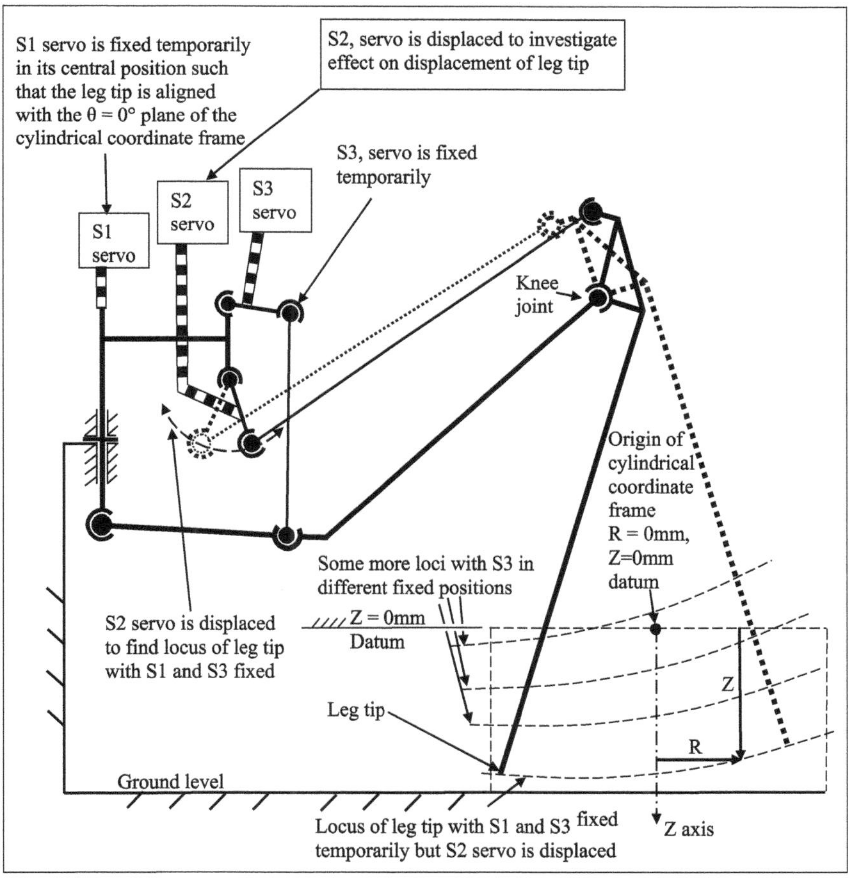

Figure 5.11 True view side elevation of 3dof leg system showing **only the S2 servo actuating the mechanical system.** The other two servos, S1 and S3 are fixed temporarily. Note that the leg tip is simultaneously displaced in R and Z coordinates but the leg system is designed such that displacement in the Z direction is kept to a minimum. It is important to realise that the individual servos, acting alone, do not displace the leg tip in the principal coordinates. We use software to accomplish that. It is called, software 'decoupling'.

Now consider the leg tip locus due to displacement of servo, S2, only. Servos, S1 and, S3 will be fixed temporarily. Figure 5.11 below shows the result where it can be seen that the leg tip simultaneously displaces in the, R, and, Z, directions. However, a plan view, shown later in figure 5.13, indicates that the leg tip also displaces in the, θ, direction except for the special case when, $\beta = \theta = 0^0$.

Now consider fixing the, S1, and, S2 servos and displacing the, S3 servo. The following diagrams, figure 5.12 and figure 5.13, show what happens to the leg tip. Displacement of servo, S3, is similar to the previous result shown in figure 5.11 but this time, S3 will displace the leg tip by a greater amount in the Z direction because the leg has been designed that way.

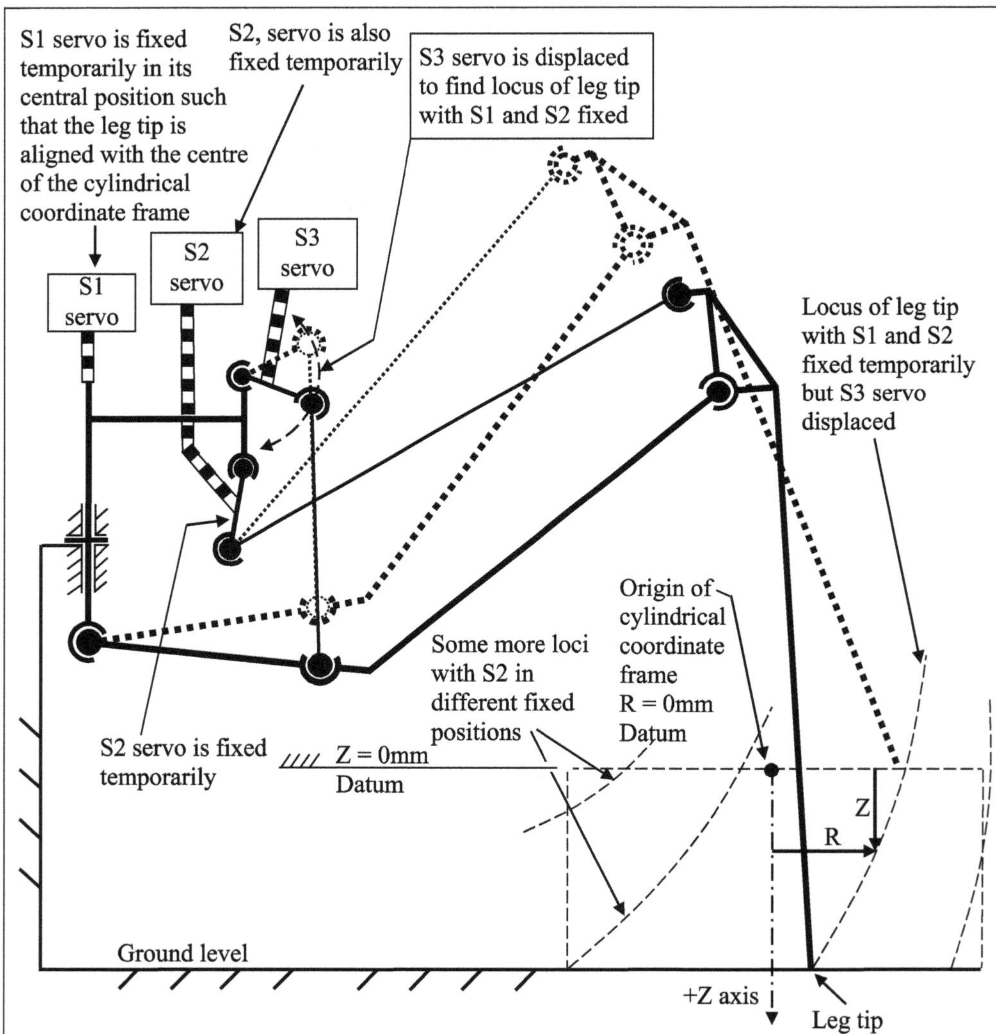

Figure 5.12 True view side elevation of 3dof leg system showing **only the S3 servo actuating the mechanical system.** The other two servos, S1 and S2 are fixed temporarily. Note that the leg tip is simultaneously displaced in R and Z coordinates but the leg system is designed such that displacement in the R direction is kept to a minimum. It is important to realise that the individual servos do not displace the leg tip in its principal coordinates, R, θ, and Z. We use software to accomplish that.

The following diagram, figure 5.13, shows a plan view of the leg system and the effect on the, R, and, θ, coordinates of displacing the, S2, and, S3, servos with, S1, fixed temporarily.

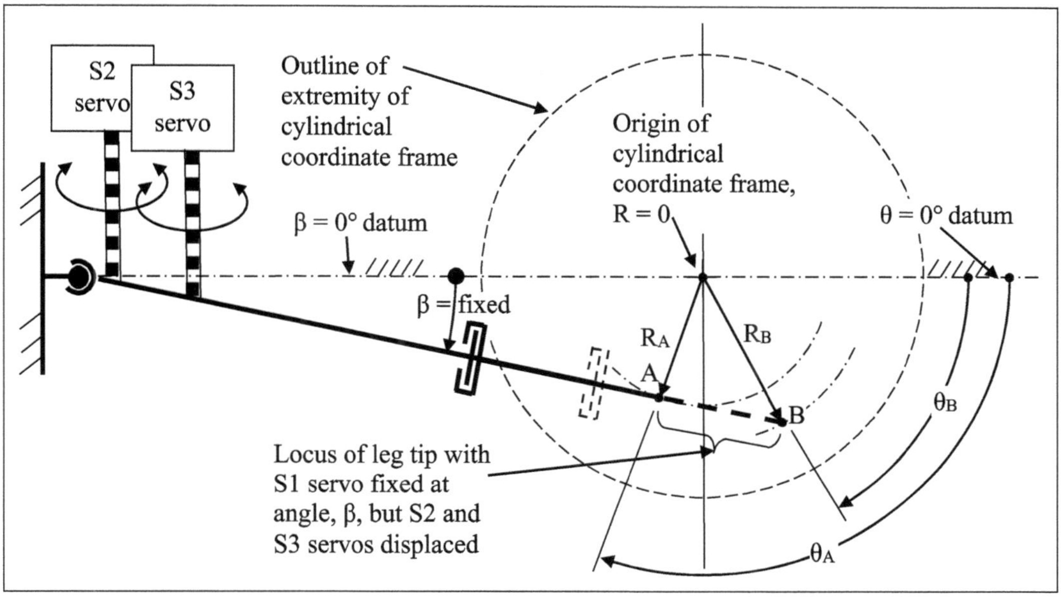

Figure 5.13 Plan view of 3dof leg system showing the, S1, servo in a fixed position at, S1 = β, and **only the S2 and S3 servos actuating the mechanical system.** Note that actuating the, S2, and, S3, servos will change, θ, and R, even when, S1, is fixed. For example, if the, S2, and, S3, servos are adjusted to position the leg tip at point A and later to point B then the leg tip will be located at, R_A, θ_A, and, R_B, θ_B, respectively. Real-time software is used to decouple the, S1, S2, and, S3, servos so as to coordinate all three servos to drive the leg tip along each principal axis, R, θ, and Z. Note that the Z-axis is perpendicular to the paper.

5.10 Decoupling the leg system

The kinematics of the leg system could be redesigned such that the S1 servo would move the leg tip in the R direction only with no displacement in, θ, or, Z, directions. Similarly, the S2 servo would displace the leg tip in the, θ, direction only with no displacement in, R, or, Z, directions and finally, S3 servo displacement would displace the leg tip in the, Z, direction only with no displacement in, R, or, θ, directions. Such a mutually exclusive mechanical system achieving these properties is likely, but not necessarily, to be heavier, more complicated and unnecessary because we have the microcomputer at our disposal. In fact herein lays the advantage of microcomputers working synergistically with servomechanisms. The great advantage of the microcomputer is that, for the first time in the history of mankind, we can dispense with levers, cams and gears that previously were used to compute mathematical equations, albeit ingeniously, and replace them with a cheaper, lightweight system with higher performance that has higher reliability. What we do is to design a simplified, lightweight uncomplicated mechanical system and then decouple the three servo displacements using real-time software. "Decoupling" means that we input values of, R, θ, Z, into a mathematical 'box', figure 5.14, which calculates inverse kinematics equations in real-time. The mathematical box works out values for, S1, S2, and, S3, angles based on the input values, R, θ, Z. In other words the mathematical box transforms the system into a principal axis system.

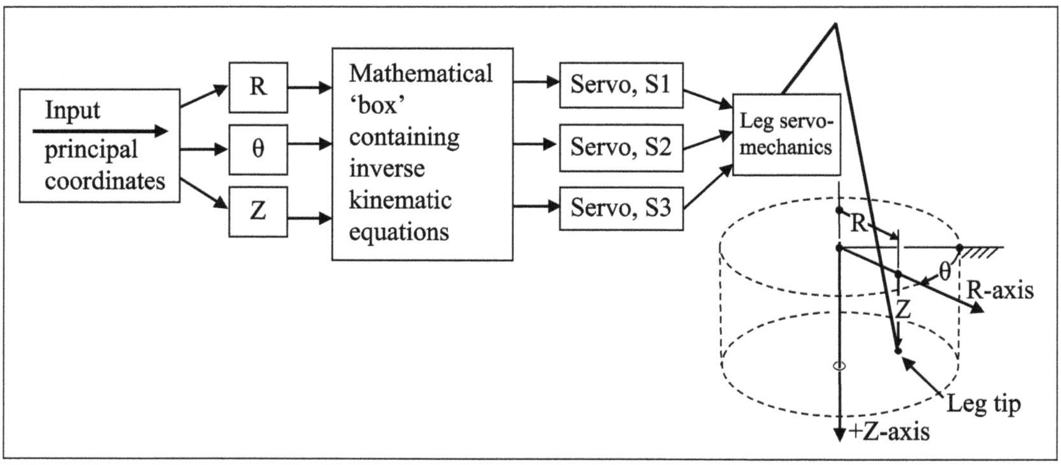

Figure 5.14 Diagram showing principal of 'decoupling' whereby we input values for the principal coordinates and the mathematical box computes values of the servomechanism angles that position the leg tip in, R, θ, Z, cylindrical coordinate space.

5.11 Deriving an inverse kinematics equation for S1= f1(R, θ)

We now analyse the simplest kinematic arrangement in the leg system. This is the S1 servomechanism system. The analysis uses traditional trigonometry. First, let us remind ourselves of the system in plan view shown in figure 5.15, which shows how the S1 servo is displaced to achieve values of, R, and, θ. The S1 servo is independent of the Z-axis which is perpendicular to the paper and so also is the S1 servo axis. The only way that the leg tip can be displaced in the Z direction is to actuate the S2 and S3 servos. These, we will talk about after the S1 servo.

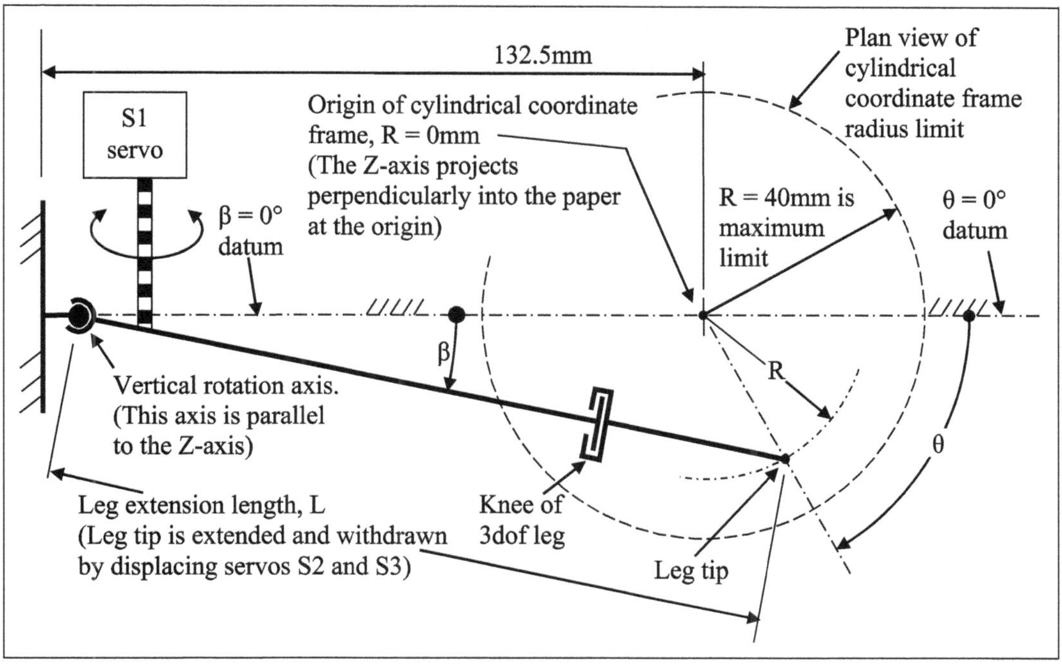

Figure 5.15 Plan view of 3dof leg system showing the leg tip accessing the coordinates, R, and, θ, in the cylindrical coordinate frame. Angle, β, is actuated directly by servo S1. The S1 servo angle is independent of, Z displacement because displacement of S1 does not displace the leg tip in the Z direction which is perpendicular to the paper.

Figure 5.16, below, is a simplified plan view of the geometry of the 3dof leg system based on figure 5.15. The S1 servo directly drives the vertical rotation axis and the S1 servo angle is represented by the angle, β.

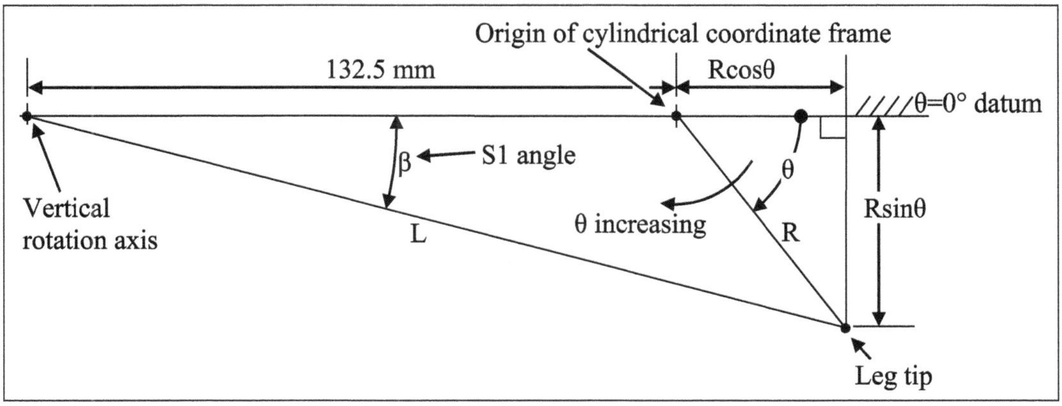

Figure 5.16 Diagram used to obtain trigonometrical relationship between, β, R, and, θ.

We now use figure 5.16 to obtain the relationship for, β, as a function of, R, and, θ. Note that β is equal to the S1 servo angle. The relationship for S1 is given as follows:

$$\text{S1 angle, } \beta(^{0}) = \text{atan}\left(\frac{R\sin\theta}{132.5 + R\cos\theta}\right) \qquad \text{Eq. 5.1}$$

Now move on to analysing the S2 servo angle

5.12 Deriving an inverse kinematics equation for S2= f2(R, θ, Z) in three stages

Referring back to figures 5.8 and 5.9 we obtain S2 in terms of, R, θ and Z in three stages.

5.12.1 1st stage is to calculate location of knee joint

Figure 5.17 below is used to obtain the P and Q coordinates of the knee joint based on given values of R, θ and Z. To find P and Q coordinates we need to calculate angle, $A_2 = A_{2a} + A_{2b}$

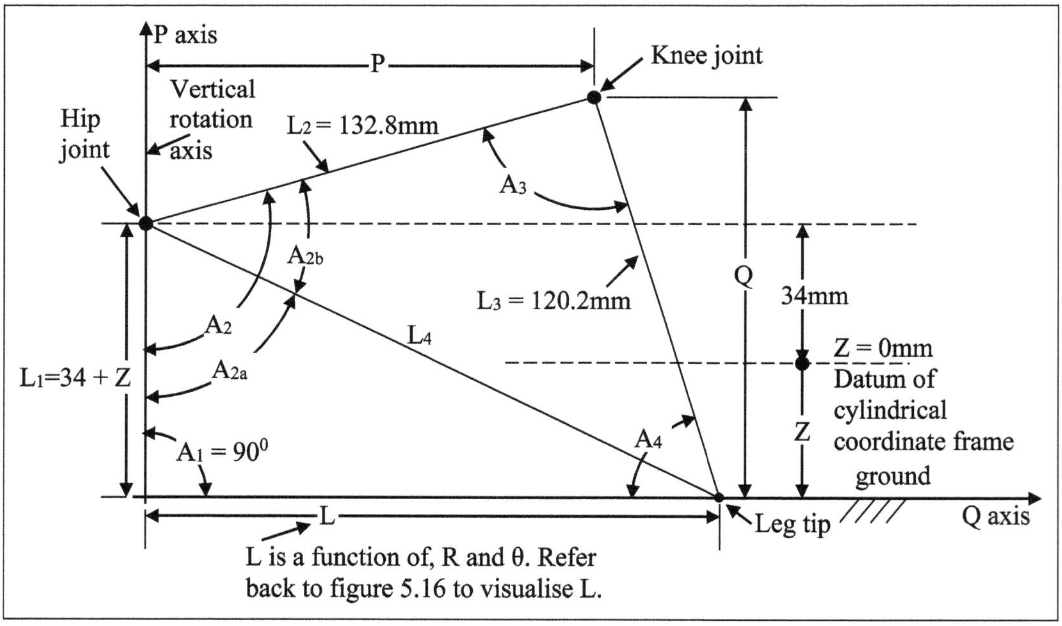

Figure 5.17 First stage diagram to obtain angle S1 as a function of R, θ and Z. Note that this diagram is a true view elevation of the plane containing the hip joint, the knee joint and the leg tip. Refer back to figures 5.8 and 5.9 for a review of the leg mechanical system design. Note that the diagram uses temporary P and Q axes.

From figure 5.16, the following equation is obtained.
Using cosine rule,

$$L = \text{sqrt}(132.5^2 + R^2 + (2 \times 132.5 \times R \times \cos\theta)) \qquad \text{Eq. 5.2}$$

Now, writing down the following trig. equations based on figure 5.17 above, (the idea is to get P and Q as functions of R, θ and Z, then later we can get S2 = function (R, θ, Z)

$$\text{angle, } A_{2a} = \text{atan}\left(\frac{L}{L_1}\right) = \text{atan}\left(\frac{L}{34+Z}\right) \qquad \text{Eq. 5.3}$$

Substituting eq. 5.2 into 53 we obtain the following;

$$\text{angle, } A_{2a} = \text{atan}\left(\frac{\text{sqrt}\left(132.5^2 + R^2 + (2 \times 132.5 \times R \times \cos\theta)\right)}{(34+Z)}\right) \qquad \text{Eq. 5.4}$$

Now moving on to obtain angle, $A_{2b;}$

$$\text{Using the cosine rule we get, angle, } A_{2b} = \text{acos}\left(\frac{L_2^{\,2} + L_4^{\,2} - L_3^{\,2}}{2 \times L_2 \times L_4}\right) \qquad \text{Eq. 5.5}$$

Also:

$$L_4^2 = L_1^2 + L^2 = (34 + Z)^2 + L^2, \text{ hence, } L_4 = \text{sqrt}((34 + Z)^2 + L^2) \qquad \text{Eq. 5.6}$$

Now substituting into eq.5.5 the values extracted from figure 5.8, i.e. L_2 = 132.8, L_3 = 120.2, and L_4 given by eq. 5.6, we obtain the following equation,

$$\text{angle, } A_{2b} = \text{acos}\left(\frac{132.8^2 + (34+Z)^2 + L^2 - 120.2^2}{2 \times 132.8 \times \text{sqrt}\left((34+Z)^2 + L^2\right)}\right) \qquad \text{Eq. 5.7}$$

Hence, we can now determine P and Q as follows,

$$P = 132.8 \times \sin(A_{2a} + A_{2b}) \qquad \text{Eq. 5.8}$$

and,

$$Q = 34 + Z - 132.8 \times \cos(A_{2a} + A_{2b}) \qquad \text{Eq. 5.9}$$

5.12.2 2nd stage is to analyse 4-bar mechanism

The analysis of a 4-bar plane mechanism as shown in figures 5.18 and 5.19 below will be used to derive the inverse kinematics equation for S2. Such a 4-bar mechanism is used by servo S2, via a push-pull tendon link, to actuate the foreleg at the knee joint, (see figure 5.11). It is assumed that the each of the four lengths of the bars, L_5, L_6, L_7, and L_8 are known values. There are two versions of the 4-bar mechanism. The first is shown in figure 5.18 and the second representing an optional mathematical solution, figure 5.19, shows two of the bars flipped over like a folded piece of paper. As a consequence, $A_6 = A_{6a} + A_{6b}$, in figure 5.18, and, $A_6 = A_{6a} - A_{6b}$, in figure 5.19.

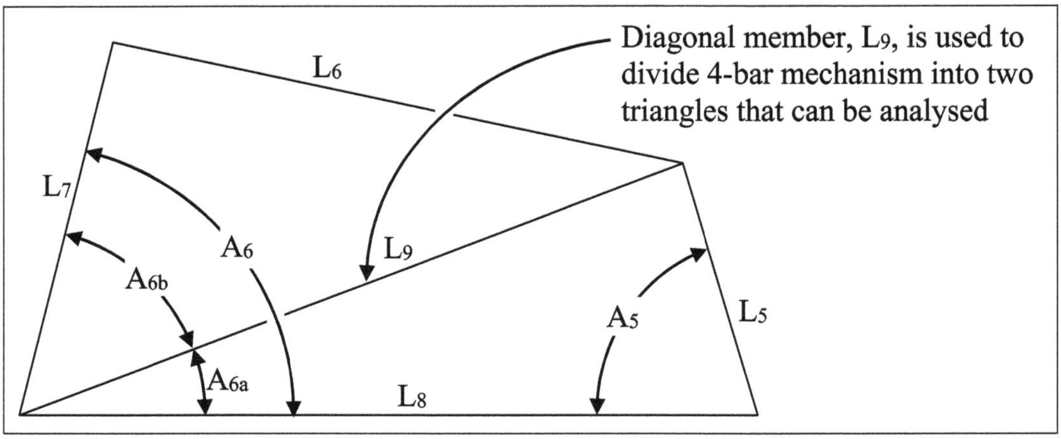

Figure 5.18 Analysis of a 4-bar linkage as a precursor to generating an inverse trigonometrical equation for computing, S2. In this figure, angle $A_6 = A_{6a} + A_{6b}$.

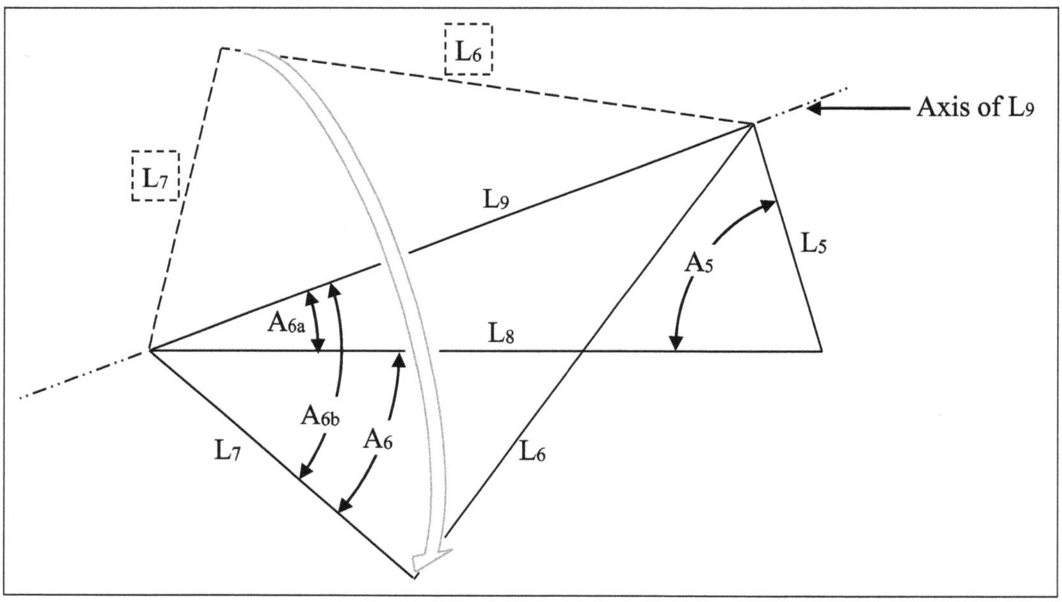

Figure 5.19 Same 4-bar mechanism as figure 5.18 above but with bars L6 and L7 flipped over, like folding a piece of paper, about axis of L_9.
In this figure, angle $A_6 = A_{6a} - A_{6b}$.

We wish to know angle A_6 as a function of A_5, because this relationship will help with the 3rd stage analysis. Now write down the following trig equations.

Using the cosine rule for figures 10.2 and 10.3:-

$$L_9^2 = L_5^2 + L_8^2 - (2 \times L_5 \times L_8 \times \cos A_5) \quad \text{Eq. 5.10}$$

Rearranging eq. 5.10 gives;

$$\text{Angle, } A_5 = \text{acos}\left(\frac{L_5^{\ 2} + L_8^{\ 2} - L_9^{\ 2}}{2 \times L_5 \times L_8}\right) \qquad \text{Eq. 5.11}$$

$$\text{Angle, } A_{6a} = \text{acos}\left(\frac{L_8^{\ 2} + L_9^{\ 2} - L_5^{\ 2}}{2 \times L_8 \times L_9}\right) \qquad \text{Eq. 5.12}$$

$$\text{Angle, } A_{6b} = \text{acos}\left(\frac{L_7^{\ 2} + L_9^{\ 2} - L_6^{\ 2}}{2 \times L_7 \times L_9}\right) \qquad \text{Eq. 5.13}$$

5.12.3 3rd stage is to apply 4-bar mechanism analysis to leg system

We are now in a position to analyse figure 5.20 below that represents a true view side elevation of the robot leg including the servo actuation links. See previous figures 5.8 and 5.9 for reference.

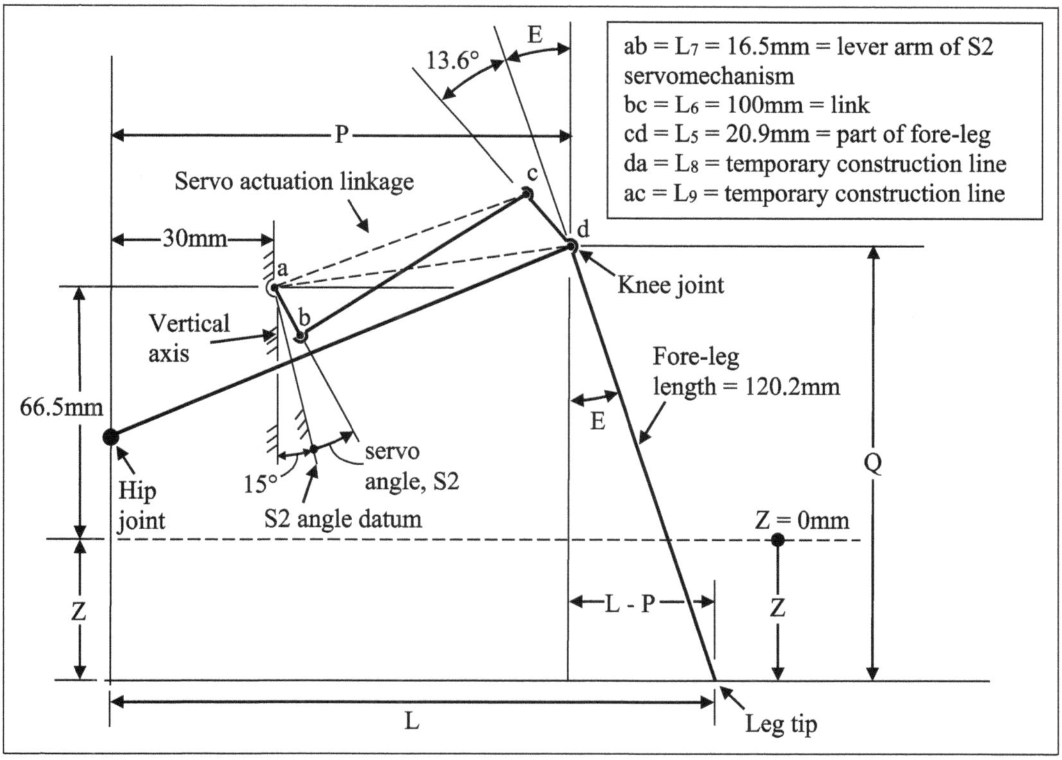

Figure 5.20 True side view of robot leg showing 3rd stage in the computation of S2 angle as a function of cylindrical coordinates, R, θ, and Z. Here we concentrate on the four-bar mechanism, abcd, where ab is the lever arm of the S2 servomechanism and, a, is the axis of rotation of the lever arm, which is perpendicular to the paper.

Now look more closely at four-bar linkage, abcd, figure 5.21 below.

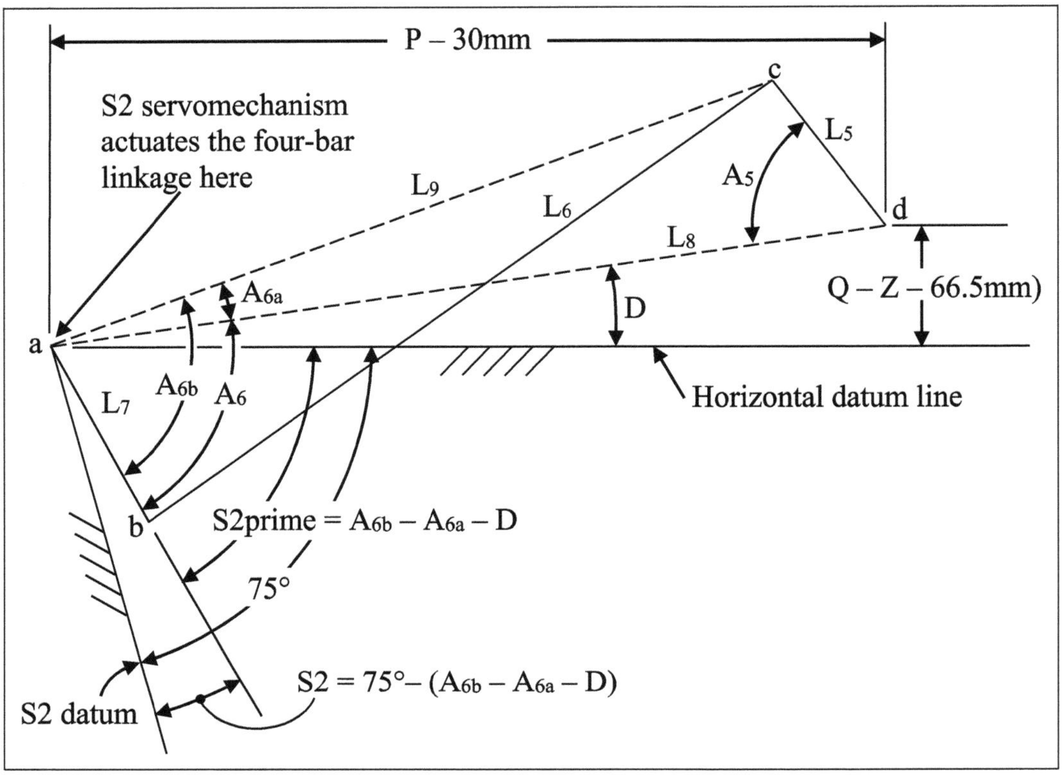

Figure 5.21 Close-up view of four-bar linkage, which is used by S2 servomechanism to actuate the leg tip. Note the location of the S2 datum.

We can now write down the following equations,

Angle, $E = \text{asin}\left(\dfrac{L - P}{120.2}\right)$ Eq. 5.14

Angle, $D = \text{atan}\left(\dfrac{Q - Z - 66.5}{P - 30}\right)$ Eq. 5.15

Angle, $A_5 = 90^0 + D - 13.6^0 - E$ Eq. 5.16

$L_8^2 = (Q - Z - 66.5)^2 + (P - 30)^2$ Eq. 5.17

and, $L_8 = \text{sqrt}\left((Q - Z - 66.5)^2 + (P - 30)^2\right)$ Eq. 5.18

$$L_9^2 = L_5^2 + L_8^2 - (2 \times L_5 \times L_8 \times \cos(A_5))$$

Substituting values,

$$L_9^2 = 20.9^2 + L_8^2 - (2 \times 20.9 \times L_8 \times \cos(A_5)) \qquad \text{Eq. 5.19}$$

$$\text{Angle, } A_{6a} = \text{acos}\left(\frac{L_8^{\,2} + L_9^{\,2} - L_5^{\,2}}{2 \times L_8 \times L_9}\right)$$

Substituting values,

$$\text{Angle, } A_{6a} = \text{acos}\left(\frac{L_8^{\,2} + L_9^{\,2} - 20.9^2}{2 \times L_8 \times L_9}\right) \qquad \text{Eq. 5.20}$$

$$\text{Angle, } A_{6b} = \text{acos}\left(\frac{L_7^{\,2} + L_9^{\,2} - L_6^{\,2}}{2 \times L_7 \times L_9}\right)$$

Substituting values,

$$\text{Angle, } A_{6b} = \text{acos}\left(\frac{16.5^2 + L_9^{\,2} - 100^2}{2 \times 16.5 \times L_9}\right) \qquad \text{Eq. 5.21}$$

...and finally, Servo angle, $S2 = 75° - (A_{6b} - A_{6a} - D)$ Eq. 5.22

The final equation, eq. 5.22, looks simple enough but no it's not; it's made up of complicated sub-equations from eq. 5.2 all the way through to eq. 5.21 inclusive. These equations are way too complicated for the Basic Stamp and unnecessarily complicated anyway, even if you have a sophisticated computer, so we'll have to think of a simpler way of computing these convoluted trigonometrical equations. This will be done in the next chapter, chapter 6. However, the next step is to derive the inverse kinematics equations for servo S3.

5.13 Deriving an inverse kinematics equation for S3= f3(R, θ, Z) in two stages

Note that we use similar symbols for the S3 analysis but they represent different quantities to the S2 analysis.

The inverse kinematics equation for S3 can be derived from figure 5.22 below which outlines the relevant dimensions of the leg system. The analysis is a two-stage process so we start with the first stage based on the geometry shown in figure 5.22.

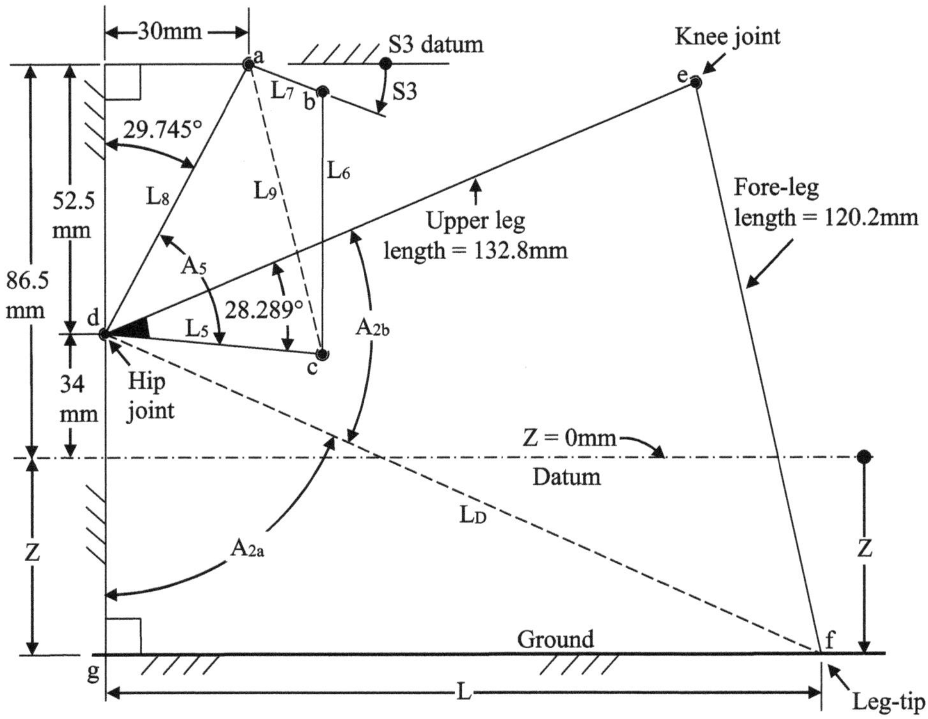

Table of dimensions for leg geometry in figure above
abcd is a 4-bar linkage
ab = L_7 = 16.5mm = lever arm of S3 servomechanism
bc = L_6 = 52.0mm = tendon link length
cd = L_5 = 44.0mm = upper leg lever arm
ad = L_8 = 60.47mm temporary construction line, fixed length and direction
ac = L_9 = temporary construction line, variable length and direction
de = 132.8mm = upper leg length, fixed length
ef = 120.2mm = fore-leg length
df = L_D = temporary construction line, variable length and direction

Figure 5.22 General layout and dimensions of leg system from which the inverse kinematics equation for servomechanism angle, S3, can be derived.

5.13.1 1st stage is to calculate angle, A_5

We set out to calculate angle, A_5, because when A_5, has been calculated then angle S3 can be calculated from the 4-bar linkage system, abcd. Analysing dimensions in figure 5.22 based on previous analysis of S2 at beginning of this chapter we recall the following equations:-

Recall Eq. 5.2 $L = \text{sqrt}(132.5^2 + R^2 + (2 \times 132.5 \times R \times \cos\theta))$ Renumber as; Eq. 5.23

Recall Eq. 5.4 angle, $A_{2a} = \text{atan}\left(\dfrac{(\text{sqrt}\left(132.5^2 + R^2 + \left(2 \text{ x } 132.5 \times R \times \cos\theta\right)\right))}{34 + Z}\right)$

Renumber as Eq. 5.24

Recall Eq. 5.7 angle, A2b $= \text{acos}\left(\dfrac{132.8^2 + (34+Z)^2 + L^2 - 120.2^2}{2 \times 132.8 \times \text{sqrt}\left((34+Z)^2 + L^2\right)}\right)$

Renumber as; Eq. 5.25

Now, angle, A_5 can be calculated as follows:-

$$A_5 = 180^0 - 29.745^0 - (A_{2a} + A_{2b}) + 28.289^0 \qquad \text{Eq. 5.26}$$

5.13.2 2nd stage is to calculate angle, S3

Figure 5.23 shows a further detailed diagram of the 4-bar linkage, abcd based on previous figure 5.22 that we now analyse based on our results so far including results from figure 5.18 and figure 5.19.

$$L_9 = \text{sqrt}(L_5^2 + L_8^2 - (2 \times L_5.L_8.\cos A_5)) \qquad \text{Eq. 5.27}$$

$$A_{6a} = \text{acos}\left(\frac{L_8^{\,2} + L_9^{\,2} - L_5^{\,2}}{2 \times L_8 \times L_9}\right) \qquad \text{Eq. 5.28}$$

$$A_{6b} = \text{acos}\left(\frac{L_7^{\,2} + L_9^{\,2} - L_6^{\,2}}{2 \times L_7 \times L_9}\right) \qquad \text{Eq. 5.29}$$

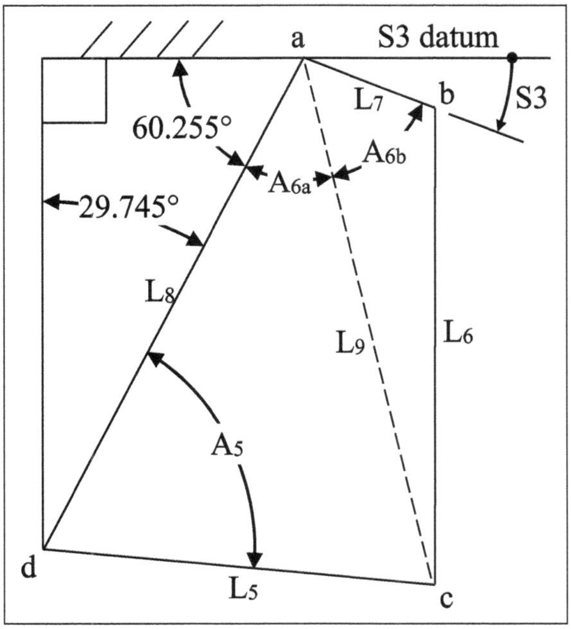

Figure 5.23 Further detailed diagram of 4-bar linkage, abcd, extracted from figure 5.22.

Substituting values from table in figure 5.22, previous equation, eq. 5.27, eq. 5.28 and eq. 5.29 become:

$$L_9 = \text{sqrt}(44.0^2 + 60.47^2 - (2 \times 44.0 \times 60.47.\cos A_5)) \qquad \text{Eq. 5.30}$$

$$A_{6a} = \text{acos}\left(\frac{60.47^2 + L_9^{\ 2} - 44.0^2}{2 \times 60.47 \times L_9}\right) \qquad \text{Eq. 5.31}$$

$$A_{6b} = \text{acos}\left(\frac{16.5^2 + L_9^{\ 2} - 52.0^2}{2 \times 16.5 \times L_9}\right) \qquad \text{Eq. 5.32}$$

...and finally, servo angle, $S3 = 180^0 - 60.255^0 - A_{6a} - A_{6b}$ Eq. 5.33

Once again the equation for servo angle S3, eq. 5.33, looks simple enough bit it's not. You can see that by referring back to the equations 5.23 to 5.32 where the equations become complicated and interconnected. A simpler method of computing S3 must be found. Before this is done, let's see what S1, S2 and S3 servo angles look like for a range of coordinates at the boundaries of the leg tip cylindrical coordinate frame.

5.14 Graphs of S1, S2 and S3 values at boundaries of the cylindrical coordinate frame

We now create 3 graphs, one for each servo, that show the servo value relationships at the boundaries of the leg tip cylindrical coordinate frame. Equations 5.1, 5.22 and 5.33 are used to compute values of S1, S2 and S3 respectively. The following table, figure 5.24, shows four sets of R, Z and θ values that represent the frame boundaries. (R and Z dimensions are mm)

Set 1.	R= 0, Z= 0, θ=0° to 360°
Set 2.	R=40, Z= 0, θ=0° to 360°
Set 3.	R= 0, Z=40, θ=0° to 360°
Set 4.	R=40, Z=40, θ=0° to 360°

Figure 5.24 Table of four sets of R, Z, θ values representing the leg tip cylindrical coordinate frame boundaries.

The relationships for the S1, S2 and S3 servo angles are shown below in pulsout units that are suitable for the Basic Stamp2SX microcomputer. The pulsout units are calculated as follows, eq. 5.34:

Servo angle, S(pulsout units) = (S(degree units) × 12) + 1875 Eq. 5.34

The symbol, S represents any one of the servos, S1, S2 or S3. The number, 12 is the nominal calibration factor in pulsout units/degree of the Futaba servo that actuates each degree of freedom of the robot legs. The number, 1875 is an offset value that sets the servo S value to give a pulse width of 1.5ms at the datum servo position. The figures below, figures, 5.25, 5.26 and 5.27, show the S1, S2 and S3 relationships as a conclusion to this chapter.

The next chapter will address how these relationships can be computed for all integer millimetre values of R and Z from 0mm to 40mm and all integer degree values of θ from 0° to 359° without relying on the complicated equations that are used to plot the graphs. A much simpler method will be concocted that is suitable for the Basic Stamp microcomputer.

5.14.1 Graph of S1 servo values

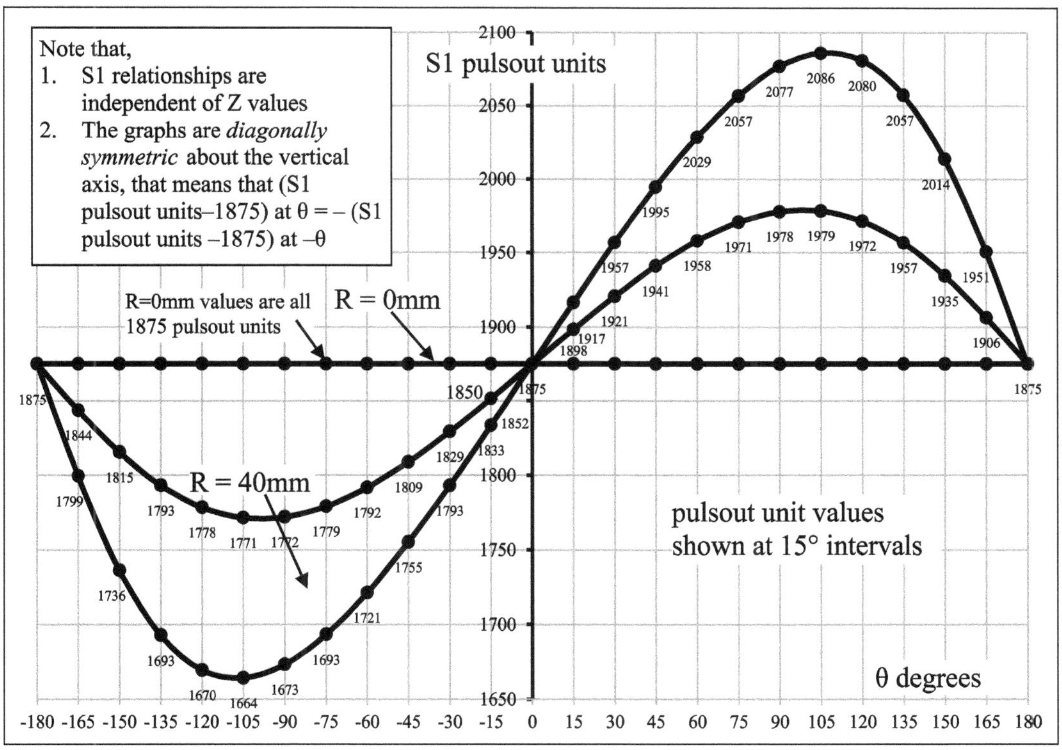

Figure 5.25 Servo S1 relationships at boundaries of leg tip cylindrical coordinate frame.

5.14.2 Graph of servo S2 values

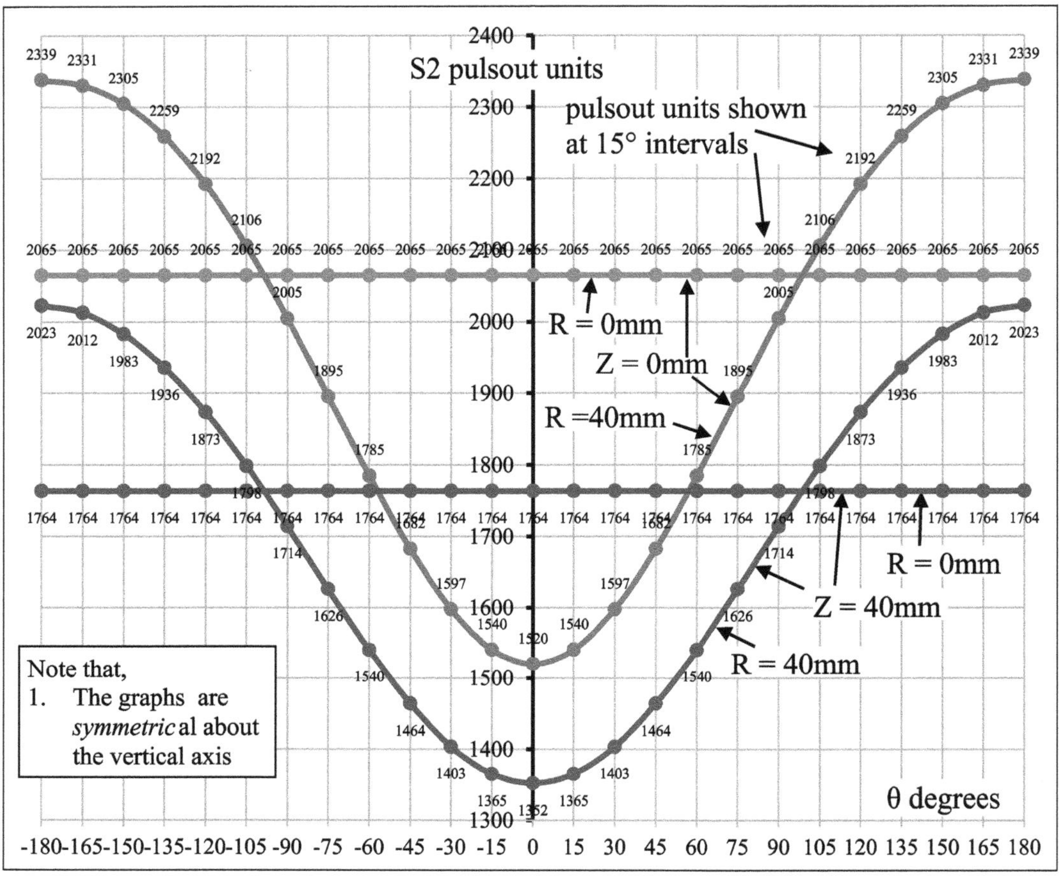

Figure 5.26 Servo S2 relationships at boundaries of leg tip cylindrical coordinate frame.

5.14.3 Graph of servo S3 values

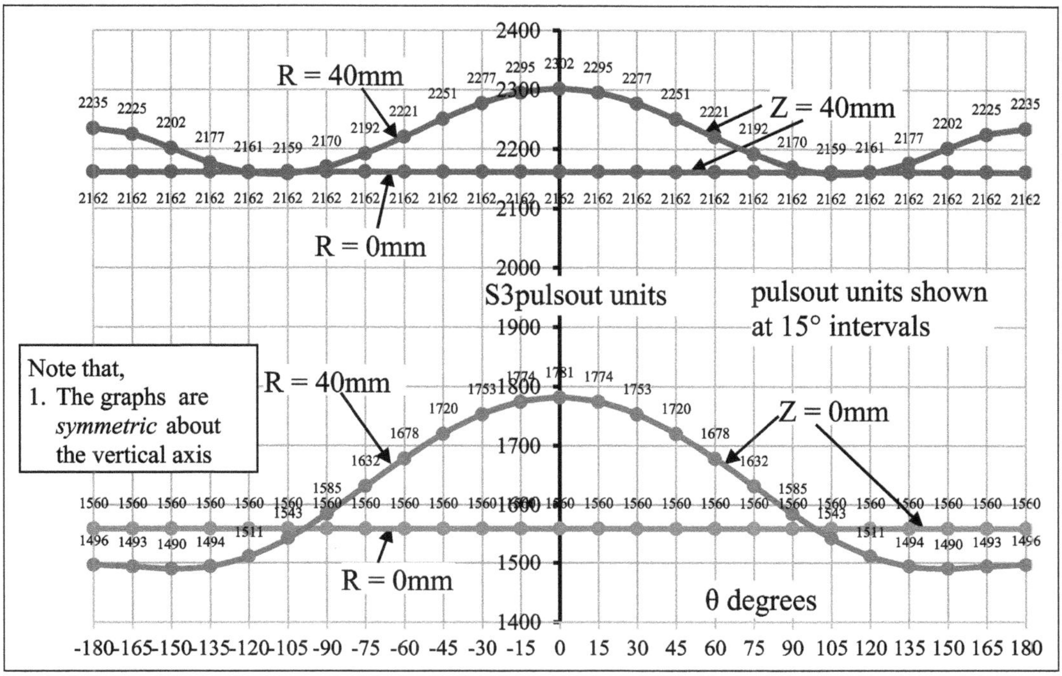

Figure 5.27 Servo S3 relationships at boundaries of leg tip cylindrical coordinate frame.

6. Inverse kinematics synthesis: computing of S1, S2, S3 from cylindrical coordinates, R, Z, θ

6.1 Introduction

The previous chapter, chapter 5, formulated the inverse kinematics equations that gave servo angles, S1, S2, S3 as functions of R, Z, θ. This chapter now formulates how to compute these functions, which are largely trigonometrical functions, with an integer-only 16-bit computer, namely the Basic Stamp microcomputer. The chapter concludes with the complete code for controlling one 3dof leg. Computation will only call upon four arithmetic instructions, which are, addition, subtraction, multiplication and division. Trigonometrical functions will not be called upon because, (i) the Basic Stamp does not possess such instructions, (ii) they are not needed and (iii) trigonometrical functions have non-unique solutions. Instead, we will compute the inverse kinematics equations using piecewise truncated Taylor series that rely solely on the ability to add, subtract, multiply and divide integer numbers as shown with equation 6.1, which is a 3-variable Taylor series equation.

$$S = \sum_{k=0}^{k=k_T} \sum_{j=0}^{j=j_T} \sum_{i=0}^{i=i_T} C_{ijk}.R^i.\theta^j.Z^k \qquad \text{Eq. 6.1}$$

..where, S is the servo angle in pulsout units; C_{ijk} are coefficients of each Taylor term

i, j, k are positive integers; R, θ, Z are the three variables

T subscript means 'truncated'

> Note that there are, $(i_T+1).(j_T+1)$ coefficients.
> E.g. $i_T=2$, $j_T=1$, then = (2+1)(1+1) = 6 coefficients.

This type of equation, because it is truncated, is usually called an 'approximate' equation. Exact equations, in engineering are of no use because engineering has to rely on practical measurement and manufacturing to acceptable tolerances. As such, these equations, eq. 6.1, are called 'accurate' equations. Otherwise, the main feature of these accurate equations is high computation efficiency which means that they, (i) require low program memory space, (ii) can be executed rapidly and (iii) give accurate solutions. By dodging and evading the temptation of using complex math functions and decimal point numbers, we dramatically increase the computing efficiency. This comes at a cost, which is the increased offline design and thinking time to formulate integer-based truncated Taylor series. However, it is well worth it in terms of computing efficiency and for student understanding, satisfaction and appreciation of computational methodology. This chapter will use eq. 6.1 to formulate three separate and independent inverse kinematics equations and computer code for servos S1, S2 and S3 as a function of variables R, θ, Z. Before that is done, we have to decide on the size and shape of the 'pieces' for the piecewise truncated Taylor series equations.

6.2 The piecewise construction of the cylindrical coordinate frame

In order to develop the piecewise truncated Taylor series inverse kinematics equations for S1, S2, S3, we need to divide up the cylindrical coordinate frame into manageable pieces as pictured in figure 6.1, which is a plan view of the cylindrical coordinate frame. The pieces are 15° radial 'cake slices' marked from T0 to T11 covering a clockwise-going semicircle starting from the cylindrical coordinate angle, $\theta=0°$ datum. The counter clockwise-going semicircle pieces are unmarked because the piecewise Taylor series equations for this semicircle are the same as the clockwise-going semicircle. You can see this in the previous chapter, figures 5.25, 5.26 and 5.27, where the relationships are either symmetric or diagonally symmetric about the servo pulsout units vertical axes. The symmetry about the $\theta=0°$ datum line means that the equations for each cake slice can be used in the mirror imaged slice on the opposite side of the cylindrical coordinate frame.

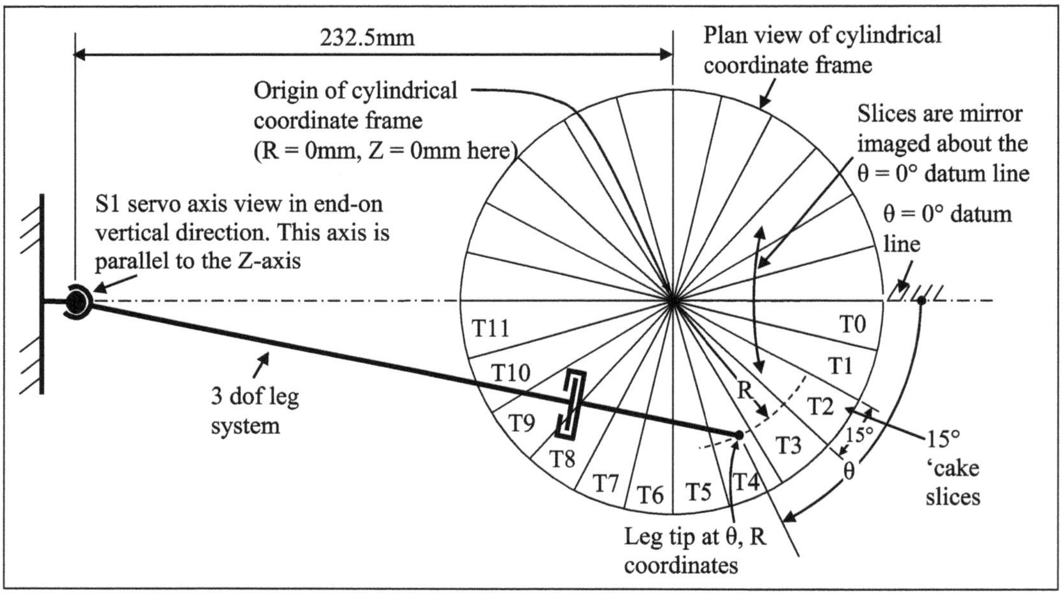

Figure 6.1 Plan view, in the +Z-axis direction, of 3dof leg system showing the cylindrical coordinate frame 'cake slice' pieces, T1 to T12. Note the R and θ directions of the cylindrical coordinate frame.

We choose 15° 'cake' slices because they are considered slim enough such that the arcuate circumference at 40mm radius is within 0.5mm of a straight line as shown in figure 6.2. A leg tip positioning tolerance of +/−0.5mm is chosen since it seems a reasonable achievable error based on the servos, the microcomputer and the mechanical system that makes up the 3dof leg system. More importantly, such a tolerance will provide suitable dead-reckoning, sensor-less accuracy when the Rickshaw and Hexapod walking robots are negotiating the ground plane table. The equations that will be derived will assume that the leg tip travels in a straight line along the arcuate circumference. Hence, 15° slices will produce a tolerable error as the leg tip travels along the outer circumference of the cylindrical coordinate frame.

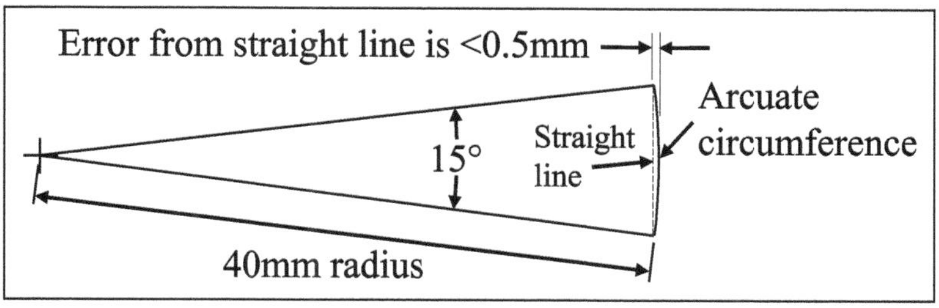

Figure 6.2 Reason for choosing 15° 'cake slice'. The arcuate circumference is within 0.5mm of a straight line.

Figure 6.1 is a 2-dimensional plan view in the direction of the Z-axis, of the cylindrical coordinate frame and is suitable for the synthesis of the S1 servo truncated Taylor series equations. The reason is because the S1 servo is independent of the Z-axis so there is no need to view the Z-axis. On the other hand the S2 and S3 servo equations *do* depend on the Z-axis. As a consequence, figure 6.3 shows the 15° cake slice pieces of figure 6.1 projected into the volume of the cylindrical coordinate frame. These volumetric cake slice pieces are used to synthesise the S2 and S3 truncated Taylor series equations. We now conclude the description of the piecewise construction of the cylindrical coordinate frame and move on to synthesising the S1, S2, S3 inverse kinematic equations.

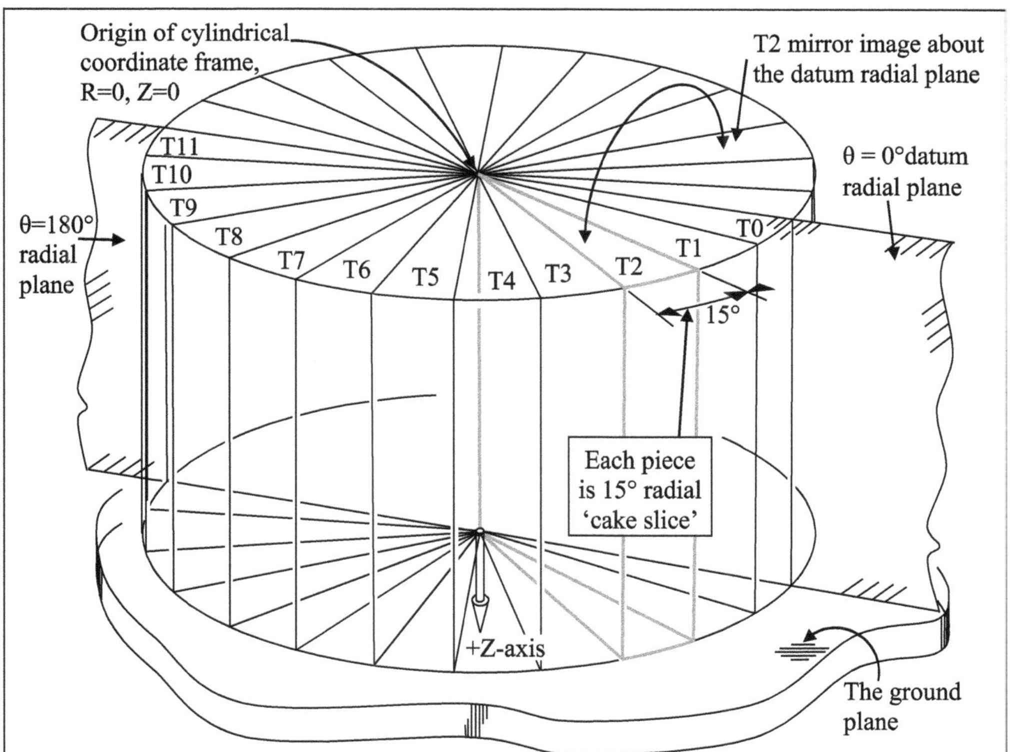

Figure 6.3 Volumetric view of the cylindrical coordinate frame shown divided into 24 pieces of 15° radial 'cake slices'. Twelve of the slices are marked T0 to T11 and the remaining twelve slices are mirror images of the T0 to T11 slices about the datum radial plane.

6.3 Computing the S1 servo angle with 1st and 2nd order equations

For the case of the S1 servo there is no dependence on the Z coordinate so eq. 6.1 is modified to eq. 6.2, as shown here. The equation is a 2-variable Taylor series.

$$S1=\sum_{j=0}^{j=j_T}\sum_{i=0}^{i=i_T} C_{ij}.R^{i}.\theta^{j} \quad \leftarrow \text{Z component removed} \qquad \text{Eq. 6.2}$$

Furthermore, we set $i_T = 2$ and $j_T = 1$ and expand out so eq. 6.2 becomes eq. 6.3, as below.

$$S1 = C_{00}R^0\theta^0 + C_{01}R^0\theta^1 + C_{10}R^1\theta^0 + C_{11}R^1\theta^1 + C_{20}R^2\theta^0 + C_{21}R^2\theta^1 \qquad \text{Eq. 6.3}$$

Further simplifying, we obtain eq. 6.4 below for which there are 12 sets of 6 coefficients to be derived for each cake slice sector.

$$S1 = C_{00} + C_{01}\theta + C_{10}R + C_{11}R\theta + C_{20}R^2 + C_{21}R^2\theta \qquad \text{Eq. 6.4}$$

6 coefficients to be derived

We have set the maximum power of R equal to 2 ($i_T = 2$) because R varies from 0mm to 40mm and if $i_T=1$ then S1 will not be computed accurately enough. However, since the cake slice is only a 15° sector and so the maximum circumferential distance at 40mm radius is approximately 10mm which is significantly less than the maximum radius and so $j_T = 1$ will give sufficient accuracy. These truncated values are the result of trial and error by the author to obtain the required tolerance of less than 0.5mm. The next step is to derive the 6 coefficients for each of the 12 sectors. This is done by choosing five points or nodes in every 15^0 slice where the servo pulsout units are S_1, S_2, s_3, S_4, S_5 at each point, see figure 6.4. For each slice the value of $\Delta\theta$ is measured CW from the beginning of the slice.

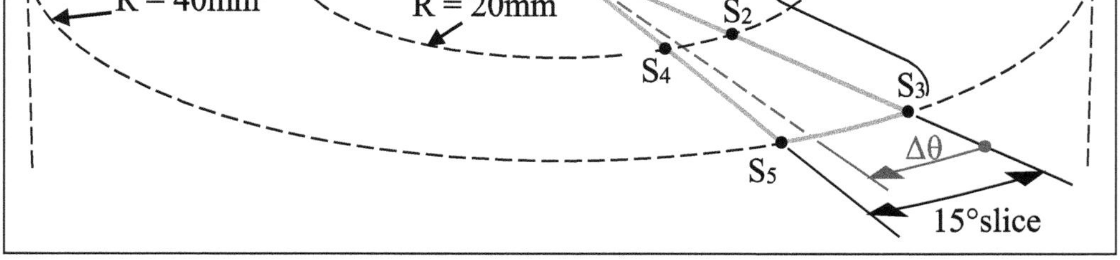

Figure 6.4 Perspective view of top surface of the cylindrical coordinate frame showing just one 15° cake slice for which an inverse kinematics equation will be derived using the 5 nodes, S_1, S_2, S_3, S_4, S_5 for computing S1 pulsout units as a function of R and θ.

We proceed as follows to obtain the 6 coefficients. First, assume second order equations as a function of R ($i_T = 2$) at the two extreme radial angles of $\Delta\theta = 0^0$, and, $\Delta\theta = 15^0$, see graphs below, figure 6.5 and relate the S_1, S_2, s_3, S_4, S_5 pulsout values to figure 6.4 above.

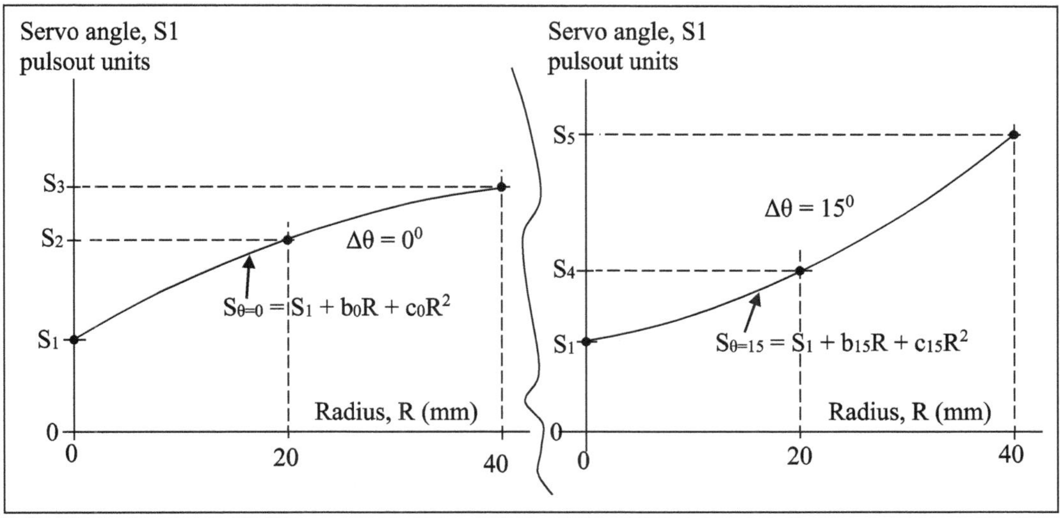

Figure 6.5 Typical graphs of S1 servo pulsout units at $\Delta\theta$=0° and $\Delta\theta$=15° at R=0mm, 20mm and 40mm.

Now write these equations as: (be careful to differentiate between S1 and S_1)

Equation for R at $\Delta\theta = 0^0$, $S1_{\Delta\theta=0} = S_1 + b_0R + c_0R^2$ Eq. 6.5

Equation for R at $\Delta\theta = 15^0$, $S1_{\Delta\theta=15} = S_1 + b_{15}R + c_{15}R^2$

...where b_0, c_0, b_{15}, and c_{15}, are coefficients. Eq. 6.6

Using Cramer's Rule or otherwise, we obtain the coefficients of the above equations as follows:

$$S1_{\Delta\theta=0} = S_1 + b_0R + c_0R^2 = S_1 + \left(\frac{-3S_1 + 4S_2 - S_3}{40}\right)R + \left(\frac{2S_1 - 4S_2 + 2S_3}{1600}\right)R^2 \quad \text{Eq. 6.7}$$

$$S1_{\Delta\theta=15} = S_1 + b_{15}R + c_{15}R^2 = S_1 + \left(\frac{-3S_1 + 4S_4 - S_5}{40}\right)R + \left(\frac{2S_1 - 4S_4 + 2S_5}{1600}\right)R^2 \quad \text{Eq. 6.8}$$

Now make the b and c coefficients become a linear function of $\Delta\theta$ as follows, eq.6.9. This equation is significant because it is used in the Basic Stamp2SX inverse kinematics code shown later in this chapter. Notably that the b and c coefficients are computed separately and both further separated into two sub components each, making four sub components, which are, B0, B1, C0, C1.

Four sub components → B0, B1 (b coefficient); C0, C1 (c coefficient)

$$S1=S_1+\left(b_0+(b_{15}-b_0)\frac{\Delta\theta}{15}\right)R+\left(c_0+(c_{15}-c_0)\frac{\Delta\theta}{15}\right)R^2 \qquad \text{Eq. 6.9}$$

Now separating out the coefficients and noting the correspondence with eq. 6.4 previously;

$$S1 = S_1 + 0 + b_0R + \left(\frac{b_{15}-b_0}{15}\right)\Delta\theta.R + c_0R^2 + \left(\frac{c_{15}-c_0}{15}\right)\Delta\theta.R^2 \qquad \text{Eq. 6.10}$$

$$S1 = C_{00} + C_{01}\theta + C_{10}R + C_{11}R\theta + C_{20}R^2 + C_{21}R^2\theta \qquad \text{Eq. 6.4}$$

Now substituting values for b_0, b_{15}, c_0 and c_{15} from eq. 6.7 and eq. 6.8 we obtain the final equation for servo S1:-

$$S1 = S_1 + \left(\frac{-3S_1+4S_2-S_3}{40}\right)R + \left(\frac{-4S_2+S_3+4S_4-S_5}{600}\right)\Delta\theta.R + \left(\frac{2S_1-4S_2+2S_3}{1600}\right)R^2 \ldots$$
$$+\left(\frac{4S_2-2S_3-4S_4+2S_5}{24000}\right)\Delta\theta.R^2$$

Eq. 6.11

Now check eq. 6.11 at the 5 points, S_1, S_2, S_3, S_4, S_5 by referring to these points in figure 6.4

R=0; θ=don't care	$S1 = S_1 + 0 + 0 + 0 + 0 = S_1$ ✓ correct
R=20; θ=0	$S1 = S_1 - 3S_1/2 + 2S_2 - S_3/2 + 0 + S_1/2 + S_3/2 - S_2 + 0 = S2$ ✓ correct
R=40; θ=0	$S1 = S_1 - 3S_1 + 4S_2 - S_3 + 0 + 2S_1 - 4S_2 + 2S_3 + 0 = S_3$ ✓ correct
R=20; θ=15	$S1 = S_1 - 3S_1/2 + 2S_2 - S_3/2 - 2S_2 + S_3/2 + 2S_4 - S_5/2 + S_1/2 + S_3/2 - S_2 + \ldots$ $+S_2 - S_3/2 - S_4 + S_5/2 = S_4$ ✓ correct
R=40; θ=15	$S1 = S_1 - 3S_1 + 4S_2 - S_3 - 4S_2 + S_3 + 4S_4 - S_5 + 2S_1 - 4S_2 + 2S_3 + \ldots$ $+ 4S_2 - 2S_3 - 4S_4 + 2S_5 = S_5$ ✓ correct

Ok good, so our equations check out correctly. Let's now formulate equations for the Basic Stamp2SX to implement inverse kinematics equations for the S1 servo. After some trial and error computation where real time execution duration and program memory size were assessed, it was decided not to use eq. 6.11 but instead to use eq. 6.9 which separates the equation into the b and c coefficients and furthermore into two components for each of the coefficients, i.e. B0, B1, C0, C1. Figure 6.6 shows a table of these coefficients given to 3 sig. fig. precision.

Slice sector	θ Coordinate range	b coefficient (3 sig.figs)		c coefficient (3 sig. figs.)		C0 coefficients all made positive by adding 0.0200
		B0	B1	C0	C1	$C0_{plus}$
T0	0°-15°	0.00	1.30	0.0000	-0.0067	0.0200
T1	15°-30°	1.30	1.24	-0.0067	-0.0055	0.0133
T2	30°-45°	2.54	1.08	-0.0122	-0.0034	0.0078
T3	45°-60°	3.62	0.86	-0.0156	-0.0003	0.0044
T4	60°-75°	4.48	0.56	-0.0159	0.0033	0.0041
T5	75°-90°	5.04	0.22	-0.0126	0.0071	0.0074
T6	90°-105°	5.26	-0.17	-0.0055	0.0100	0.0145
T7	105°-120°	5.09	-0.57	0.0045	0.0108	0.0245
T8	120°-135°	4.52	-0.90	0.0153	0.0081	0.0353
T9	135°-150°	3.62	-1.13	0.0234	0.0010	0.0434
T10	150°-165°	2.49	-1.24	0.0244	-0.0085	0.0444
T11	165°-180°	1.25	-1.25	0.0159	-0.0159	0.0359

Figure 6.6 Table of inverse coefficients for computing the S1 servo inverse kinematics equations.

Using these coefficients together with eq. 6.9, we now show how the coefficients are used to compute the S1 servo angle in pulsout units in a few examples as follows.

Assume in all examples that S_1 =1875 pulsout units which gives a pulse width of 1.5ms for servo S1.

Example 1 Assume $\theta = 108°$ and R = 40mm
Since $\theta = 108°$, slice is T7 and $\Delta\theta = 108° - (7 \times 15°) = 3°$
$S1 = 1875 + (B0 + B1.\Delta\theta/15)R + (C0 + C1.\Delta\theta/15)$
Thus; $S1 = 1875 + (5.09 - 0.57.\Delta\theta/15)R + (0.0045 - 0.0108.\Delta\theta/15)R^2$
Substituting values for $\Delta\theta$ and R:
$S1 = 1875 + (5.09 - (0.57 \times 3/15)) \times 40 + (0.0045 + (0.0108 \times 3/15) \times 40^2$
Thus S1 = 1875 + 210
So, servo S1 = 2085 pulsout units (Check and confirm, figure 5.25 in previous chapter)

Example 2 Assume $\theta = 252°$ ($= -108°$) and R = 40mm
Slice is in mirror image of T7 slice and so using data in example 1 above we get;
S1 = 1875 – 210 (NOT S1 = 1875 + 210 due to diagonal symmetry)
So servo S1 = 1665 pulsout units (Check and confirm, figure 5.25 in previous chapter)

It is necessary to explain the furthest right column in figure 6.6 above that is headed, 'C0 coefficients all made positive by adding 0.200'. The reason for this column is due to some of the C0 coefficients being negative and this may cause trouble with the Basic Stamp during the multiplication of negative numbers. So to overcome this potential trouble, we add 0.0200 to all the C0 coefficients to make them all positive and rename the positive a coefficients as $C0_{plus}$. Thus the Basic Stamp equation for computing S1 servo values is shown as follows: The equation to be formulated in Basic Stamp format is:

This term brings the $C0_{plus}$ coefficient back to its correct value

$$S1 = 1875 + (B0 + B1.\Delta\theta/15)R + (C0_{plus} + C1.\Delta\theta/15)R^2 - 0.0200R^2 \quad \text{Eq. 6.12}$$

We do this as follows with an example with $\theta = 108°$ and R = 40mm

For this example, slice = T7, B0=5.09, B1= −0.57, $C0_{plus}$=0.0245, C1=0.0108

Let,

b coefficient = twelvesects
c coefficient = thetemp

Don't worry about the strange named variables; they are recycled from another application. The complete programme code is shown at end of the chapter

Just before writing the code, we introduce a gain factor that accounts for servo calibration. All the coefficients in the table, figure 6.6 have been worked out for a servo calibration of 12.0 pulsout units/° but what about is servos have a different calibration? For example, suppose experiments show that the servo calibration is in fact 12.5 pulsout units/°; how do we modify our equations to accommodate this calibration shift? We do it by multiplying the S1 values by a gain factor called, 'S1gain' where S1gain is in integer percentage units. So the integer percentage gain for a 12.5pulsoutunits/° calibration value is integer number value of 12.5/12.0 × 100% = 104. Now it is important to realise that we multiply the pulsout unit value by the gain that is above or below the mid range 1875 value and not to multiply the 1875 value by the gain because servos are accurately calibrated to give the mid-range position as 1875 pulsout units, i.e. 1875 × 0.8μs = 1.5ms pulse width. So the 1875 value does not have a gain factor associated with it but it does have a positive or negative integer offset value added to it because the servo actuating horn or lever cannot be accurately located at the correct angle aligned with its 0° datum due to practical difficulties. The way to solve this problem is to use a software offset, called 'S1off'.

Ok, enough said, let's get back to the task in hand and write some sample Basic Stamp, PBasic code for eq. 6.12 as follows.

Note that here we get an integer result

```
S1gain =104          'assume a servo calibration gain of 104%
S1off= −136          'assume a servo offset to be added to the 1875 midrange position
Δθ = θ − ((θ/15)*15)          'Δθ=108−((108/15)*15)=108−((7)*15)=108−105=3
twelvesects = (509 + 57*Δθ/15)     'coefficients are multiplied 100 to give integers
thetemp = (245 + 108*Δθ/15)        'coefficients are multiplied 10000 to give integers
S1 =(twelvesects*R/100)+(thetemp*R/20*R/500) − (200*R/20*R/500)*S1gain/100
S1 = 1875 + S1off + S1
```

This is midrange servo angle value (1.5ms pulse width) and is midrange because of diagonal symmetry

Basic Stamp works from left to right and completes calculations in brackets so putting S1gain/100 at the right hand end multiplies the whole result by the gain factor

Now suppose the leg tip has to be positioned at $\theta = 252°$ ($= -108°$) which is the mirror image of the T7 slice. The S1 function involving twelvesects and thetemp is now subtracted from the S1 function in the final line. The final line of the PBasic programme is now:

```
S1 = 1875 + S1off – S1
```

This time, due to diagonal symmetry, S1 is subtracted from the function

So there you have it. That's the synthesis of the S1 servo inverse kinematics computation completed in its general form. The full programme code is shown at the end of the chapter. Now we move on to explain the S2 and S3 servo computation.

6.4 Computing the S2 and S3 servos with 1st order equations

The S2 and S3 servo equations are more complex than the S1 servo because the inverse kinematics equations concern three variables rather than two. The S2 and S3 equations are a function of R, θ, Z. Because the maximum ranges of R and Z are both 40mm then the equations involving R and Z should be 2nd order, i.e. truncated values set equal to, $i_T=2$, $k_T=2$. The θ variable, having a maximum range of approximately 10mm, can still stay as, $j_T=1$. So with $i_T=2$, $j_T=1$, $k_T=2$, the number of coefficients, see eq. 6.1, is $(2+1)(1+1)(2+1)=18$ which is a large number and three times the number of coefficients of the S1 servo. The Basic Stamp can compute these complex equations using its 8 pages of 2kbytes/page but we make a difficult choice and decide to simplify matters at the expense of accuracy by using just one 8kbyte page and use $i_T= j_T= k_T=1$, i.e. 1st order equations throughout, thus generating $(1+1)(1+1)(1+1) =8$ coefficients. Thus we set out to solve equations for S2 and S3 using the generic eq. 6.13.

$$S=\sum_{k=0}^{k=1}\sum_{j=0}^{j=1}\sum_{i=0}^{i=1}C_{ijk}.R^{i}.\theta^{j}.Z^{k} \qquad \text{Eq. 6.13}$$

Eq. 6.13 will produce the following expanded equation, eq. 6.14, with 8 coefficients.

$$S = C_{000} + C_{001}Z + C_{010}\theta + C_{100}R + C_{011}\theta Z+ C_{101}RZ + C_{110}R\theta + C_{111}R\theta Z \qquad \text{Eq. 6.14}$$

8 coefficients to be derived

There will be a just a small amount of programme page space in our one page memory to add corrections and this will be shown later. Ok, let's start working out these 1st order 3-variable equations for S2 and S3 using generic eq. 6.14. We will use generic values for servo pulsout units to generate an equation that can be used for both S2 and S3 servos. Figure 6.7 is a more detailed view of the cylindrical coordinate frame, previous figure 6.3.

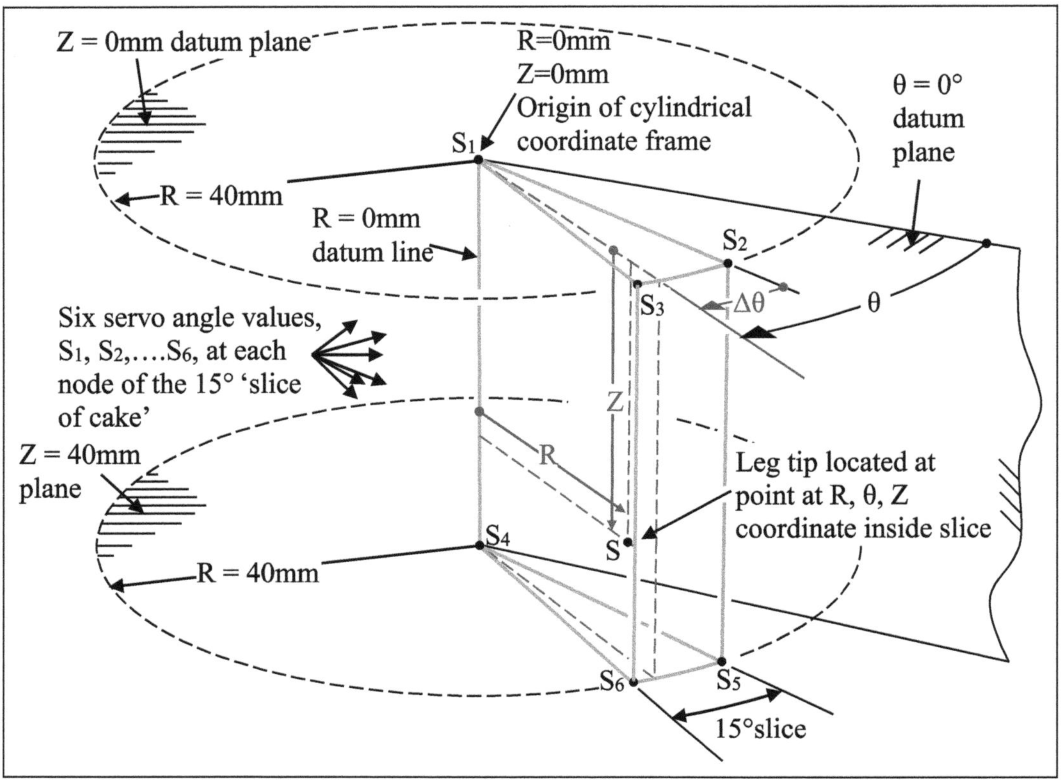

Figure 6.7 Diagram illustrating 15^0 slice method for computing leg tip servo angle, S_P, located at cylindrical coordinates, R, Δθ, Z.

Using figure 6.7, let the servo angle values at each node of the slice be equal to, S_1, S_2, S_3, S_4, S_5, and S_6. We wish to derive an 8-coefficient, 3-variable Taylor series according to eq. 6.14. We proceed as follows using figure 6.7 above. Start first at the Z=0mm plane using the S_1, S_2 and S_3 values.

Equations at; $\Delta\theta = 0$, $Z = 0$

$$S = S_1 + \frac{(S_2 - S_1)R}{40} \qquad \text{eq. 6.15}$$

Equations at; $\Delta\theta = 15$, $Z = 0$

$$S = S_1 + \frac{(S_3 - S_1)R}{40} \qquad \text{eq. 6.16}$$

Equations at; $\Delta\theta = 0 \text{ to} 15$, $Z = 0$

Coefficient =0

$$S = \left\{S_1 + \frac{(S_3 - S_1)}{15}.\Delta\theta\right\} + \left\{\frac{(S_2 - S_1)}{40} + \left(\frac{(S_3 - S_1) - (S_2 - S_1)}{40 \times 15}\right).\Delta\theta\right\}R$$

$$S = S_1 + \frac{(S_2 - S_1)}{40}.R + \frac{(S_3 - S_2)}{600}.\Delta\theta.R \qquad \text{Eq. 6.17}$$

A similar procedure for the Z = 40mm plane, using S_4, S_5 and S_6 values, produces the following equation:-

Equation for the range ; $\Delta\theta = 0$ to15
and $Z = 40$

$$S = S_4 + \frac{(S_5 - S_4).R}{40} + \frac{(S_6 - S_5).\Delta\theta.R}{600} \qquad \text{Eq. 6.18}$$

Now linearising as a function of Z between equations 6.17 and 6.18 we obtain;

$$S = S_1 + \frac{(S_4 - S_1)}{40}.Z + \frac{(S_2 - S_1)}{40}.R + \left(\frac{(S_5\text{-}S_4) - (S_2 - S_1)}{40\ \ 40}\right)Z.R + \frac{(S_3 - S_2)}{600}.\Delta\theta.R + \left(\frac{(S_6 - S_5) - (S_3 - S_2)}{600\ \ 40}\right)Z.\Delta\theta.R$$

Simplifying:

$$\underbrace{S = S_1 + \frac{(S_4 - S_1)}{40}.Z + \frac{(S_2 - S_1)}{40}.R + \left(\frac{S_5 - S_4 - S_2 + S_1}{1600}\right).Z.R + \frac{(S_3 - S_2)}{600}.\Delta\theta.R + \left(\frac{S_6 - S_5 - S_3 + S_2}{24000}\right).Z.\Delta\theta.R}$$

Eq. 6.18

Only 6 terms, 2 terms missing! Which ones? They are, $C_{010}\theta$ and $C_{011}\theta.Z$. Why are they missing? Because the two radial planes at $\Delta\theta=0°$ and $\Delta\theta=15°$ intersect producing two equal servo values at S1 and S4.

To make eq. 6.18 suitable for the Basic Stamp programme code, we modify it by extracting $(a + b\Delta\theta/15)$ factor components to form five coefficients as shown in the following equation, eq. 6.19.

Five coefficients → A0; First $(a + b\Delta\theta/15)$ component: B0, B1; Second $(a + b\Delta\theta/15)$ component: C0, C1

$$S = S + \overbrace{\left(\frac{S_4 \quad S_1}{40}\right)}^{A0}. + \left(\overbrace{\left(\frac{S_2 \quad S_1}{40}\right)}^{B0} + \overbrace{\left(\frac{S_3 - S_2}{40}\right)}^{B1}.\frac{\Delta\theta}{15}\right)R + \left(\overbrace{\left(\frac{S_5 - S_4 - S_2 + S_1}{1600}\right)}^{C0} + \overbrace{\left(\frac{S_6 - S_5 - S_3 \quad S_2}{1600}\right)}^{C1}.\frac{\Delta\theta}{15}\right)RZ \qquad \text{Eq. 6.19}$$

Equation 6.19 is similar to the equation for servo S1, eq. 6.9. The next step is to calculate the five coefficients, A0, B0, B1, C0, C1 for each slice. These values are shown in the table in figure 6.8. Note that the S2 coefficients, B0 and C0, have negative values so constant values are added to them to equal $B0_{plus}$ and $C0_{plus}$ respectively so that the coefficients are always positive. Once again, this is to aid the basic Stamp computation. Computation of the S2 servo pulsout value is as follows.

slice sector	θ coordinate range	S2 Coefficients (3 sig figs based on $B0_{plus}$ and $C0_{plus}$)					B0 coefficients all made positive by adding 15.0 $B0_{plus}$ (3sig figs)	C0 coefficients all made positive by adding 0.020 $C0_{plus}$ (3sig figs)
		A0	B0	B1	C0	C1		
T0	0°-15°	-7.53	-13.6	0.5	0.0831	-0.004	1.4	0.103
T1	15°-30°	-7.53	-13.1	1.4	0.0788	-0.012	1.9	0.099
T2	30°-45°	-7.53	-11.7	2.1	0.0669	-0.015	3.3	0.087
T3	45°-60°	-7.53	-9.6	2.6	0.0519	-0.017	5.4	0.072
T4	60°-75°	-7.53	-7.0	2.7	0.0350	-0.015	8.0	0.055
T5	75°-90°	-7.53	-4.3	2.8	0.0200	-0.014	10.7	0.040
T6	90°-105°	-7.53	-1.5	2.5	0.0063	-0.011	13.5	0.026
T7	105°-120°	-7.53	1.0	2.2	-0.0044	-0.007	16.0	0.016
T8	120°-135°	-7.53	3.2	1.7	-0.0113	-0.003	18.2	0.009
T9	135°-150°	-7.53	4.9	1.1	-0.0138	0.001	19.9	0.006
T10	150°-165°	-7.53	6.0	0.7	-0.0131	0.002	21.0	0.007
T11	165°-180°	-7.53	6.7	0.2	-0.0113	0.002	21.7	0.009

Figure 6.8 Table of inverse kinematics coefficients for computing S2 servo angle.

6.5 Example computing of S2 in T7 sector

Assume that the R,θ, Z values are, 40mm, 108°, 40mm. The angle of 108° puts the cylindrical slice in T7 sector so the coefficients, A0, $B0_{plus}$, B1, $C0_{plus}$, C1 are taken from figure 6.8 slice sector T7. The equation to compute S2 is eq. 6.20:

Don't worry about the strange variable names ⟶ twelvesects, thetemp

$$S2 = S_1 + A0.Z + \overbrace{(B0_{plus} + B1.\Delta\theta/15)}^{\text{twelvesects}} R + \overbrace{(C0_{plus} + C1.\Delta\theta/15)}^{\text{thetemp}} R.Z \quad \text{Eq. 6.20}$$

The sample code is as follows;

```
Δθ = θ – ((θ/15)*15)            'Δθ=108–((108/15)*15)=108–((7)*15)=108–105=3
twelvesects = (160 + 22*Δθ/15)  'coefficients are multiplied by 10 to give integers
thetemp = ( 16 – 7*Δθ/15)       'coefficients are multiplied by 1000 to give integers
S2 = 0–(753*Z/100)+(twelvesects*R/10)+(thetemp*R/25*Z/40)–(15*R)–(R*Z*20/1000)
S2 = 1875 + S1off ± S1          'include offset values
```

This value is set to position the full operating range of the S3 values to within the max and min values of the servo, i.e. 1250 to 2500 pulsout units

This sign changes to negative for mirror image sector slices

These terms bring the $B0_{plus}$ and $C0_{plus}$ coefficients back to B0 and C0

Ok that's the S2 inverse kinematics computation sorted out now we move on to the S3 computation which is the same in principle as for S2.

6.6 Servo S3 computation

We follow the same procedure as for the S2 servo by utilising the data in figure 5.27 of the previous chapter, together with eq. 6.19 to produce the table of S3 coefficients in figure 6.9.

Cylindrical slice	θ coordinate range	S3 Coefficients (3 sig figs based on A0, $B0_{plus}$ and $C0_{plus}$ coefficients)					B0 coefficients all made positive by adding 2.00	C0 coefficients all made positive by adding 0.060
		A0	B0	B1	C0	C1	$B0_{plus}$ (3sig figs)	$C0_{plus}$ (3sig figs)
T0	0°-15°	15.1	5.53	-0.18	-0.0506	0.000	7.53	0.009
T1	15°-30°	15.1	5.35	-0.52	-0.0506	0.002	7.35	0.009
T2	30°-45°	15.1	4.83	-0.83	-0.0488	0.004	6.83	0.011
T3	45°-60°	15.1	4.00	-1.05	-0.0444	0.008	6.00	0.016
T4	60°-75°	15.1	2.95	-1.15	-0.0369	0.011	4.95	0.023
T5	75°-90°	15.1	1.80	-1.17	-0.0263	0.016	3.80	0.034
T6	90°-105°	15.1	0.63	-1.06	-0.0106	0.019	2.63	0.049
T7	105°-120°	15.1	-0.43	-0.80	0.0088	0.021	1.57	0.069
T8	120°-135°	15.1	-1.23	-0.42	0.0300	0.021	0.77	0.090
T9	135°-150°	15.1	-1.65	-0.10	0.0506	0.018	0.35	0.111
T10	150°-165°	15.1	-1.75	0.07	0.0688	0.013	0.25	0.129
T11	165°-180°	15.1	-1.68	0.08	0.0813	0.004	0.32	0.141

Figure 6.9 Table of inverse kinematics coefficients for computing S3 servo angle.

Note in figure 6.9 that some of the B0 and C0 coefficients are negative so this may cause problems once again with the Basic Stamp programme. To solve the problem, similar to the S2 servo, the coefficients are made positive by adding constant values to create new coefficients, $B0+2.00=B0_{plus}$ and $C0+0.060=C0_{plus}$, respectively.

And now, we take the A, $B0_{plus}$, B1, $C0_{plus}$, C1 coefficients in figure 6.9 to produce our equation, eq.6.21, for computing S3.

Once again, don't worry about the strange variable names → bp1 bp2

$$S3 = S_1 + A0.Z + \overbrace{(B0_{plus} + B1.\Delta\theta/15)}^{bp1} R + \overbrace{(C0_{plus} + C1.\Delta\theta/15)}^{bp2} R.Z \qquad \text{Eq. 6.21}$$

..and once again we use a similar algorithm to compute S3 with Basic Stamp code as follows using, as an example, R,θ, Z values equal to, 40mm, 108°, 40mm respectively.

```
Δθ = θ – ((θ/15)*15)        'Δθ=108–((108/15)*15)=108–((7)*15)=108–105=3
bp1 = (157 – 80*Δθ/15)      'coefficients are multiplied by 100 to give integers
bp2 = ( 69 + 21*Δθ/15)      'coefficients are multiplied by 1000 to give integers
S3 =  (151*Z/10)+(bp1*R/100)+(bp2*R/25*Z/40) – (2*R) – (R*Z*6/100)
S3 = 1875 + S1off + S1      'include offset values
```

This value is set to position the full operating range of the S3 values to within the max and min values of the servo, i.e. 1250 to 2500 pulsout units	**This sign changes to negative for mirror image sector slices**	These terms bring the $B0_{plus}$ and $C0_{plus}$ coefficients back to B0 and C0

So that's it, we're done. It has been shown how all servo values are calculated, albeit lacking some accuracy due to the 1st order polynomial equations, using Taylor series and furthermore how the series are computed with the Basic Stamp microcomputer. The complete Basic Stamp code for one 3dof leg is given in chapter 8 where the inverse kinematics algorithm described in this chapter is given in detail. The code has many comments describing the algorithm and the computation method. A more detailed and highly accurate derivation of the inverse kinematics equations using higher order Taylor series is shown in a paper entitled 'A Tabular Format for Computing Inverse Kinematic Equations for a 3DOF Robot Leg' authored by F. Nickols and published in the open access journal, the International Journal of Advanced Robotic Systems, Vol. 6, No. 3 (2009) ISSN 1729–8806, pp. 177–200. You can download a free copy at this website, https://journals.sagepub.com/doi/pdf/10.5772/7234

The complete Basic Stamp code of chapter 8 includes some corrections to the errors caused by using 1st order Taylor series. The correction method we will now describe.

6.7 Corrections

The inverse kinematics equations for the S2 and S3 servos are 1st order Taylor equations which produce accurate leg tip positioning at the nodes, S_1, S_2, S_3, S_4, S_5, S_6, of the piecewise cake slice of the leg tip cylindrical coordinate frame but otherwise produce warping of the slice; see figure 6.10, 'before corrections'. To correct for these errors, the complete Basic Stamp code for one degree of freedom leg, chapter 8, has a short subroutine at the end of the code to correct for the S2 and S3 servos. The S1 servo is accurate since it has been formulated in 2nd order Taylor equations. The short correction code is as follows;

```
corrections:        S2=S2+(twelvesects*R/100)-(thetemp*R/25*R/400)
                    S3=S3-(        bp1*R/100)+(     bp2*R/25*R/400)
                    RETURN
```

This section of code corrects the S2 and S3 servos for the ground plane at Z=40mm which is the height that the leg tips are mainly set for walking and serves as some degree of correction for the remaining warped slice; see figure 6.10, 'after corrections'. The corrections are a function of radius, R only and so the errors in the slice largely remain uncorrected between the top and ground planes of the slice.

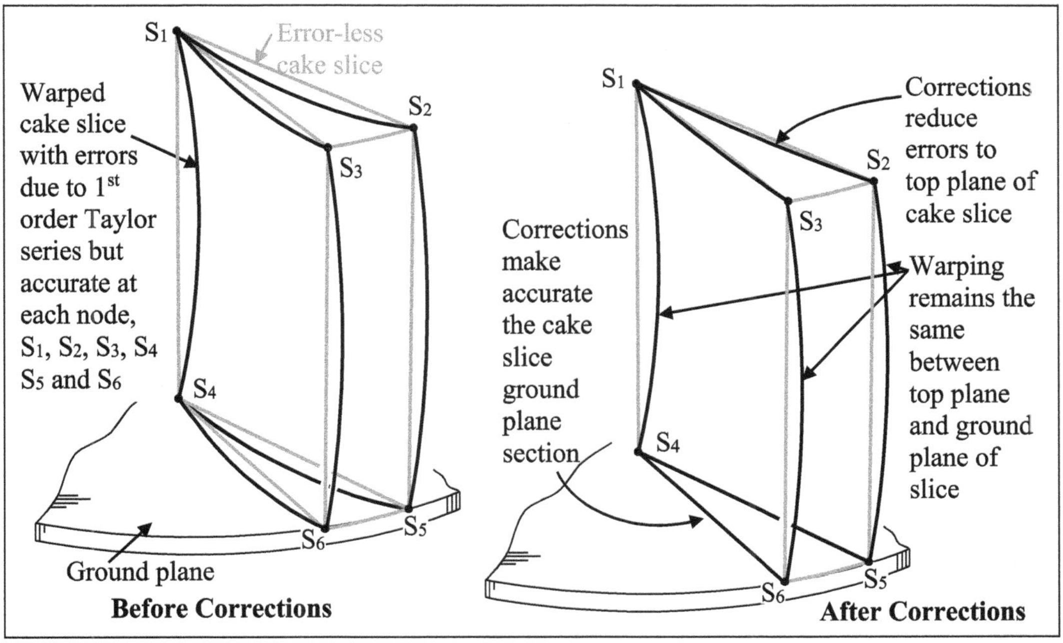

Figure 6.10 Before and after cake slice shapes due to corrections being applied to it.

7. Inverse kinematics using the ServoTeach module: a project-based, hands-on solution using experimentation

7.1 Introduction

This chapter is written in the style of a project-based, hands-on teaching and learning experiment based tutorial. As such it is very helpful in understanding the inverse kinematic Taylor series equations as defined in chapter 6, eq. 6.1. It is a stand-alone chapter in that it can be used by students to derive an elementary robot leg walking algorithm together with its real-time code.

In chapter 6 the robot leg inverse kinematics equations were solved using conventional trigonometry. Such a theoretical approach is fine and is an excellent study, albeit tedious, for students to derive the equations for a given robot design. However, there is an alternative practical approach to originate the inverse kinematics equations that uses experimentation allied with the ServoTeach Module; see experimentation system, figure 7.1. But what about advantages and disadvantages of the two methods? Well, the trigonometrical method takes significant development time to derive the equations, so that is a disadvantage, but is intrinsically highly accurate, so that's an advantage. It also needs calibration due to manufacturing tolerances in the mechanical system, the microcomputer system and the servos, so that's another disadvantage. On the other hand, the experimental method is also accurate and naturally includes calibration of manufacturing tolerances, (advantage), but as soon as any part of the system is changed then calibration has to be repeated, (disadvantage). But then, so also, does the theoretical approach require recalibration when a part is changed.

Interestingly, there is a third hybrid method which is to enlist the help of a highly accurate computer-aided-design and draughting software package that can be used to *do the experiments virtually* that measures servo angles. This hybrid method is left as an idea for students to explore based on the ideas for doing inverse kinematics presented in this book.

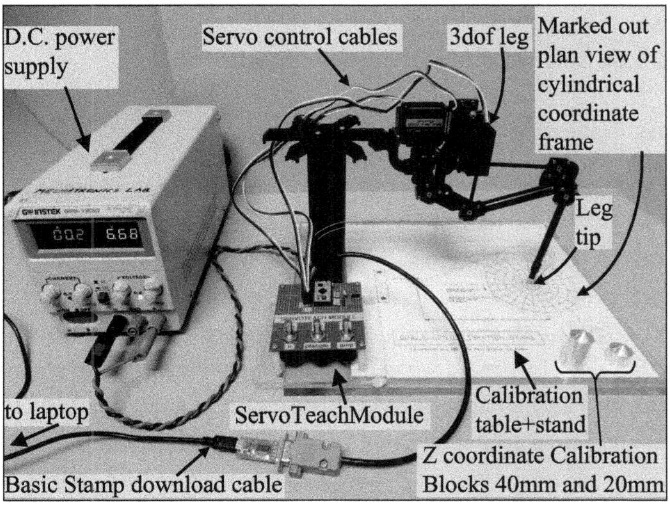

Figure 7.1 Experimentation system. The laptop computer is out of sight but connected to the download cable.

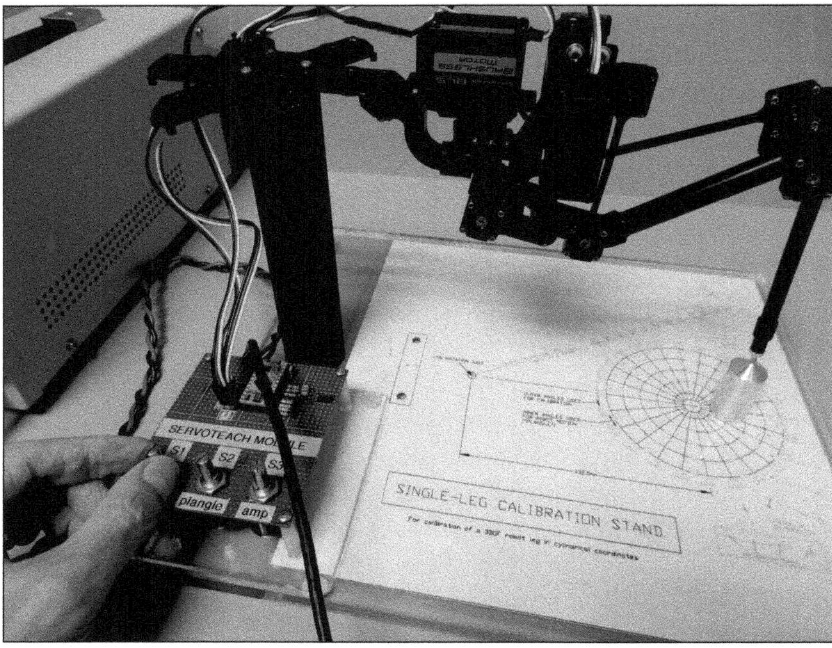

Figure 7.2 Closer view showing ServoTeach Module being manually adjusted to place the leg tip just touching the calibration block that represents a node on a sector slice.

Otherwise, this chapter describes an inverse kinematics method using the ServoTeach Module to measure the servo angles at each node of the cylindrical coordinate sector slices. (Recall sector slices, figure 6.7). The servo angles are measured by mounting a 3 degree-of-freedom robot leg on a calibration stand plus table and then manually operating the ServoTeach Module, figure 7.2, to place the leg tip at each node of the twelve slices of the cylindrical coordinate frame. A plan view of the cylindrical coordinate frame is marked out on the calibration table that shows R and θ coordinates and height calibrated aluminium blocks are used for the Z coordinate.

Ok, enough of the introduction, let's now set about doing the experiment.

7.2 Description of apparatus - overview

For the experiment hardware we will use a different version of the 3 dof leg and its calibration stand shown in figures 7.1 and 7.2, which show a black anodised aluminium version with the geometry as detailed in the drawings of chapter 5. The black anodising, whilst looking classy and professional, does not show up clearly in photographs. As such, for the forthcoming experiment description, we will use a trimmed down, 'skinny', non-anodised version of the leg, figure 7.3, which shows up clearly in photographs. The skinny leg was used as an early prototype version of the hexapod robot. It has identical geometry to the black version so serves as a perfectly valid device for the experimentation.

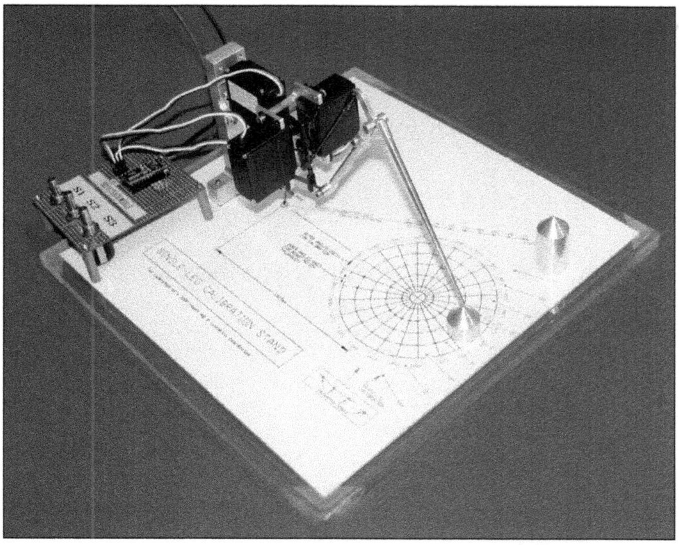

Figure 7.3 A skinny version of the 3dof leg used in the forthcoming experimentation.

7.3 Description of apparatus - detailed

Figure 7.4 shows a close-up view of the servoteach module which is part of the experimentation system shown in figure 7.1, adjacent. Note the components;

1. Basic Stamp2SX microcomputer.
2. Basic Stamp programming download cable.
3. Three variable resistor knobs that will be used manually to move the robot leg tip in 3-d cylindrical coordinate space.
4. Nominal 6V d.c. electrical power supply
5. Three servo connectors

The servoteach module was described in detail in chapter 2 which also lists the module code. This code is to be downloaded into the servoteach module in order to do the experimentation in this chapter.

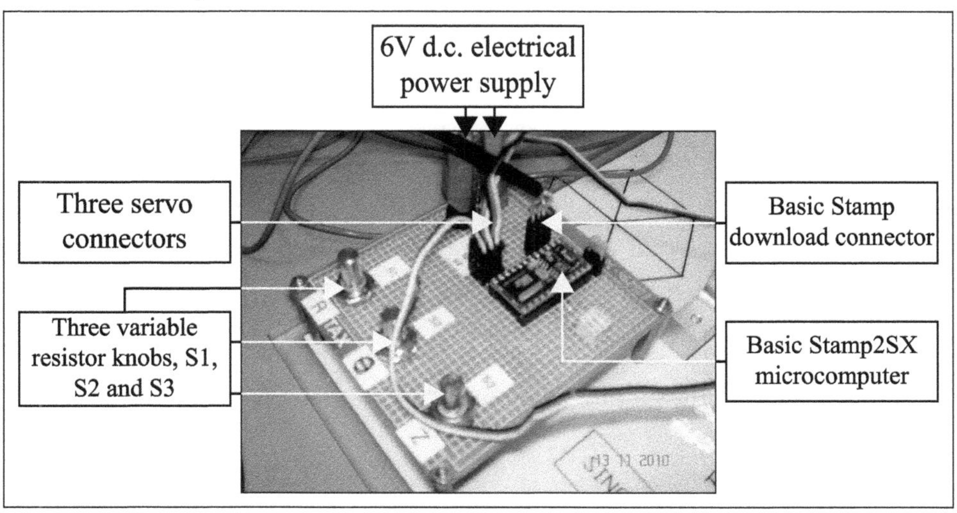

Figure 7.4 The ServoTeach Module set up to control servos S1, S2 and S3.

Figure 7.5 shows a close-up view of the robot leg tip and the calibration tools which consist of three types as follows:

1. Z blocks (2 pieces, 20mm and 40mm high) which are used to calibrate the leg tip in the Z vertical direction for any R, θ coordinate..
2. A Cartesian calibration plane which is used to determine leg tip errors in an R Z vertical plane for any θ.
3. A cylindrical calibration plane which is used to calibrate the leg tip in an R θ horizontal plane for Z=40mm.

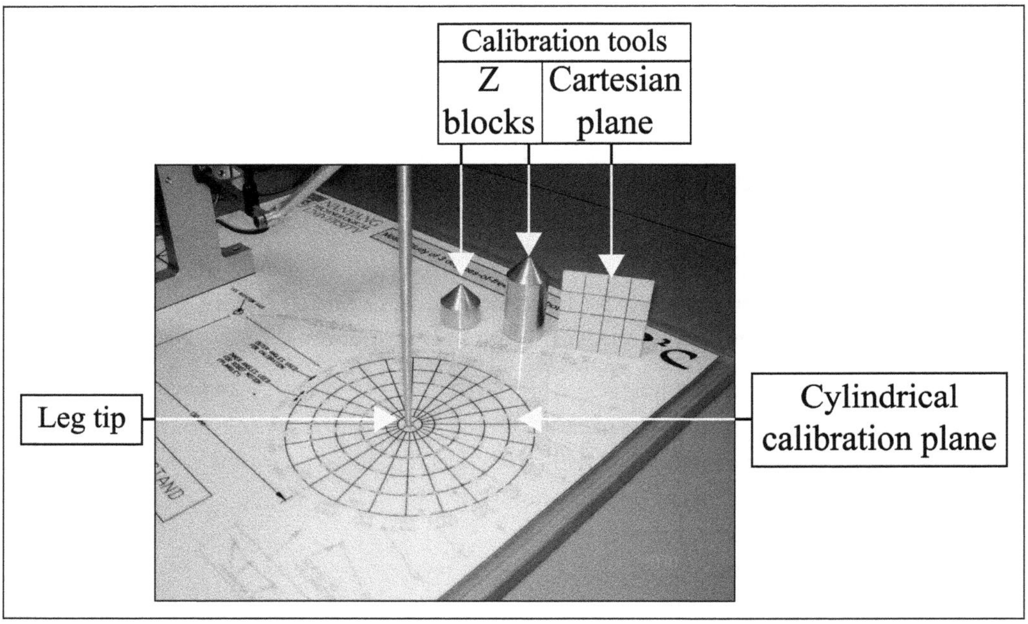

Figure 7.5 The Cylindrical Coordinate Calibration space and tools.

7.4 Manual control of S1, S2 and S3 servomechanisms

Let's do a test to see the robot leg moving under manual control using the ServoTeach Module, figure 7.6. Switch on power to the ServoTeach Module then open the Basic Stamp editor on your personal computer or laptop then copy and paste the BS2SX program, *servoteach with backlash filtering.bsx, from chapter 2* into the Basic Stamp editor then run the program by downloading it to the servoteach/programming module.

Almost immediately the servos kick into action and the leg will move in the directions according to your adjustments of the three variable resistor knobs, marked, S1, S2 and S3, see figure 7.7. A debug window will open on your computer screen and show you the three angle control numbers, S1, S2 and S3 being sent to servos, like this:

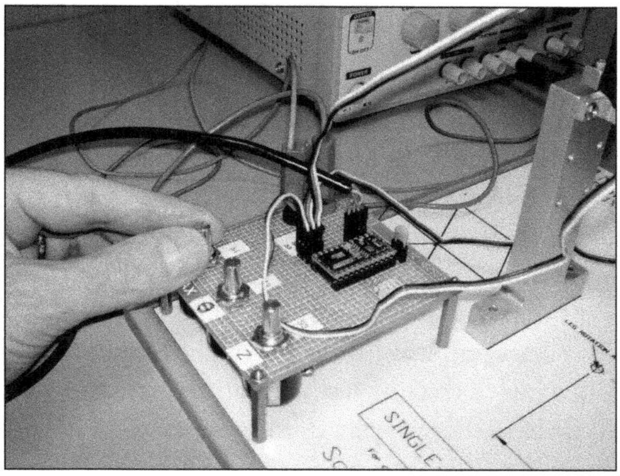

Figure 7.6 Manual adjustment of the servos, S1, S2 and S3 using the ServoTeach Module.

S1=2050 S2=1981 S3=1609

The servo numbers, S1, S2 and S3 are called *pulsout numbers*. They are integers; they do not have a decimal point!

Now play with the ServoTeach Module knobs to move the leg tip around. You will notice that:

1. S1 knob moves the leg tip in a left/right arc.
2. S2 knob moves the leg tip in a forward/backward arc.
3. S3 knob moves the leg tip up in an up/down arc.

Our problem now is how to programme the servos to control the leg tip in R, θ, Z cylindrical coordinates. Let's start simple by programming the leg tip to travel in a straight line along the R-axis.

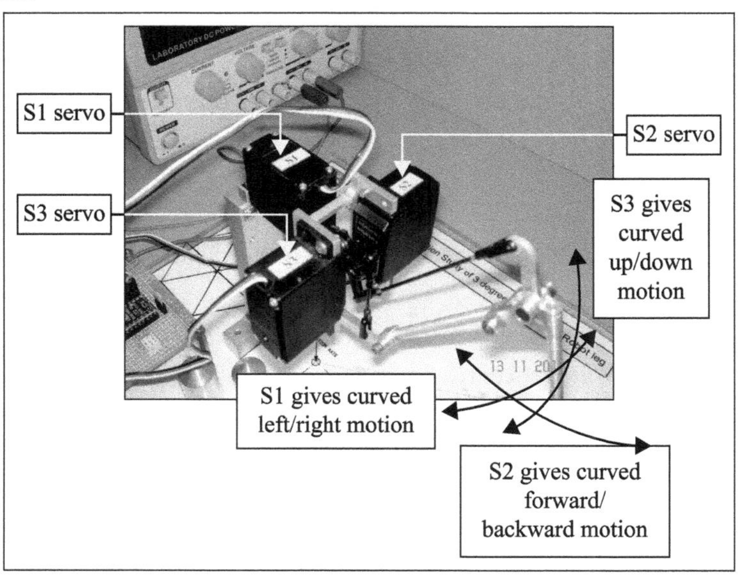

Figure 7.7 Servo number annotation and adjustment of actuation directions using the ServoTeach Module.

7.5 Steps in programming leg tip straight line motion along R-axis at θ=0°, Z=0mm

We will now develop an inverse kinematics equation that will control the leg tip to travel along the R-axis in figure 7.8. This line is located at Z = 0mm and θ = 0°. In order for the leg tip to travel along this line, the value of radius, R is programmed to increase from R = 0mm to R = 40mm with the S1 knob being used as the R input controller. We will write a program that will move the leg tip in 1mm steps controlled by the S1 knob. This means that the value of R will be 0, 1, 2, 3,.....38, 39, 40. If we use a basic 1st order Taylor series equation then a straight line will not be possible. Instead the leg tip will travel in a curve thus deviating from its straight line but will start accurately at R=0mm and finish accurately at R=40mm. Later a 2nd order equation will be used to straighten the curve. Ok, let's get started with the experimentation; we start with step 1.

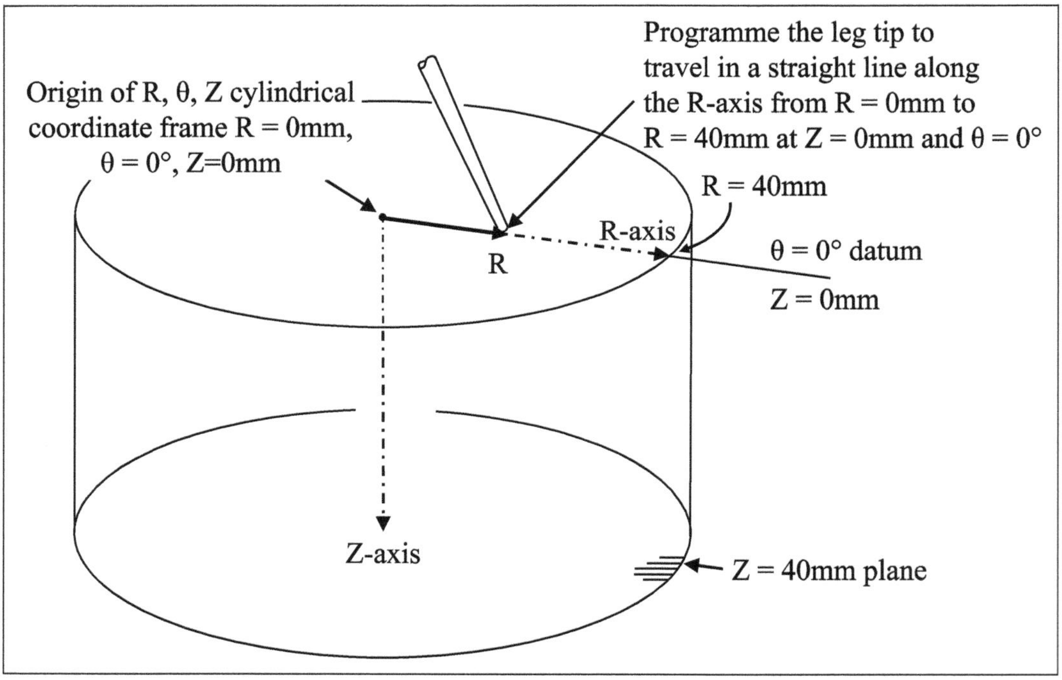

Figure 7.8 Programming of leg tip to travel in a straight line.

7.5.1 Step 1 Calibrate coordinates, R=0mm, Z=0mm, θ = 0°

Place the 40mm Z block at the centre position of the cylindrical coordinate frame, figure 7.9 and adjust the three ServoTeach Module knobs to place the leg tip at the top point of the Z block. The leg tip is now located at the origin of the cylindrical coordinate frame at R = 0mm, θ = 0°, Z = 0mm. Read and record the values of S1, S2 and S3 from the debug window. The values shown in figure 7.10 are the values recorded by the author, and are shown as an example.

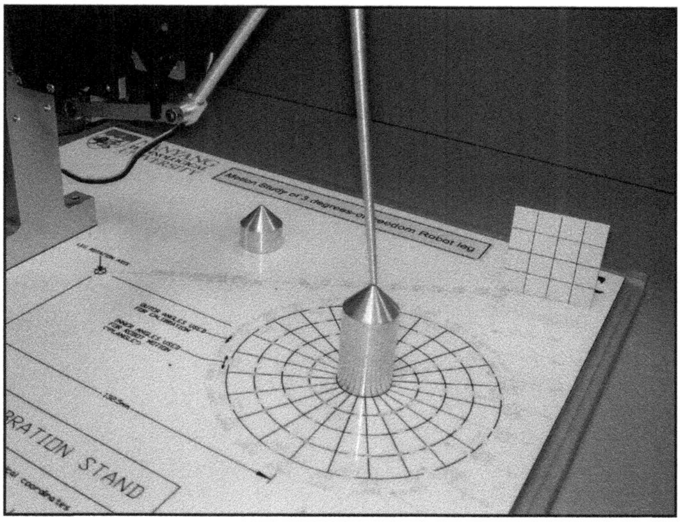

Figure 7.9 Leg tip at origin, R = 0mm, θ = 0°, Z = 0mm.

S1 = 1911 S2 = 2081 S3 = 1269

Figure 7.10 Servo values at Z=0mm, θ=0°, R=0mm.

Your values will be different and could be up to a few hundred. Don't worry, this is what calibration is all about…everybody will get different values. The numbers, which represent the servo angles, will be called *pulsout* numbers and the angle units will be expressed as *pulsout units*.

7.5.2 Step 2 Calibrate coordinates, R=40mm, Z=0mm, θ = 0°

For the 2nd step we choose to calibrate the leg tip at the point, R=40mm, θ=0° Z=0mm. Figure 7.11 shows the 40mm Z block moved to give this point. Make sure you accurately place the Z block on the calibration table according to figure 7.12.

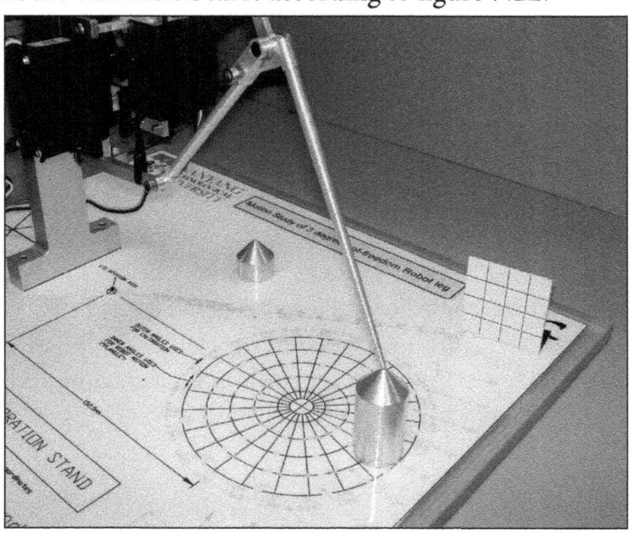

Figure 7.11 Leg tip at R=40mm, θ=0°, Z=0mm.

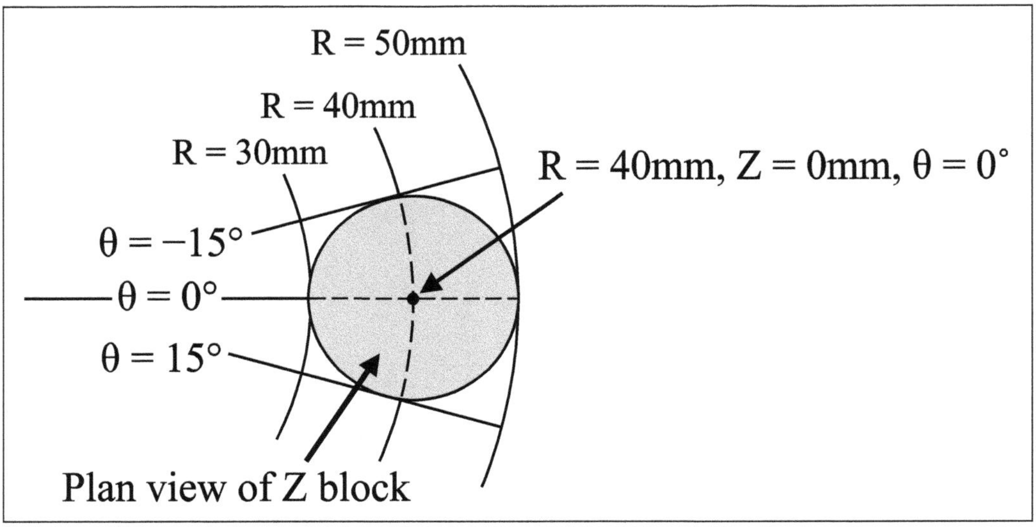

Figure 7.12 Ensure the Z block is placed correctly for locating the R=40mm, θ=0°, Z=0mm point.

7.5.3 Step 3 Record servo values

Now record the servo values, figure 7.13. Notice that the S1 value hasn't changed because no rotation by S1 was necessary because the cylindrical coordinate frame radial datum line, θ=0°, intersects the leg rotation axis.

S1 = 1911	**S2 = 1472**	**S3 = 1508**

Figure 7.13 Servo values at Z=0mm, θ=0°, R=40mm.

7.5.4 Step 4 Derive 1st order inverse kinematics equations

Now we have enough numbers to derive 1st order inverse kinematics equations. Put your results in an Excel spreadsheet and generate three graphs and their linear trendline equations shown on the graphs, figure 7.14. Note that the coefficients of R have been set to 3 sig. fig. precision based on the largest coefficient in its set. For example, this means that the largest coefficient of R is 15.2 which has been set to 3 sig. fig. precision. This number has 1 dec. pl. and so the decimal places of the other two coefficients must be set to 1 dec. pl. However, we must check the error of working to 1 dec. pl. with the R coefficient. The error is +/−0.05 (the error of 1 dec. pl.) multiplied by the largest value of R which is 40 thus giving an error of +/−2 pulsout units. Let us work to a total computed error not exceeding 3 pulsout units which is an error that was found to give leg tip positioning errors not exceeding 0.5mm.

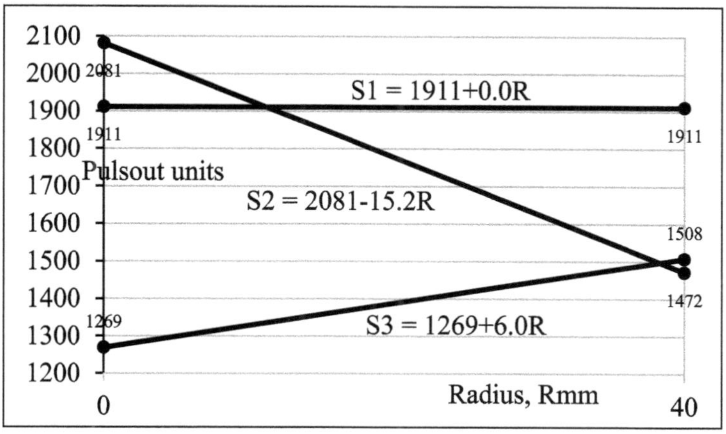

$S1 = a_1 + b_1R$
$= 1911 + 0.00R$

$S2 = a_2 + b_2R$
$= 2081 - 15.2R$

$S3 = a_3 + b_3R$
$= 1269 + 6.0R$

The 1st order coefficients and trendline equations

Figure 7.14 Graph of S1, S2, S3 pulsout numbers against radius, R for θ=0° and Z=0mm. Also shown are 1st order linear trendline equations.

7.5.5 Step 5 Formulate Basic Stamp code using 1st order equations

Now use the equations in figure 7.14 to formulate a Basic Stamp2SX program. First of all the program should read the value of the S1 knob which is the 1st knob of the ServoTeach Module and transform this knob value into an R value which will rise in integers from 0, 1, 2, 3,....up to 39, 40. We will then expect the leg tip to travel along the, θ = 0°, Z = 0mm, radial line in 1mm steps from R = 0mm, 1mm, 2mm,...up to 39mm, 40mm. Ok let's try it with the following program, figure 7.15 below.

Enter your own a, b and c coefficients for S1, S2 and S3 into the equations in figure 7.15 adjacent. The integers that you modify are highlighted in yellow. Now run this program in the ServoTeach Module, thus over writing the *servoteach with backlash filtering.bsx* programme. If you hear any disturbing noises coming from your servos then you have made a mistake in entering the coefficients. What you do now is to do a debug operation by activating the two lines just before the pulsout instructions. You can now adjust the S1 knob and check what numbers are wrong and why. As soon as you have solved the problem then deactivate those two lines of code before the pulsout instructions and try again.

Note that the Basic Stamp cannot handle decimal places. This is a blessing in disguise because you will develop a greater understanding of efficient precision computation.

In fact all digital computers that are in common use cannot handle the decimal point at machine code and assembler level. Computers can only do decimal point calculations due to software algorithms. *Significant figures* and *integers*, and *not decimal point numbers*, are the key ingredients to understanding accuracy and precision of computation.

```
' {$STAMP BS2sx}
' 1st order inverse kinematics line motion
' 1-dimensional inverse kinematics equation
'for driving leg tip along the datum axis, i.e.
'the radial line, theta=0deg, Z = 0mm from R=0 to
'R=40mm in 1mm steps

R        VAR Word
S1       VAR Word
S2       VAR Word
S3       VAR Word

again:   HIGH 15
         PAUSE 1
         RCTIME 15,1,R
         R=R–1/10 MAX 40
         DEBUG DEC R,CR
         S1=1911 + (  0*R/100)   'a1+ b1R
         S2=2081 – (152*R/ 10)   'a2+ b2R
         S3=1269 + (600*R/100)   'a3+ b3R
'DEBUG DEC S1," ",DEC S2," ",DEC S3,CR
'GOTO again
         PULSOUT 5,S1
         PULSOUT 6,S2
         PULSOUT 7,S3
         GOTO again
```

Measure, set limits and display the ServoTeach Module S1 knob demand value that varies from 0 to 40 in steps of 1 (annotation to lines HIGH 15 … DEBUG DEC R,CR)

Figure 7.15 Program to use S1 knob to position leg tip along a straight line at $\theta=0°$, Z=0mm.

7.5.6 Step 6 Assess precision of programming using 1st order equations

We now use the Cartesian calibration plane, figure 7.16, to assess the accuracy of the inverse kinematics program. By adjusting S1 knob we can run the leg tip in 1mm steps back and forth across the top edge of the calibration plane and at the same time read the value of R on the debug screen. We note that the leg tip positioning is very good at R = 0mm and R = 40mm but there is a significant error at R = 20mm. Here we note that there is approximately 4.5mm of error in the +R direction. The error in the Z direction is much smaller, being less than 0.5mm. Let us now decide on what accuracy we want. Usually +/−1% of full scale deflection is considered good, i.e. +/−1% of 40mm = +/−0.4mm.

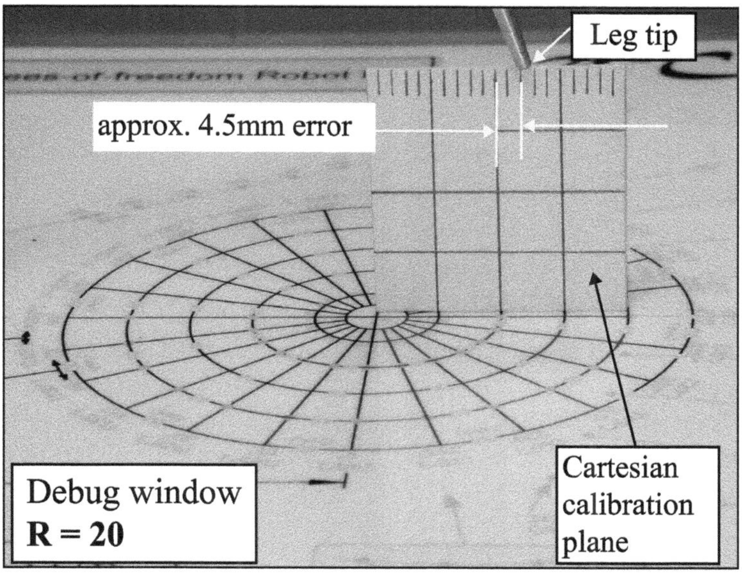

Figure 7.16 Error along radial axis at R = 20mm, θ = 0°, Z = 0mm. You can see approximately 4.5mm of error in the +R direction but negligible error in the Z direction.

The robot leg has noticeable backlash (looseness) in its joints and servos and the servos themselves have some compliance, meaning that the output shaft rotates slightly under load, so we cannot expect our errors to be less than 1%. So let's try for a maximum error of +/−1mm which represents +/−2.5% of 40mm which is our full scale deflection. Actually this will be an *unforced* error of +/−1mm where *unforced* means that we measure errors without any force or load being attached to the leg tip. We now define what is accurate and inaccurate as follows

From now on if the leg tip has a positioning error of greater than +/−1mm then we consider the robot leg system to be *inaccurate*.

From now on if the leg tip has a positioning error of less than or equal to +/−1mm then we consider the robot leg system to be *accurate*.

So the conclusion is that our present straight line system using 1st order inverse kinematics equations is inaccurate!

Don't forget that a system is only accurate or inaccurate if we have previously worked out and specified a maximum error, i.e. a tolerance.

So it is back to the drawing board to figure out a way of making our robot leg accurate over the straight line. The strategy will be to use 2nd order equations instead of a linear (1st order) equation.

7.6 Steps in deriving an accurate inverse kinematics equation using 2nd order Taylor series equation

Ok let's try 2nd order Taylor series equations to reduce the positioning error. Second order equations take the form; $y = a + bx + cx^2$. Three coefficients require 3 points to solve for their values. We already have S1, S2 and S3 values at R = 0mm and 40mm. A good place for the 3rd point is the mid-range point, R = 20mm.

7.6.1 Step 1 Increase accuracy by adding mid-range coordinate

Place the 40mm Z block at the R = 20mm, θ = 0° position, taking care to locate the Z block as shown in figure 7.17. The tip of the Z block will now indicate the coordinate, R = 20mm, θ = 0°, Z = 40mm which is the point where we had 4.5mm error so this will be the point where the error must be reduced.

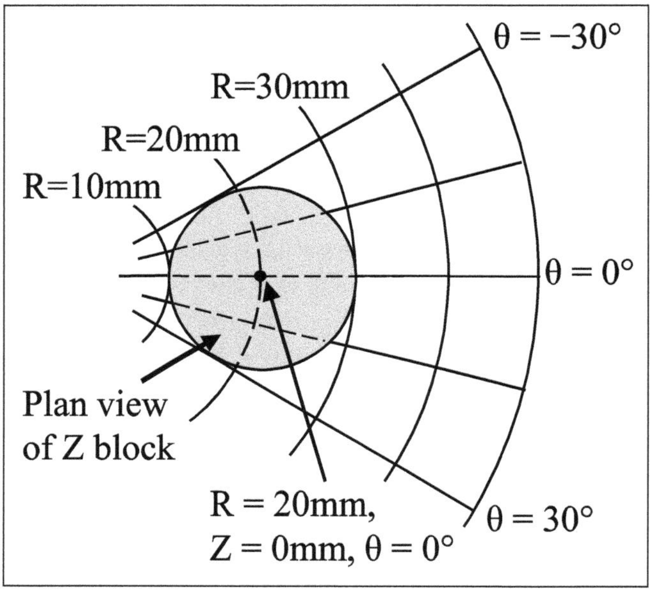

Figure 7.17 Ensure the Z block is placed correctly for locating the Z = 0mm, R = 20mm point.

7.6.2 Step 2 Derive accurate 2nd order inverse kinematics equation

Reload and run the program, *servoteach with backlash filtering.bsx* Adjust the S1, S2 and S3 knobs to place the leg tip at the top point of the Z block, figure 7.18. Now record the S1, S2 and S3 values in Excel spreadsheet. The S1 number should be the same. If it isn't, you should repeat measurements of S1, S2 and S3, using the 40mm Z block at R = 0, 20 and 40mm. We now have a new set of values from which 2nd order equations can be fitted to the three points of each servo. Now plot these values on Excel spreadsheet together with their trendline equations as in figure 7.19.

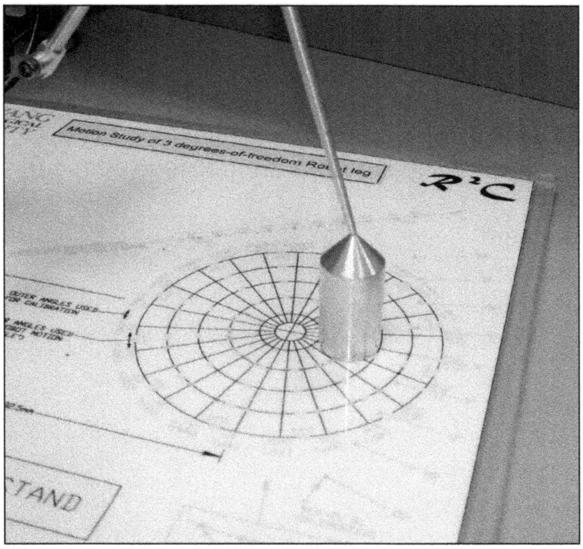

Figure 7.18 Calibration of servos at point, R=20mm, θ=0°, Z=0mm.

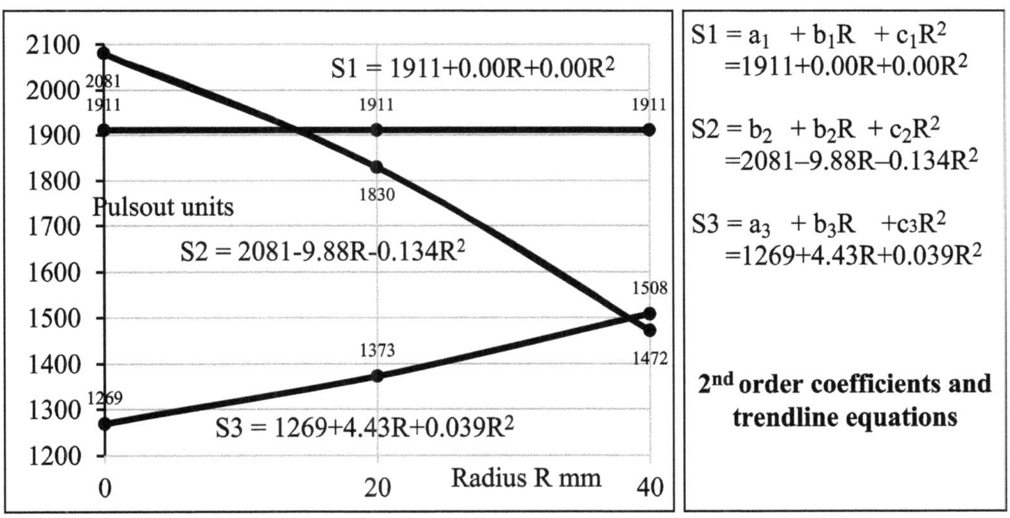

Figure 7.19 Graph of S1, S2, S3 pulsout numbers against radius, R for θ=0° and Z=0mm. Also shown are 2nd order linear trendline equations.

7.6.3 Step 3 Formulate Basic Stamp code using 2nd order equations

This step implements the 2nd order equations in the Basic Stamp, figure 7.20 adjacent, utilizing the coefficients of figure 7.19. Remember to be careful with the coefficients, for example; the maximum sized coefficient of R is –9.88 to 3 sig. fig precision. This number has 2 dec. pl. so the other coefficients of R must also have 2 dec. pl. Similarly with the coefficients of R^2, the largest value is 0.134 to 3 sig. figs. and this number has 3 dec. pl. so that fixes the number of dec. pl. for the remaining coefficients of R^2. Also note that any computation must not exceed the value of 65535 ($=2^{16}-1$). Thus the R^2 numbers are divided by 1000 in two stages, first by 25 then finally by 40

```
' {$STAMP BS2sx}
'2nd order inverse kinematics line motion
' 1-dimensional inverse kinematics equation
'for accurately driving leg tip along the datum axis, i.e.
'the radial line, theta=0deg, Z = 0mm from R=0 to 40mm in 1mm steps

R       VAR Word
S1      VAR Word
S2      VAR Word
S3      VAR Word

again:  HIGH 15
        PAUSE 1
        RCTIME 15,1,R
        R=R-1/10 MAX 40
        DEBUG DEC R,CR
        S1=1911 + (  0*R/100) + (  0*R/25*R/40)   'a1 + b1R + c1R2
        S2=2081 - (988*R/100) - (134*R/25*R/40)   'a2 + b2R + c2R2
        S3=1269 + (443*R/100) + ( 39*R/25*R/40)   'a3 + b3R + c3R2
        'DEBUG DEC S1," ",DEC S2," ",DEC S3,CR
        'GOTO again
        PULSOUT 5,S1
        PULSOUT 6,S2
        PULSOUT 7,S3
        GOTO again
```

Figure 7.20 Accurate programme that uses S1 knob to position leg tip along a straight line at $\theta=0°$, Z=0mm in 1mm increments.

7.6.4 Step 4 Assess precision of programming using 2nd order equations

Write the programme of figure 7.20 into your Basic Stamp editor and download it into the ServoTeach Module then do tests to ratify the servo pulsout values. Activate the pulsout values then adjust S1 knob to see if the leg tip positioning performance is now accurate meaning that it is within tolerance of +/−1mm for the range R=0 to 40mm.

The leg tip is, indeed, accurate over this range and in particular at R=20mm where it can be seen, figure 7.21, that the error has dropped from 4.5mm to less than 0.5mm which is well within our tolerance of +/−1mm. You should also check the errors at R=10mm and R=30mm.

The resolution of the programme is 1mm, meaning that we can only control the R values to the nearest 1mm. How about increasing resolution to 0.1mm so that we can control R values from 0 to 40mm in 0.1mm increments, i.e. 0.0, 0.1, 0.2... 39.9, 40.0mm? Let's do that now.

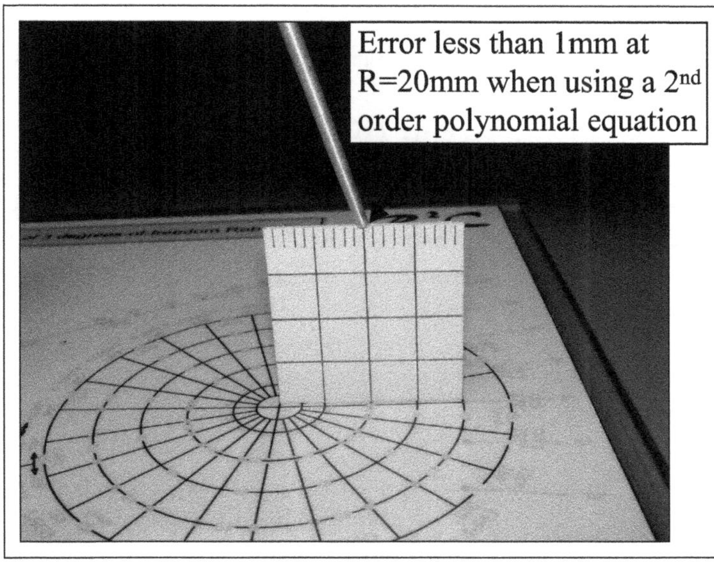

Figure 7.21 Program to use S1 knob to position accurately leg tip along a straight line at Z=0mm, θ=0°.

7.7 Increasing the positioning resolution of the leg tip

We are concerned at the moment in positioning the leg tip along a straight line. Currently you will notice that the leg tip moves in 1mm steps thus resulting in quite jumpy steps. The size of the steps is called the *positioning resolution*, i.e. the positioning resolution of the leg tip is 1mm. Positioning or displacement resolution is defined as the smallest increment that the leg tip can be positioned. Ok so what if we want to increase the resolution. *Increase* means that we want to *decrease* the size of the 1mm steps. Let's say we want 10 times more resolution. In other words, we want a positioning resolution of 0.1mm. How can we do this? The answer is to return to the equations in figure 7.19 and rewrite them as follows, figure 7.22:

```
R=0 to 400 meaning that R increases from 0*0.1mm=0.1mm to 400*0.1mm=40mm
S1=1911 + ( 0*R/100) + (0*R/1000) +( 0*R/250*R/40) – (0*R/25*R/4000)
S2=2081 – (98*R/100) – (8*R/1000) – (13*R/250*R/40) – (4*R/25*R/4000)
S3=1269 + (44*R/100) + (3*R/1000) + ( 3*R/250*R/40) + (9*R/25*R/4000)
```

This coefficient cannot exceed 65535/400 = 163 but for convenience we say it should not exceed 2 sig.figs, i.e. 99

Divide by .100 not 10

Coefficients need to work to 3 sig. figs. So add another component with 1 sig. fig.

Divide by .1000 not 100

Work to 2 sig. figs., i.e. max is 99

Divide by (99*400*400) /65535=244 so round up to 250

Divide by total of 10^4 =250*40

Divide by (9*400*400) /65535=24 so round up to 25

Divide by total of 10^5 = 25*4000

Figure 7.22 Rescaling the 2nd order Taylor series inverse kinematics equations to give 0.1mm resolution and to suit the Basic Stamp. Notes are given for each term.

The RCtime instruction has also to be rewritten so that it is scaled from 0 to 400 like this:
R=R–1/1 MAX 400
Furthermore the DEBUG statement that displays the demanded R value in mm is rewritten as:
DEBUG "R=",DEC R/10,".",DEC R–(R/10*10),"mm",CR

It will show a decimal place, for example, R=6.3mm. Remember the Basic Stamp, as with all digital computers, will only recognize integer numbers. This is a really important concept for science and engineering students to appreciate. Let's say that we rotate the S1 knob to obtain an R value of 6.3mm. This means that the RCtime value will be 63 integer units. The debug line works like this:

R/10 = 63/10 which gives a result of 6 and this is printed to the screen as 6

Then the debug instruction "." prints a full stop which means a decimal point.

Then R–(R/10*10) = 63–(63/10*10)...working from left to right inside the brackets gives 63–(6*10) = 3

..so the number 3 is now printed to the screen then mm is printed to the screen and finally what we see on the screen is, R=6.3mm

The complete programme is shown in figure 7.23 below.

```
' {$STAMP BS2sx}
'2nd order inverse kinematics line motion with 0.1mm  resolution
' 1-dimensional inverse kinematics equation
'for accurately driving leg tip along the datum axis, i.e.
'the radial line, theta=0deg, Z = 0mm from R=0 to 40mm in 0.1mm steps

R      VAR Word
S1     VAR Word
S2     VAR Word
S3     VAR Word

again: HIGH 15
       PAUSE 1
       RCTIME 15,1,R
       R=R–1/1      MAX 400

       DEBUG "R=",DEC R/10,".",DEC R– (R/10*10),"mm",CR

       S1=1911 + (  0*R/100) + (0*R/1000) +(  0*R/250*R/40) – (0*R/25*R/4000)
       S2=2081 – (98*R/100) – (8*R/1000) – (13*R/250*R/40) – (4*R/25*R/4000)
       S3=1269 + (44*R/100) + (3*R/1000) + ( 3*R/250*R/40) + (9*R/25*R/4000)
       PULSOUT 5,S1
       PULSOUT 6,S2
       PULSOUT 7,S3
       GOTO again
```

Figure 7.23 2nd order equation program to give 0.1mm resolution in straight line from R=0 to R=40mm.

After downloading the programme and running it you will notice how much smoother is the motion of the leg tip. The smoothness has the advantage of inputting much less mechanical stepping and shaking noise into the robot. However, there is a disadvantage of fine resolution control that is the maximum leg tip speed is reduced to 0.1mm per 20ms update rate period = 5mm/s maximum speed. If the resolution is decreased to 1mm then the maximum speed is 50mm/s. You can jump steps such that the resolution is decreased to 8mm and in this case the maximum leg tip speed is 8mm/20ms=400mm/s. Don't forget though that the maximum leg tip speed is limited not by the resolution but by the maximum slew speed of the servo that has to be found by experiment. The author has found that it is about 400mm/s which relates to the maximum speed of the robot. The overall diameter of the hexapod robot from leg tip to diametrically opposite leg tip is 465mm so the normalised maximum speed of the robot is 400mm/s divided by 465mm = 0.86 body lengths/s

Since our inverse kinematics equations are getting more complex, especially the next section on planar motion, it is helpful at this point to list some rules, tips and procedures for doing integer math computation with the Basic Stamp.

7.8 Basic Stamp microcomputer maths computation. Some rules, procedures and tips

1. The Basic Stamp carries out calculations from left to right so: 3 × 2+6 will compute 12 whereas 3+2 × 6 will compute 30. Multiplications are not given precedence as in hand held calculators.
2. If you divide integers you do not get fractions or decimal point numbers. So, 3/2=1, the 0.5 is lost. Other examples; 32/2=16; 33/2=16; 36/3=12; 35/3=11; 56/57=0; 2/3=0; 57/56=1; 95/10=9
3. Be careful of over-running (=exceeding) 65,535; for example, the calculation, 95 × 40 × 40/10 is calculated in steps from the furthest left number as firstly 95 × 40 = 3800 which is less than 65,535 so that's ok, then, 3800 × 40 = 152,000 which is greater than 65,535 so that's not ok since we have over-run. The solution is to change the order of multiplication and division like this; 95 × 40/10 × 40 which gives the sequence of number calculations as; 95 × 40=3800 then 3800/10=380 then 380 × 40 =15,200.
4. Do multiplication first and division last. For example, 95 × 40/10=380 but 95/10 × 40=360 which is wrong. This is because the sequence of computation is; 95/10=9 then 9 × 40=360.
5. For improved accuracy multiply the biggest numbers first. For example, 95 × 93 × 41/10. The correct answer is 36,223.5, but here are the three combinations of doing Basic Stamp computations:

95 × 93/10 × 41=883 × 41=36,203	error = 36,223.5 – 36,203=20.5
95 × 41/10 × 93=389 × 93=36,177	error = 36,223.5 – 36,177 = 46.5 (why the greatest error?)
41 × 93/10 × 95=381 × 95=36,195	error = 36223.5 – 36,195 = 28.5

Hence the least error is produced by multiplying the biggest numbers first, with the proviso that 65,535 is not exceeded, then divide later in stages.

We now do one more experiment and that is to programme the leg tip to go to an R, Z coordinate in the θ=0° plane. This means we can give planar motion to the leg tip.

7.9 Leg tip planar motion: programming the leg tip to travel to any coordinate, R, Z, in the, θ = 0° plane

Now let's move on from straight line motion to planar motion so that the leg tip can be positioned in a plane. We seek a set of 2-variable, R, Z, inverse kinematics equations, figure 7.24, that will position the leg tip in a plane; the θ = 0° plane is chosen for convenience, figure 7.25.

$$S1=f_1(R,Z)$$
$$S2=f_2(R,Z)$$
$$S3=f_3(R,Z)$$

Figure 7.24 Set of 2-variable equations for planar motion.

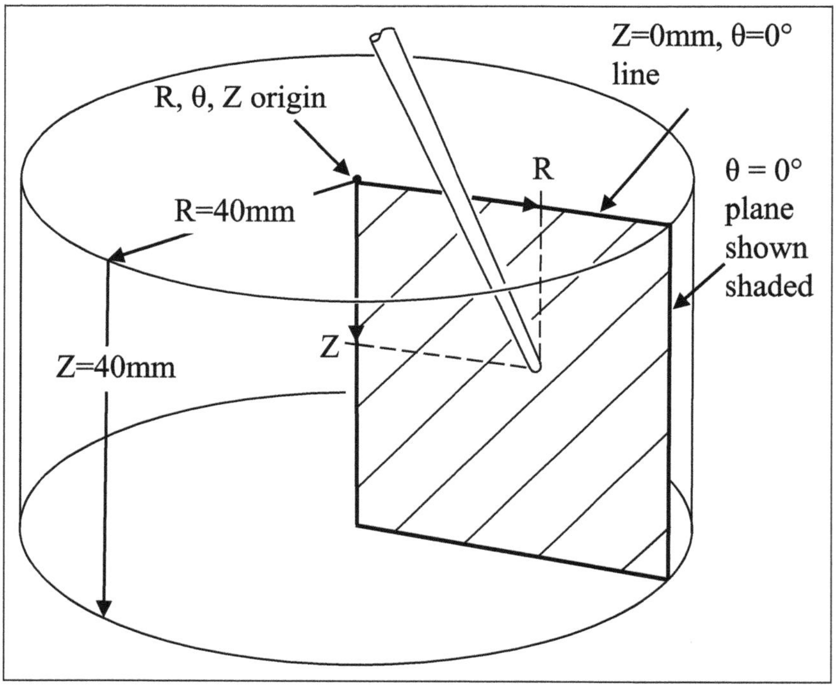

Figure 7.25 Programming the leg tip to any coordinate R, Z in the θ = 0° plane.

In order to derive the set of accurate 2-variable equations for planar motion we use the ServoTeach Module, calibration table and Z blocks to obtain servo pulsout numbers at nine equispaced points in the θ=0° plane as shown in figure 7.26. These nine points are at R=0mm, R=20mm and R=40mm and repeated for Z=0mm, Z=20mm and Z=40mm.

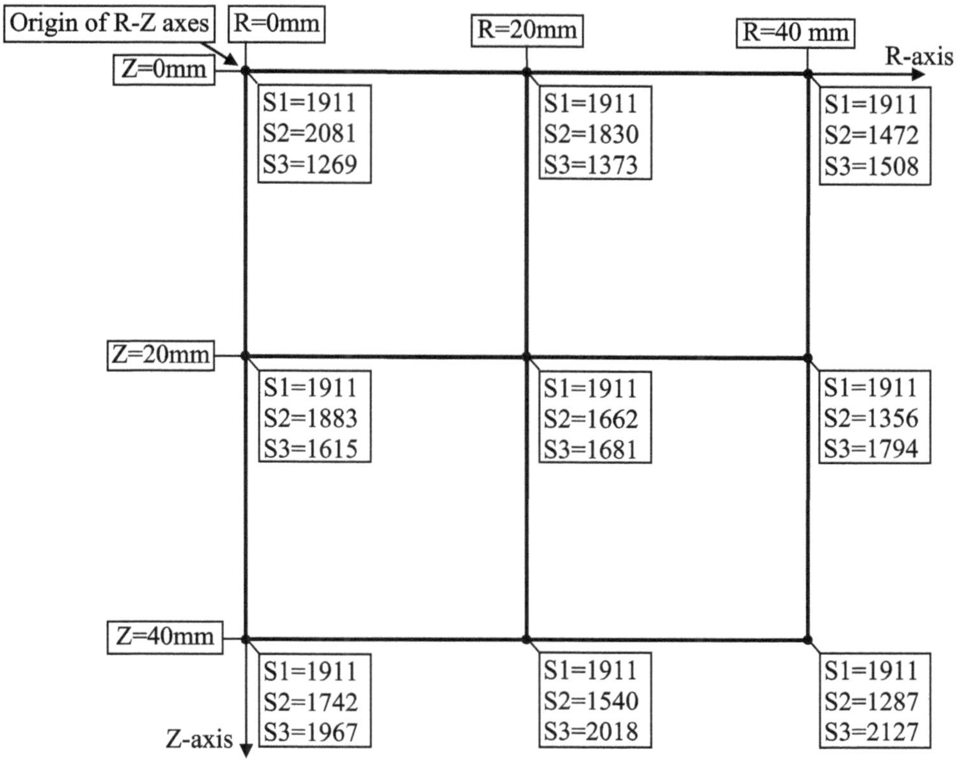

Figure 7.26 S1, S2 and S3 pulsout numbers for all the nine calibration coordinate points in the $\theta = 0°$ plane.

Figures 7.9, 7.11, and 7.18 have already shown how the three Z=0mm servo values have been measured. Figures 7.27 to 7.32 inclusive show, together with the servo pulsout numbers, how the remaining six sets of servo values are measured at Z=20mm and Z=40mm. Note that the positive Z direction is downwards so the Z=40mm coordinate is the plane of the calibration table so there is no need to use a Z-block, figures 7.30 to 7.32 inclusive.

We now have all the data to derive 2nd order inverse kinematics equations of the form shown in figure 7.24 but how do we go about it? Well, first of all plot the servo pulsout numbers as a function of R for Z=0mm, Z=20mm and Z=40mm. We start with servo S3 data, figure 7.33. Note importantly that the precision of all calculated coefficients is 3 sig. figs.

S1=1911 S2=1883 S3=1615

Figure 7.27 Leg tip position R, Z(0,20).

S1=1911 S2=1662 S3=1681

Figure 7.28 Leg tip position R, Z(20,20).

S1=1911 S2=1356 S3=1794

Figure 7.29 Leg tip position R, Z(40,20).

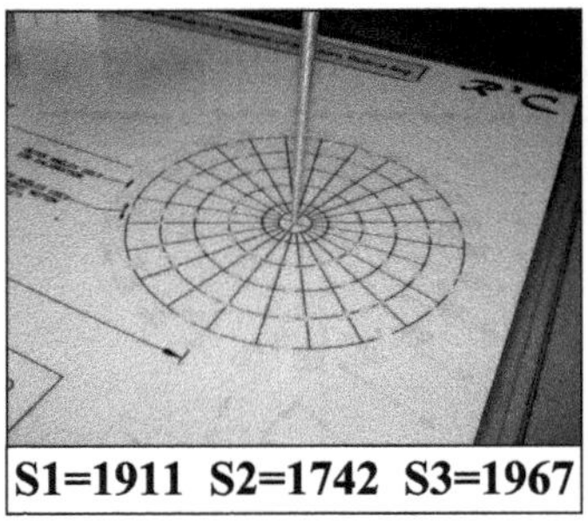

Figure 7.30 Leg tip position R, Z(0,40).

Figure 7.31 Leg tip position R, Z(20,40).

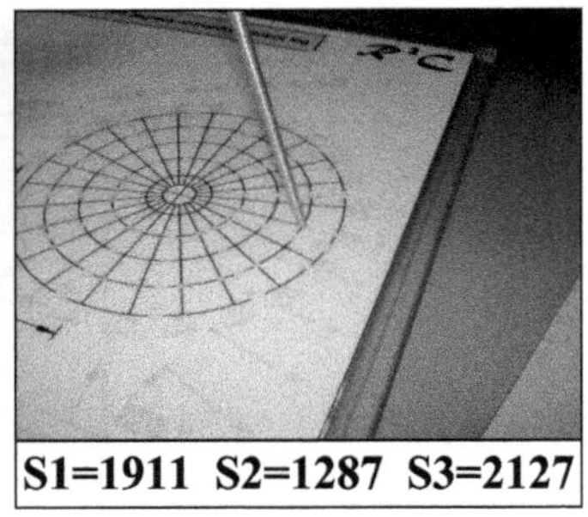

Figure 7.32 Leg tip position R, Z(40,40).

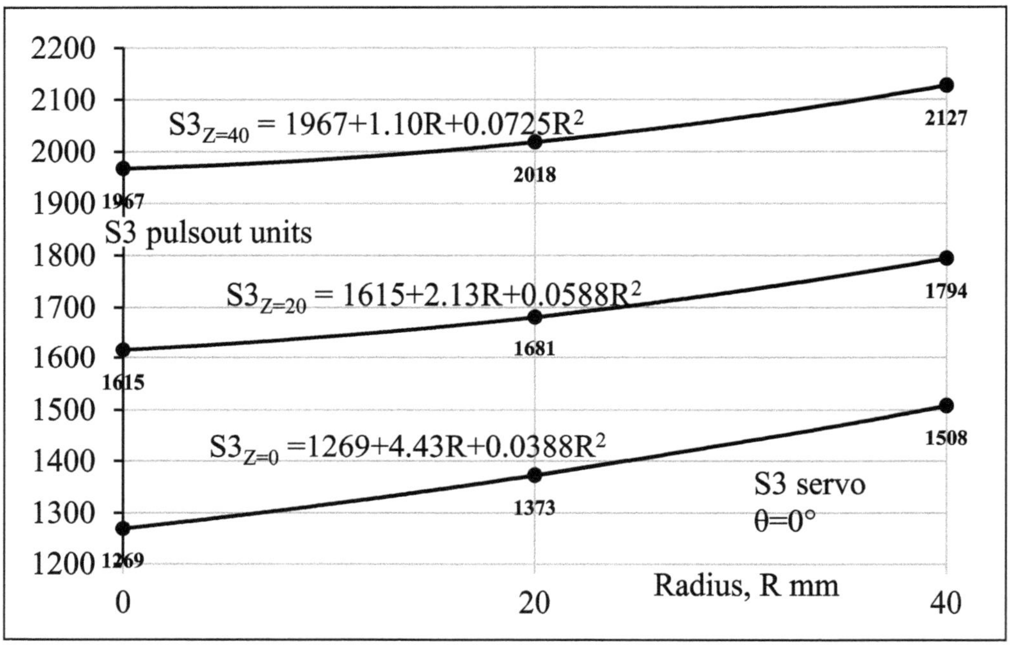

Figure 7.33 Three 2nd order equations as a function of R for servo S3 at Z=0mm, Z=20mm Z=40mm.

The equations are, figure 7.34:

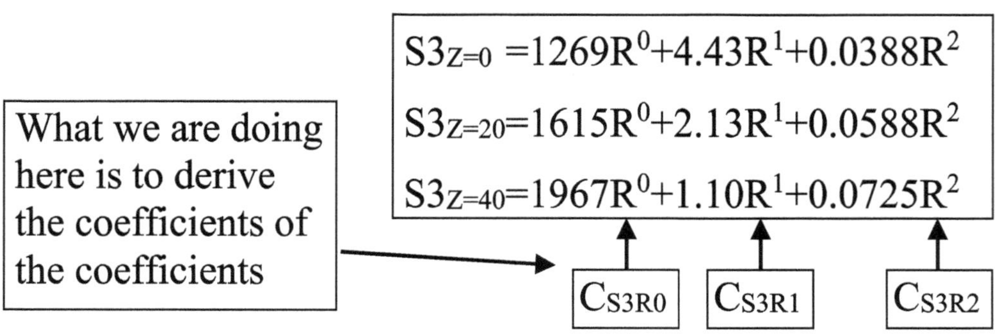

Figure 7.34 2nd order equations as function of R for S3 at different Z values.

Now derive 2nd order function for the coefficients as a function of Z. See figures 7.35, 7.36 and 7.37

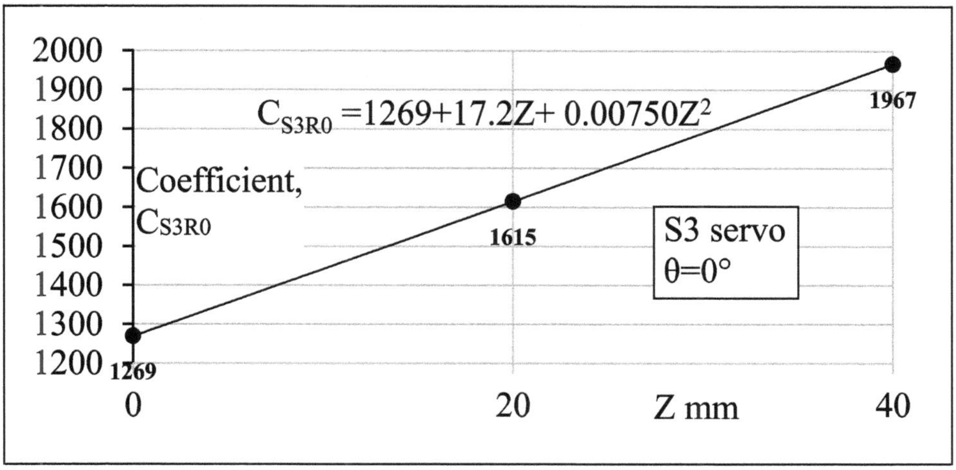

Figure 7.35 The coefficient of the R^0 coefficient as a function of Z.

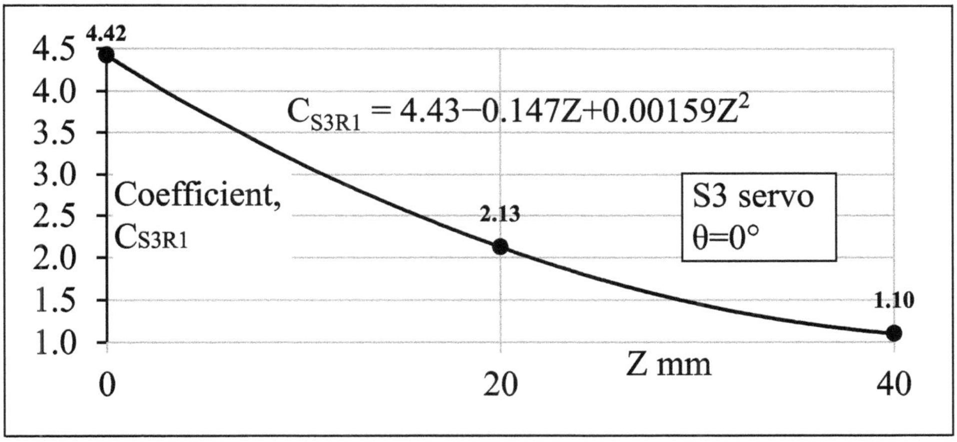

Figure 7.36 The coefficient of the R^1 coefficient as a function of Z.

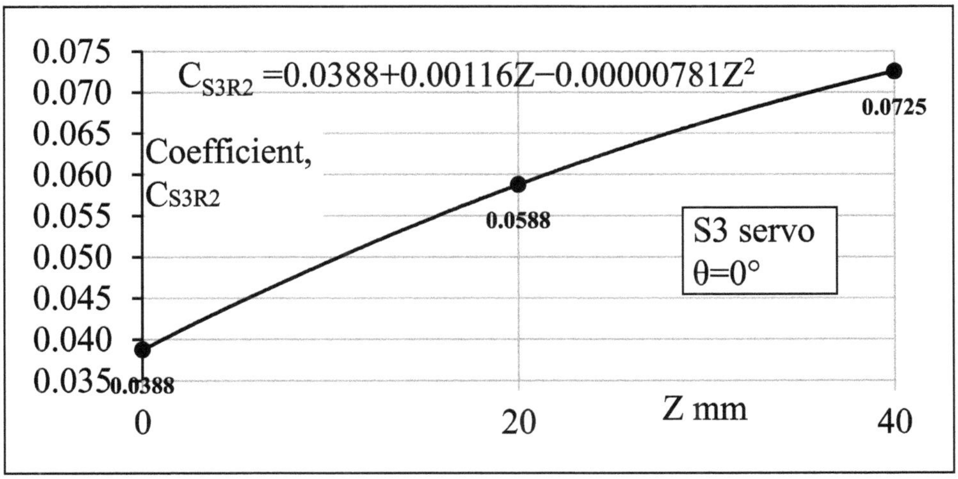

Figure 7.37 The coefficient of the R^2 coefficient as a function of Z.

Taking data from the preceding figures 7.35, 7.36 and 7.37, the coefficients of the coefficients for S3 servo are, figure 7.38:

$$C_{S3R0}=1269+17.2Z+0.00750Z^2$$
$$C_{S3R1}=4.43-0.147+0.00159Z^2$$
$$C_{S3R2}=0.0388+0.00116Z+0.00000781Z^2$$

Figure 7.38 Coefficients of coefficients for S3 servo. Note 3 sig.fig. precision.

Now merging equations from figure 7.38 into equations from figure 7.34, we obtain $S3= C_{S3R0}R^0 + C_{S3R1}R^1 + C_{S3R2}R^2$ which expands to the equation in figure 7.39 below.

$$\begin{aligned} S3= \quad & +(1269+17.2Z+0.00750Z^2)R^0 \\ & +(4.43-0.147Z+0.00159Z^2)R^1 \\ & +(0.0388+0.00116Z-0.00000781Z^2)R^2 \end{aligned}$$

Figure 7.39 The 2-variable 2nd order Taylor series equation for servo S3.

Ok, that's S3 done. Now let's move on to S2 where its data are shown in graphs in figures 7.40, 7.41, 7.42 and 7.43. We use the same technique as for S3 and, using the data for S2 from its graphs, produce the equations in figure 7.44 (page 7-190).

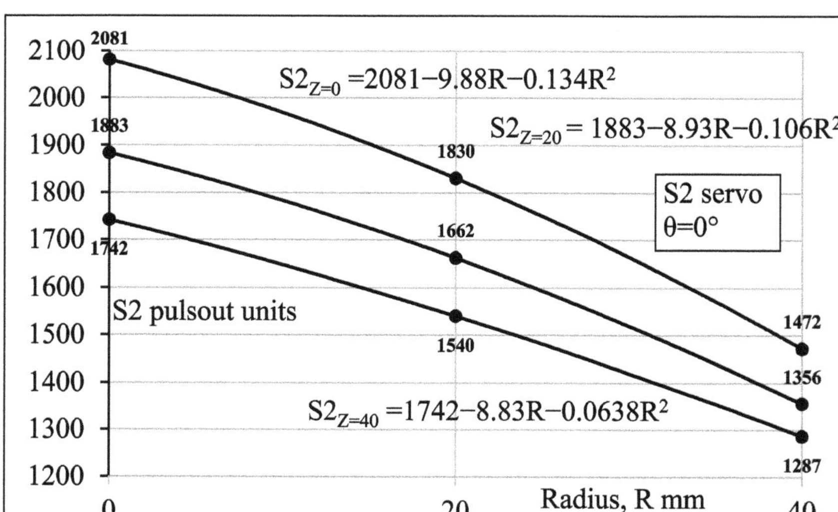

Figure 7.40 Three 2nd order equations as a function of R for servo S2 at Z=0mm, Z=20mm Z=40mm.

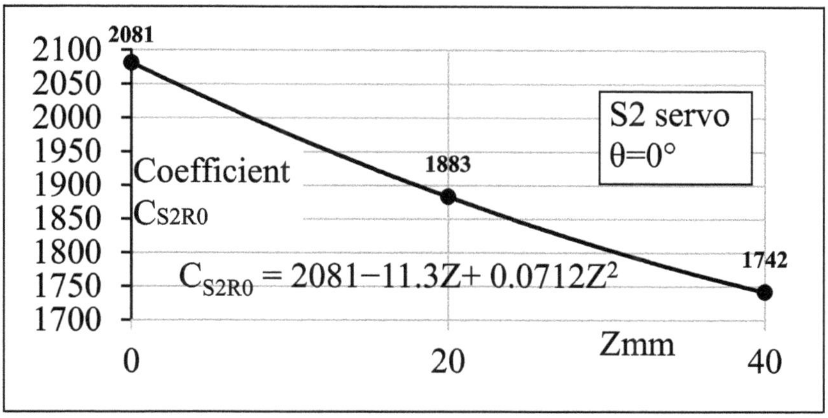

Figure 7.41 The coefficient of the R^0 coefficient as a function of Z.

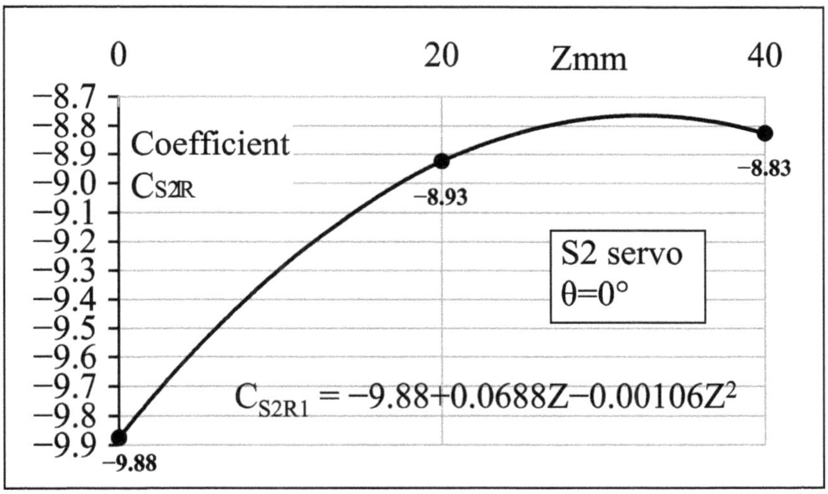

Figure 7.42 The coefficient of the R^1 coefficient as a function of Z.

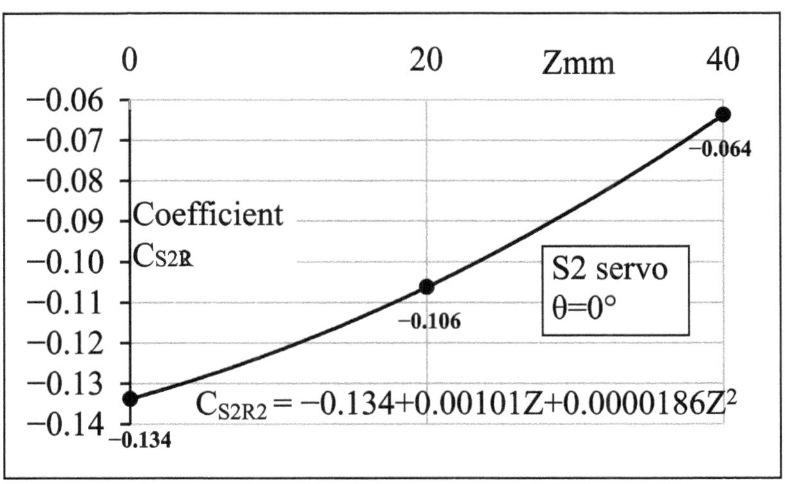

Figure 7.43 The coefficient of the R^1 coefficient as a function of Z.

$$C_{S2R0}=2081-11.3Z+0.0712Z^2$$
$$C_{S2R1}=-9.88+0.0688Z-0.00106Z^2$$
$$C_{S2R2}=-0.134+0.00101Z+0.0000186Z^2$$

Figure 7.44 Coefficients of coefficients for S2 servo. Note 3sig.fig. precision.

Now merging figure 7.44 equations into equations from figure 7.34 but for the S2 servo, we obtain; $S2= C_{S2R0}R^0 + C_{S2R1}R^1 + C_{S2R2}R^2$ which expands to the equation in figure 7.45 below.

$$\begin{aligned} S2= \quad & +(2081-11.3Z+0.0712Z^2)R^0 \\ & +(-9.88+0.0688Z-0.00106Z^2)R^1 \\ & +(-0.134+0.00101Z+0.0000186Z^2)R^2 \end{aligned}$$

Figure 7.45 The 2-variable 2nd order Taylor series equation for servo S2.

Moving on to servo S1, its graph is shown in figure 7.46. The S1 servo on for the θ=0° plane is a special case and its pulsout number stays constant

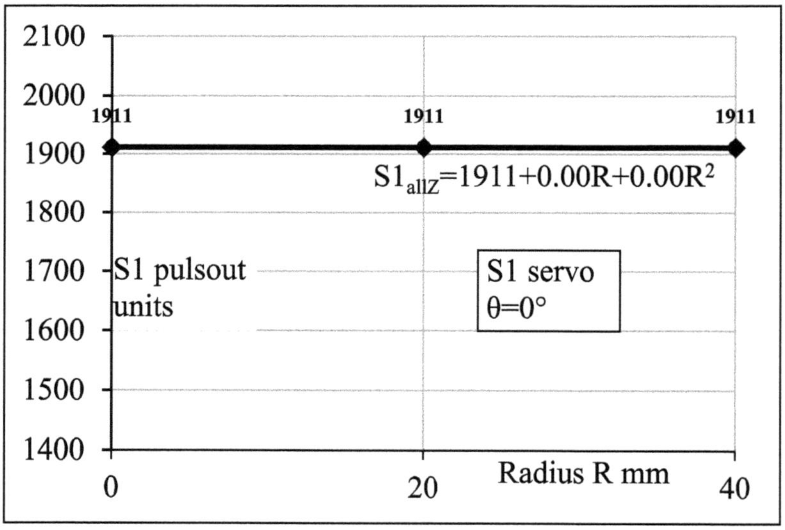

Figure 7.46 Three 2nd order equations as a function of R for servo S1 at Z=0mm, Z=20mm Z=40mm. Note that all the graphs coincide and are constant.

As a consequence, all the coefficients of the coefficients are zero and the final 2-variable 2nd order equation for S1 is the equation in figure 7.47.

$$\begin{aligned} S1= \quad & +(1911+0.00Z+0.00Z^2)R^0 \\ & +(\ 0.00+0.00Z+0.00Z^2)R^1 \\ & +(\ 0.00+0.00Z+0.00Z^2)R^2 \end{aligned}$$

Figure 7.47 The 2-variable 2nd order Taylor series equation for servo S2.

It is important to note that the form of all the 2-variable 2nd order Taylor series equations for the three servos are of the generalised form, figure7.48.

$$S=C_{00}R^0Z^0+C_{01}R^0Z^1+C_{02}R^0Z^2+C_{10}R^1Z^0+C_{11}R^1Z^1+C_{12}R^1Z^2+C_{20}R^2Z^0+C_{21}R^2Z^1+C_{22}R^2Z^2$$

Figure 7.48 Generalised form for the servo inverse kinematics equations.

Note that there are nine terms in the generalised form. The next step is to calculate to what precision the coefficients should be specified. We have to ask ourselves the question; "What error can we tolerate?" Let's say the answer is that we can only tolerate 2 pulsout units error for each term then at a worse case, since there are 6 terms with coefficients, we could have an error of 12 pulsout units which would be an unacceptable error. However, the signs of each term are not always positive or negative so there will be some cancellation of errors. Anyway, figure 7.49 below shows the precision of each term based on 2 pulsout units error per term.

Term	calculate 1 unit precision			nearest lowest decimal place	
$C_{00}R^0Z^0$	specify to nearest integer				
$C_{01}R^0Z^1$	2/40 =	0.05	(1.sig.fig.)	0.1	(1 dec.pl.)
$C_{02}R^0Z^2$	2/(40*40)	0.001	(1 sig.fig.)	0.001	(3 dec.pl.)
$C_{10}R^1Z^0$	2/40	0.05	(1 sig.fig.)	0.1	(1 dec.pl.)
$C_{11}R^1Z^1$	2/(40*40)	0.001	(1 sig.fig.)	0.001	(3 dec.pl.)
$C_{12}R^1Z^2$	2/(40*40*40)	0.00003	(1 sig.fig.)	0.00001	(5 dec.pl.)
$C_{20}R^2Z^0$	2/(40*40)	0.001	(1 sig.fig.)	0.001	(3 dec.pl.)
$C_{21}R^2Z^1$	2/(40*40*40)	0.00003	(1 sig.fig.)	0.00001	(5 dec.pl.)
$C_{22}R^2Z^2$	2/(40*40*40*40)	0.0000008	(1 sig.fig.)	0.000001	(6 dec.pl.)

Figure 7.49 Calculating lowest decimal place coefficient precision for each of the 9 terms in the servo equations.

Now rewrite the servo equations with the coefficients specified to the number of decimal places as calculated in figure 7.49 *and* specified to 3 sig. figs.

First the S1 servo;

$$\begin{aligned} S1= \quad & +(1911+\underline{00.0}Z+0.\underline{000}Z^2)R^0 \\ & +(\ \underline{00.0}+0.\underline{000}Z+0.00\underline{000}Z^2)R^1 \\ & +(\ 0.\underline{000}+0.00\underline{000}Z+0.000\underline{000}Z^2)R^2 \end{aligned}$$

Note that the underlined numbers are the 3 sig. figs. required

Figure 7.50 Specifying the S1 equation to the appropriate precision to give approx. less than 1 pulsout unit error per term.

…and the S2 servo;

$$
\begin{aligned}
S2= \; & +(2081+\underline{11.3}Z+0.\underline{071}Z^2)R^0 \\
& +(-\underline{09.9}+0.\underline{069}Z-0.00\underline{106}Z^2)R^1 \\
& +(-0.\underline{134}+0.00\underline{101}Z+0.000\underline{019}Z^2)R^2
\end{aligned}
$$

Note that the underlined numbers are the 3 sig. figs. required

Figure 7.51 Specifying the S2 equation to the appropriate precision to give approx. less than 1 pulsout unit error per term.

…and the S3 servo;

$$
\begin{aligned}
S3= \; & +(1269+17.2Z+0.\underline{008}Z^2)R^0 \\
& +(\underline{4.4}-0.\underline{147}Z+0.00\underline{159}Z^2)R^1 \\
& +(0.\underline{039}+0.00\underline{116}Z-0.000\underline{008}Z^2)R^2
\end{aligned}
$$

Note that the underlined numbers are the 3 sig. figs. required

Figure 7.52 Specifying the S3 equation to the appropriate precision to give approx. less than 1 pulsout unit error.

The above equations are interesting because not one of the coefficients of R or Z exceeds 3 sig figs. This is not only coincidental but also results from specifying the error budget of 2 pulsout units per term. A smaller error budget would have resulted in the requirement for higher precision in the coefficients and a consequent increase of significant figures.

Ok, the final step is to create Basic Stamp programmes for these equations, so here we go, see figure 7.53.

```
S1 = 1911     -(  0*Z/10) + (  0*Z/25*Z/40)
S1 = S1       -(  0*R/10) + (  0*Z /25*R/40) - (  0*Z/25*Z/40*R/100)
S1 = S1       -(  0*R/25*R/40) + (  0*Z/25*R/40*R/100) + (  0*Z/25*Z/40*R/40*R/25)

S2 = 2081     -(113*Z/10) + ( 71*Z/25*Z/40)
S2 = S2       -( 99*R/10) + ( 69*Z /25*R/40) - (106*Z/25*Z/40*R/100)
S2 = S2       -(134*R/25*R/40) + (101*Z/25*R/40*R/100) + ( 19*Z/25*Z/40*R/40*R/25)

S3 = 1269     +(172*Z/10) + (  8*Z/25*Z/40)
S3= S3        +( 44*R/10) - (147*Z /25*R/40) + (159*Z/25*Z/40*R/100)
S3= S3        +( 39*R/25*R/40) + (116*Z/25*R/40*R/100) - (  8*Z/25*Z/40*R/40*R/25)
```

Figure 7.53 The Basic Stamp PBasic inverse kinematics equations programme to compute the servo pulsout units as a function of R,Z in the θ=0°plane. Note that each equation is split into 3 stages because there is a limit to the length of each instruction line.

These equations are now loaded into the ServoTeach Module together with a programme, figure 7.54 that uses the S1 and S3 knobs to control the R and Z values respectively.

7.10 Programme code listing for knob controlled R, Z positioning of leg tip in θ=0°plane

```
' {$STAMP BS2sx}
'program using two-variable 2nd order Taylor series to position
'leg tip in R, Z coordinates in the theta=0deg plane controlled by the ServoTeach Module
'S1 and S3 knobs respectively
R       VAR   Word
Z       VAR   Word
S1      VAR   Word
S2      VAR   Word
S3      VAR   Word
again:  HIGH 13
        HIGH 15
        PAUSE 1
        RCTIME 13,1,R        'S1 knob controls the R value'
        RCTIME 15,1,Z        'S3 knob controls the Z value
        R=R−1/10 MAX 40
        Z=Z−1/10 MAX 40
        DEBUG "R=",DEC R," Z=",DEC Z,CR

plane0:
S1 = 1911     −(  0*Z/10) + (  0*Z/25*Z/40)
S1 = S1       −(  0*R/10) + (  0*Z /25*R/40) − (  0*Z/25*Z/40*R/100)
S1 = S1       −(  0*R/25*R/40) + (  0*Z/25*R/40*R/100) + (  0*Z/25*Z/40*R/40*R/25)

S2 = 2081     −(113*Z/10) + ( 71*Z/25*Z/40)
S2 = S2       −( 99*R/10) + ( 69*Z /25*R/40) − (106*Z/25*Z/40*R/100)
S2 = S2       −(134*R/25*R/40) + (101*Z/25*R/40*R/100) +
(19*Z/25*Z/40*R/40*R/25)

S3 = 1269     +(172*Z/10) + (  8*Z/25*Z/40)
S3= S3        +( 44*R/10) − (147*Z /25*R/40) + (159*Z/25*Z/40*R/100)
S3= S3        +( 39*R/25*R/40) + (116*Z/25*R/40*R/100) −
(8*Z/25*Z/40*R/40*R/25)

pout:   PULSOUT 5,S1
        PULSOUT 6,S2
        PULSOUT 7,S3
        GOTO again
```

Figure 7.54 Program that positions leg tip accurately with 1mm resolution in the θ=0° plane with coordinates R, Z.

One final note is to say that these Taylors series, 2-variable inverse kinematics equations are extremely useful since, being in the form of a sum of integer power polynomials, they are easily differentiated with respect to time. This is very useful to derive the speed of the servos given the coordinates of the leg tip and the required speeds in the R, θ and Z directions.

We now conclude the chapters related to inverse kinematics and move on to the precision programming of roving robots at the higher level now that the lower level inverse kinematic systems are in place.

8 Walking gaits and motion control

8.1 Introduction

An omni-directional walking robot requires legs whose leg-tips can independently access 3-dimensional space. 'Omni-directional' means the ability to walk in all directions such that the walking robot is able to walk sideways left, sideways right, forwards and backwards, and turn left and right. Furthermore, possession of 3dof legs enables the robot to turn while translating, which is similar to ballroom dancers when they dance the Viennese waltz as already discussed in chapter 1. In fact other complex manoeuvres are possible that include maintaining the body in a level attitude as the legs negotiate uneven terrain. Such motion can be analysed as the body rotating about any Instantaneous Axis of Rotation (note, 'axis' as compared to 'centre') which is a three dimensional vector as compared to a two dimensional point. (This interesting problem serves as a mathematical challenge to students). The leg tip motion and the leg tip locus have been analysed in detail in chapter 4 so we now consider how leg tip motion can achieve walking gaits and motion control.

8.2 Review of the leg tip locus

Figure 8.1 reviews the shape and important points of the leg tip digital locus. An effective way of analysing and synthesising walking gaits is to graph the horizontal component of the leg tip position against time. It is not necessary to show the vertical component of the leg tip position. Instead it is only necessary to indicate when the leg tip is in the air and when it is on the ground. The following diagram, figure 8.1, illustrates the idea.

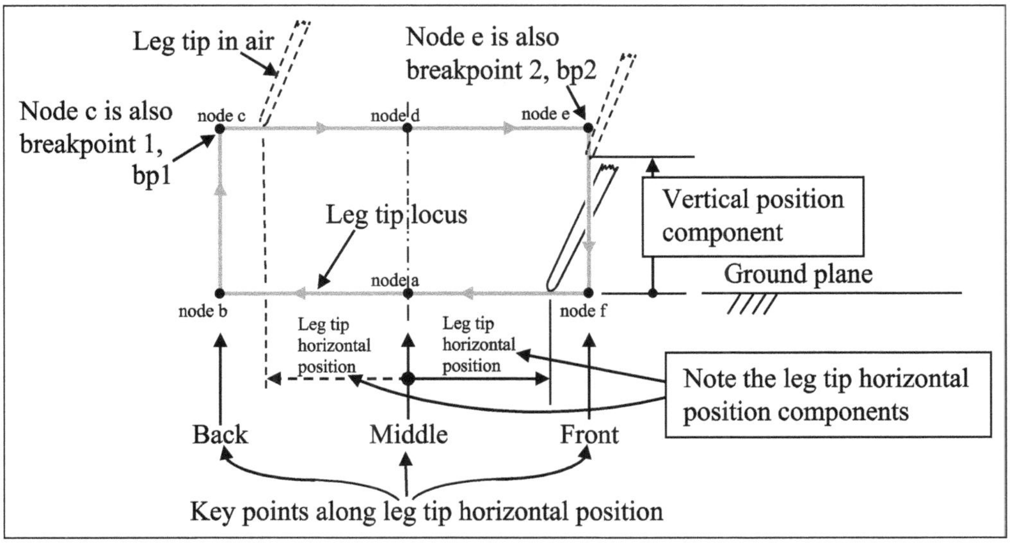

Figure 8.1. Review of the shape of the leg tip displacement during walking, i.e. the leg tip locus. The diagram indicates the leg tip horizontal position components. Future diagrams in this chapter will not show the vertical position component.

Note that we are not concerned with the location of the two breakpoints, bp1, and, bp2, because they are not significant with the gait. They only depend on the values of zag and amp.

The different walking gaits of legged systems are analogous to driving in different gears for a wheeled vehicle, e.g. an automobile. The horse is a good example of a legged system capable of different walking gaits that are used as the horse gathers speed, e.g. walk, trot, canter and gallop. (It would appear, then, that the horse has a 4-speed gearbox). The concept of walking gaits, also leads to the problem of the detailed leg motion that is required in order to transfer smoothly to a new gait. (More interesting challenges for students). Automobile gearboxes solve the problem of gear changing, i.e. gait changing, by easing off on the torque being transmitted before disengaging a pair of mating gears then bringing a new set of mating gears to an equal tangential velocity before meshing.

Let's look at a gait that is analogous to first gear in an automobile. The gait always has 5 leg tips in contact with the ground. It gives the slowest locomotion speed but with the highest traction force and can support the highest pay load on its back because of 5 legs all combining to support the load and also to provide traction, albeit maximum traction is dependent on body weight and friction. It is called the "5-on" gait and is now described. (Actually six legs on the ground will support the highest payload but then it cannot walk).

8.3 The "5-on" gait

The gait is shown in figure 8.2 below. One gait cycle time, i.e. the time for one circuit of the digital locus, T_G, is divided into six equal sub-cycles of time duration, $T_G/6$. Each leg tip will spend 5 of these six sub-cycles on the ground giving locomotion to the robot and the remaining sub-cycle quickly returning the leg tip, through the air, to its front starting point on the ground.

Each leg tip carries out the same motion except that the phase of each leg is shifted by one sub-cycle. This means that there are always 5 legs on the ground supporting the robot except for an instant in time, at the junction of two sub-cycles when one leg leaves the ground and another touches down on the ground.

Since each leg tip spends 5 times longer on the ground than along the air return path then the leg tip speed along the air return path will be at least 5 times faster than when on the ground. Don't forget the speeds we are talking about are relative to the robot body and not compared to relative to the ground.

The question now arises how to get the leg tips to follow the waveforms in figure 8.2 and the answer is to do it with an, *n, versus time waveform*. For example, the leg tip horizontal position versus time waveform of the LF leg, (Left Front leg), is extracted from the above figure 8.2 and in figure 8.3. Note that the figure indicates a few notable values of, n, relating to the path of the leg tip along the digital locus. Also note that the leg tip is on the ground for 128 digital increments and 5 time sub-cycles, whereas it is in the air also for 128 digital increments but these take place in just 1 sub-cycle time.

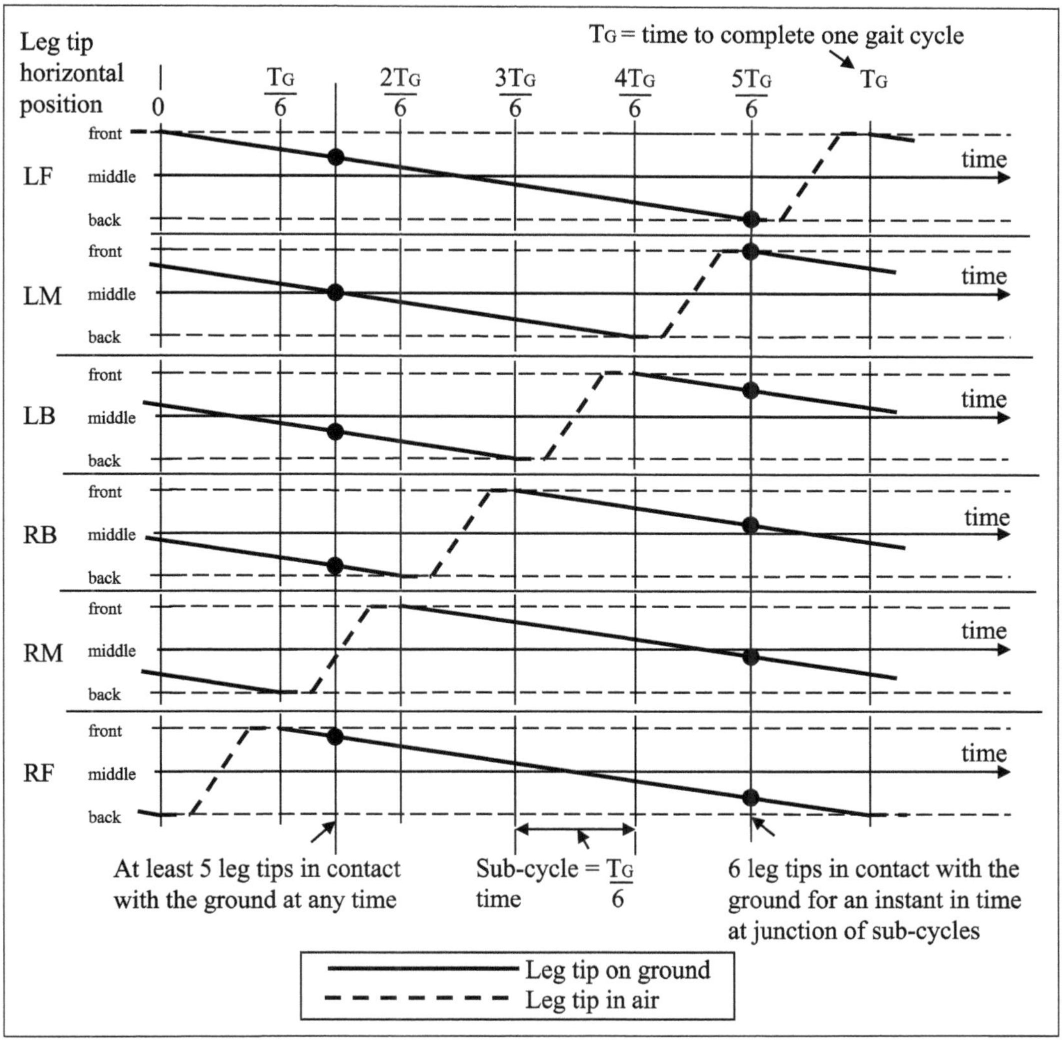

Figure 8.2. Horizontal motion for each of the six legs during 5-on gait.

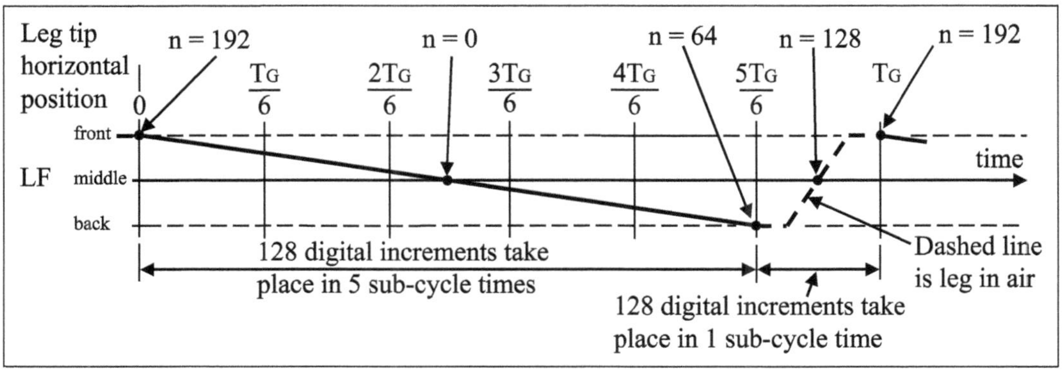

Figure 8.3. Using the n-value to obtain the correct leg motion in 5-on gait using the LF leg as an example.

Experiments carried out to measure the maximum speed of response of the Futaba S9450 servomechanisms that are used to actuate the 3dof leg show that not less than six 20ms clock pulses, (= 0.120sec), should be used for the duration of 1 time sub-cycle. Hence, using 6 clock pulses for 1 time sub-cycle will produce the maximum leg tip speed. Thus, the minimum gait cycle time, $T_G = 6 \times 0.120\text{sec} = 0.72\text{sec}$. A suitable range of locomotion speed is from speed 1 to speed 10 where speed 1 is given by 60 clock pulses for 1 sub-cycle time and speed 10 is given by 6 clock pulses for 1 sub-cycle time.

Now transform the leg tip horizontal position axis into values of, n, and the time axis into clock pulse counts as follows, figure 8.4, using the left front leg as an example and using the slowest speed, speed 1, that produces a gait cycle period of 7.2sec.

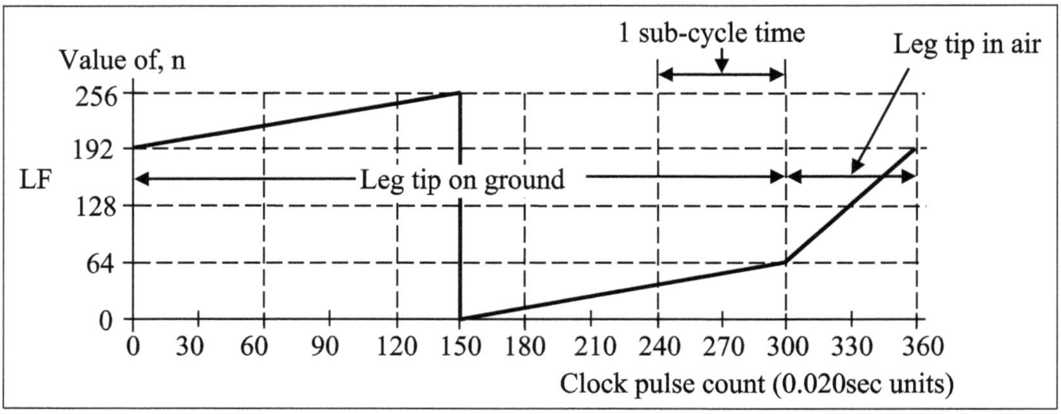

Figure 8.4 Detailed n-values versus time for the 5-on gait at the slowest speed, (speed 1), 7.20sec/gait cycle.

The waveform shape in the diagram above of, n, versus, clock pulse count is the same for each leg tip except that the phase is shifted by 1 sub-cycle which is 60 clock pulses. All the waveforms for each leg can be shown on one diagram, figure 8.5.

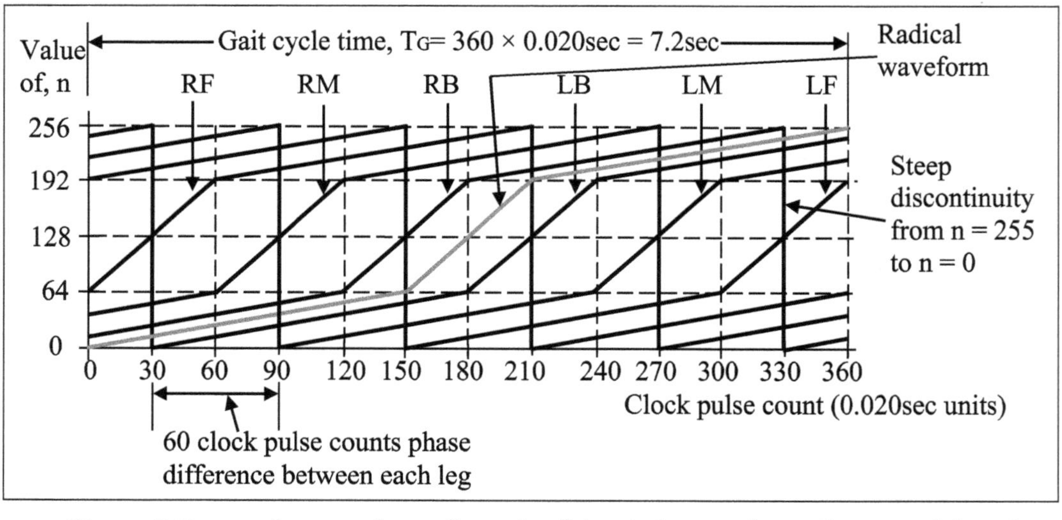

Figure 8.5 n-value waveforms for each of the six legs to obtain 5-on gait. Note the phase difference between each of the six waveforms.

Figure 8.5 also shows a radical waveform which is explained in detail in the next section together with a computer programme to implement it.

8.4 Microcomputer real-time program for the 5-on gait

Now is a suitable time to write a computer program that will produce the six waveforms, one for each leg, as shown in figure 8.5 above. A method that we will use is to create the radical waveform, figure 8.5, and then shift it left or right by the correct number of clock pulse counts to obtain the phase shift relevant to each leg tip. The radical waveform is shown separately below, figure 8.6 alone for clarity. This shape is chosen because it can be specified with only three equations and it eliminates the steep discontinuity from n = 255 to n = 0, by placing it at the end of the radical cycle.

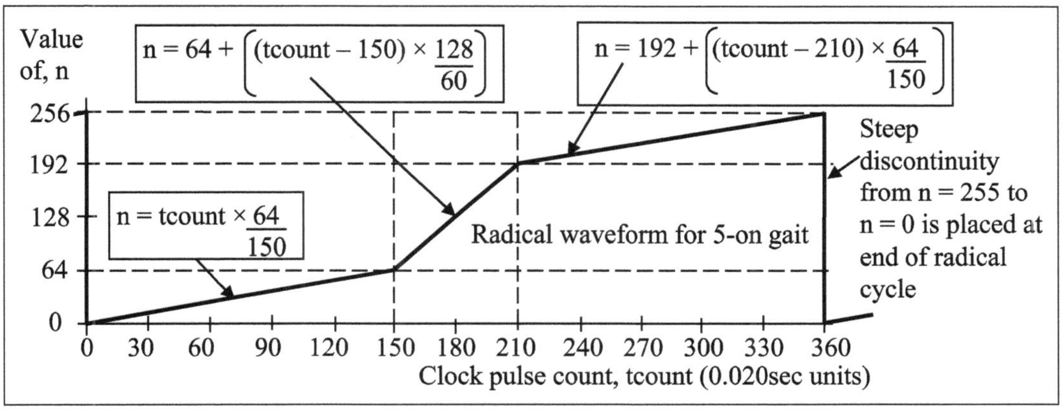

Figure 8.6. n-value radical waveform for the 5-on gait. Note the three piece-wise linear equations shown in boxes that specify the radical cycle shape.

Figure 8.6 shows three equations, one for each of the three, piece-wise, linear parts of the radical waveform. A Basic Stamp2SX program is now written using these equations to compute the values of, n, for each of the six leg tips in order to carry out the 5-on gait. Note that a 50Hz, 20ms clock pulse signal is sent to pin 15 of the BS2SX microcomputer that implements the 5-on gait algorithm. This clock signal serves as a 50Hz polled interrupt signal that is recognised by the instruction line: pulsin 15,1,j

```
' BS2SX program to carry out the 5-on gait
n         VAR   byte    'n value of radical waveform
j         VAR   nib     'used for the pulsin instruction
nlf       VAR   byte    'n value of LF leg tip
nlm       VAR   byte    'n value of LM leg tip
nlb       VAR   byte    'n value of LB leg tip
nrb       VAR   byte    'n value of RB leg tip
nrm       VAR   byte    'n value of RM leg tip
nrf       VAR   byte    'n value of RF leg tip
pcount    VAR   word    'clock pulse count number
tcount    VAR   word    'temporary clock pulse count number
speed     VAR   nib     'locomotion speed
```

```
                    speed=1                           'set to minimum speed
again:              for pcount= 0 TO 359 STEP speed          'step through one gait cycle
                    pulsin 15,1, j            'wait for polled interrupt 20msec period
                                              clock pulse
LFleg:              tcount=pcount+150                 'calculate n for LF leg tip
                    GOSUB redo
                    GOSUB calcn
                    nlf=n
LMleg:              tcount=pcount+90                  'calculate n for LM leg tip
                    GOSUB redo
                    GOSUB calcn
                    nlm=n
LBleg:              tcount=pcount+30                  'calculate n for LB leg tip
                    GOSUB redo
                    GOSUB calcn
                    nlb=n
RBleg:              tcount=pcount+330                 'calculate n for RB leg tip
                    GOSUB redo
                    GOSUB calcn
                    nrb=n
RMleg:              tcount=pcount+270                 'calculate n for RM leg tip
                    GOSUB redo
                    GOSUB calcn
                    nrm=n
RFleg:              tcount=pcount+210                 'calculate n for RF leg tip
                    GOSUB redo
                    GOSUB calcn
                    nrf=n
        '*********************************************************************

'display results
DEBUG DEC nlf," ",DEC nlm," ",DEC nlb," ",DEC nrb," ",DEC nrm," ",DEC nrf,CR
'you should disable debug instruction as soon as results are satisfactory to resume real time
ndone:              NEXT

again1:             GOTO again1               'infinite loop, stop after one gait cycle
        '*********************************************************************

redo:               IF tcount>359 THEN skip1    'subroutine to reset the clock pulse
                    GOTO skip2                  'count number. See text for
                                                explanation
skip1:              tcount=tcount-360
skip2:              RETURN
        '*********************************************************************
```

```
calcn:        IF tcount>210 THEN section3        'subroutine uses radical
              IF tcount>150 THEN section2        'waveform to calculate n
section1:     n=tcount*64/150
              GOTO skip0
section2:     n=64+((tcount-150)*128/60)
              GOTO skip0
section3:     n=192+((tcount-210)*64/150)
skip0:        RETURN
'*********************************************************************
```

8.5 Wave or 'ripple' action of leg tip motion during walking

It is interesting to see the robot walking with the 5-on gait. The immediate visible feature is that of each leg tip, one after the other, returning quickly through the air to its starting position on the ground. It appears like a constantly rotating ripple or a "Mexican wave" in a football stadium travelling CCW around the robot as shown in this diagram.

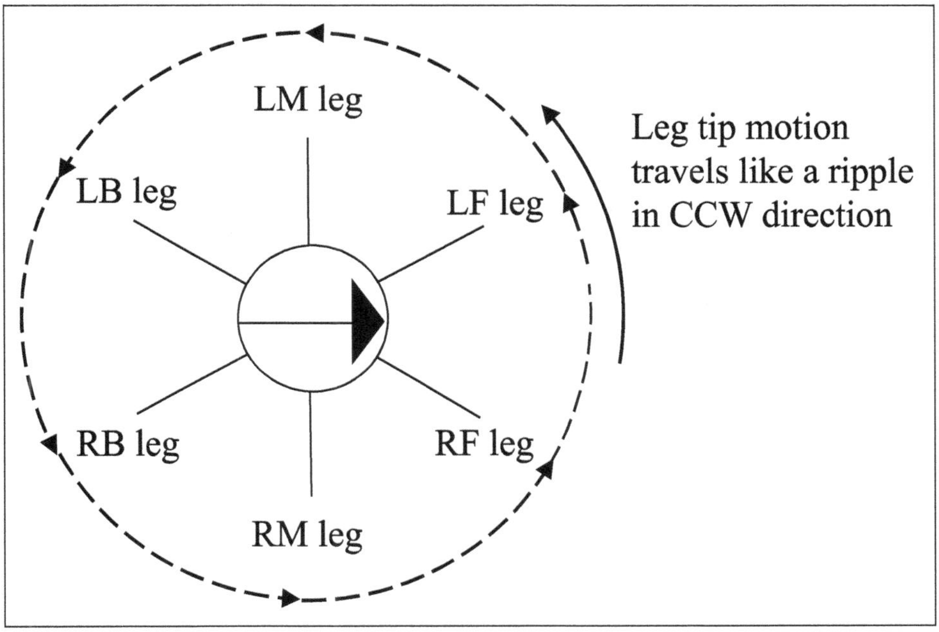

Figure 8.7. Constantly rotating wave or ripple action during 5-on gait.

There are many combinations in the order that the leg tips can be actuated. For example, to make the leg actuation style more symmetrical, the ripple can travel down the left hand side from front to back and then travel down the right hand side from front to back like this, figure 7.8. You can do this by changing the phase of each leg in the BS2SX programme

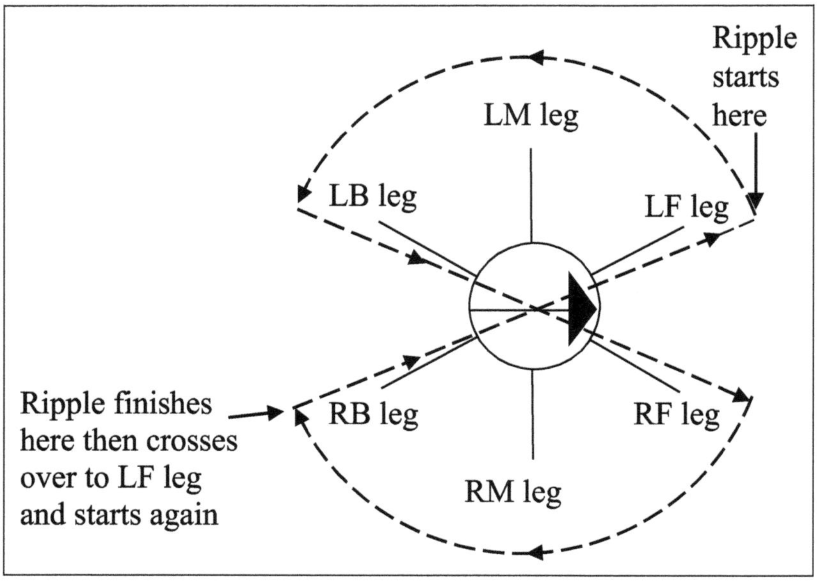

Figure 8.8. Variation of the leg tip ripple action for the 5-on gait.

Another combination with further increased leg actuation symmetry is a zig-zag ripple like this, figure 7.9.

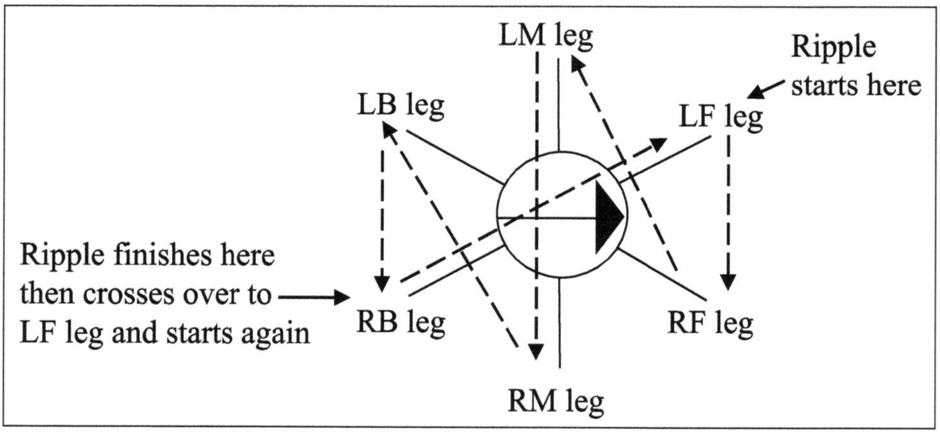

Figure 8.9. Zig-Zag ripple with 5-on gait.

Ok, that's enough on the 5-on gait. Now let's move on to the 4-on gait.

8.6 The "4-on" gait

The 4-on gait is equivalent to second gear and means there are at least 4 leg tips supporting the robot and 2 leg tips in the air on the return stroke. The leg tips are grouped into three pairs which are diametrically opposite, i.e. LF/RB, LM/RM, and LB/RF. The leg tips in each pair move together with each pair moving consecutively in the following order, LF/RB, LM/RM, LB/RF and then repeat. As a result a continuous CCW ripple is set up in the robot legs.

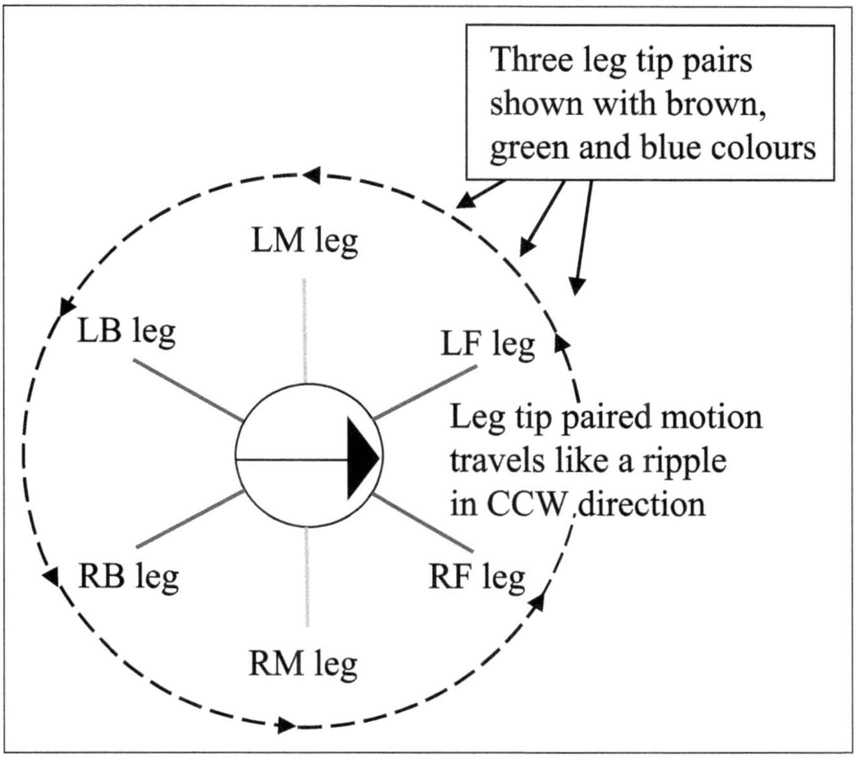

Figure 8.10 Leg tips moving in pairs demonstrating the 4-on gait.

The leg tip motion for each of the three pairs is as follows, figure 8.11.

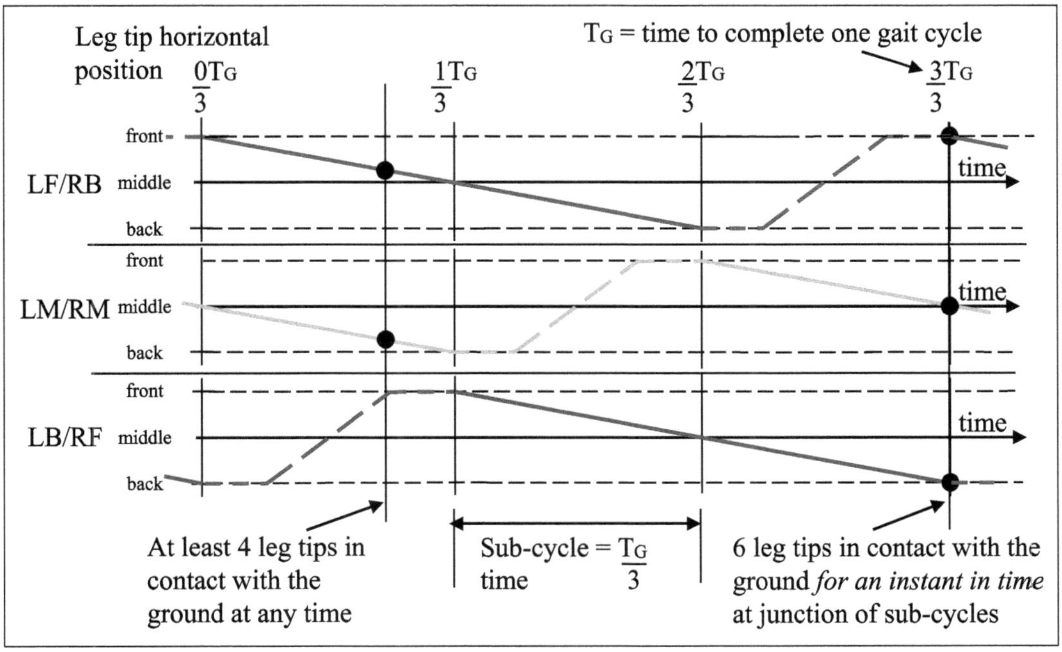

Figure 8.11 Horizontal motion for three pairs of legs during 4-on gait.

You can see from the above diagram, figure 8.11, that the three pairs of leg tips cause one gait cycle to be divided up into three sub-cycle times. Each leg tip spends twice the time in contact with the ground as compared to the time spent in the air. Also note that at least two pairs of leg tips are in contact with the ground at any time and three pairs are in contact for an instant in time at the junction of sub-cycles.

To achieve the 4-on gait the following radical waveform is formed, figure 8.12.

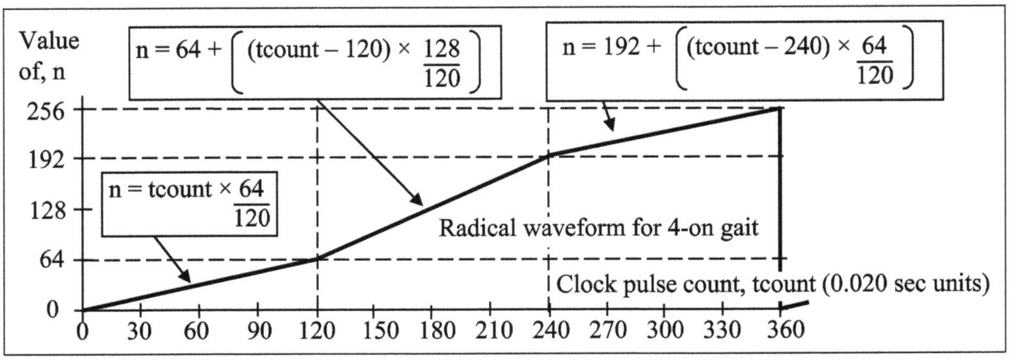

Figure 8.12 n-value radical waveform for the 4-on gait. Note the three piece-wise linear equations shown in boxes that specify the radical cycle shape.

The 4-on radical waveform uses 360 clock pulse counts for one gait cycle time which is the same as the 5-on gait. Also 120 clock pulses are spent with the leg tip in the air and 240 clock pulses are spent with the leg tip on the ground.

Utilising the radical waveform, the following waveforms, figure 8.13 represent the 4-on gait.

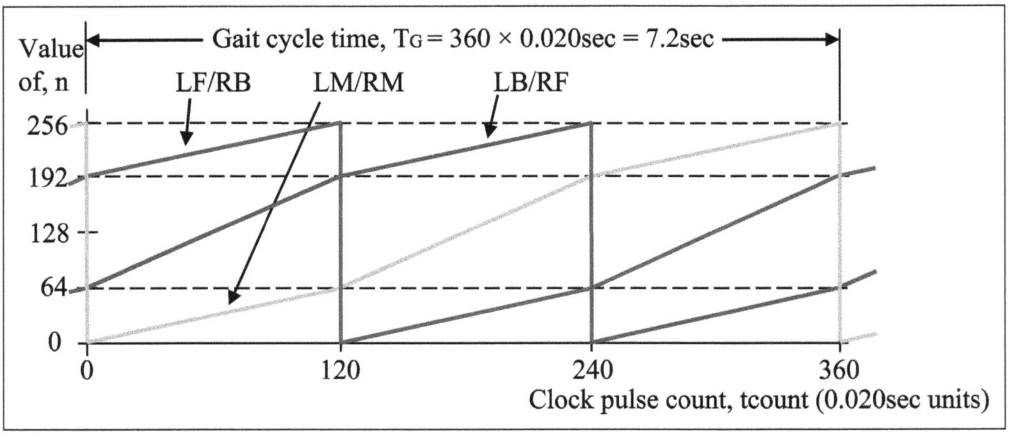

Figure 8.13 n-value waveforms for each of the three pairs of legs to obtain 4-on gait.

It is straightforward to modify the 5-on gait computer program shown in the previous section to carry out the waveforms for the 4-on gait. First the Now let's move on to the gait that produces the maximum speed which uses the minimum number of leg tips in contact with the ground for static stability. The number of leg tips is three and is similar in principle to the static stability provided by a three-legged stool. The gait is called "3-on" gait or "double tripod" gait.

8.7 The "3-on" or "double tripod" gait

The double tripod gait is probably equivalent to the canter gait of the horse in so far that it is one gear before a dynamic gait that causes the all legs of the horse not to be in contact with the ground for a short period of time. At present the low power-to-weight ratio of most man-made servomechanisms presents problems in carrying out a dynamic gait for our six-legged robot. In other words the servomechanisms just don't have enough mechanical power output to make the robot jump off the ground. If we program the legs to do this then very quickly the servomechanism gearboxes and/or electronic drives are stripped or burn out respectively and this is not a good idea because the servomechanisms are very expensive.

The left and right tripods are indicated by blue and red lines respectively, figure 8.14.

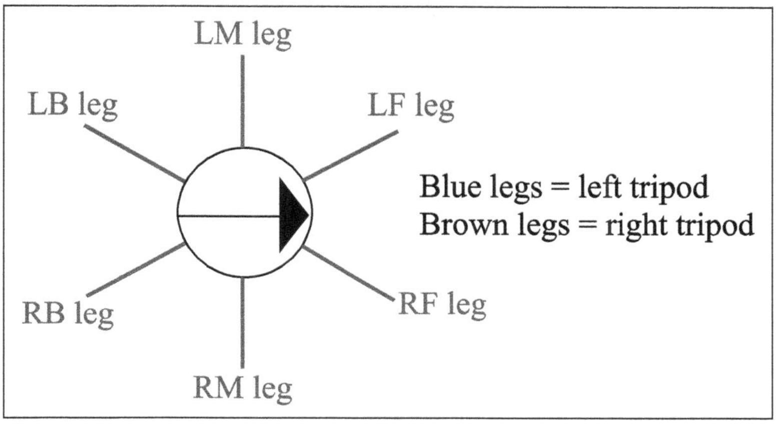

Figure 8.14 Leg tips moving in tripods demonstrating the 3-on gait, otherwise known as the 'double tripod gait'.

The LF, LB, and RM leg tip horizontal position waveforms are mutually identical, figure 8.15, as are the LM, RF, and RB leg tip waveforms since these two sets represent the left tripod and right tripod respectively. The two tripods move in antiphase such that the left tripod spends half the gait cycle time on the ground and half in the air and vice versa for the right tripod.

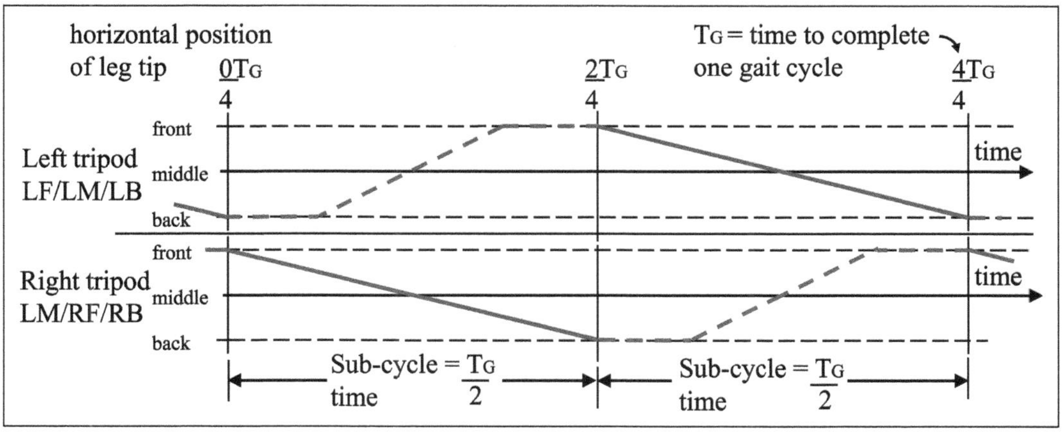

Figure 8.15 Horizontal motion for each of the two tripods legs during 3-on, double tripod gait.

The n-value radical waveform for the double tripod gait is shown in figure 8.16 below. We keep the cycle time clock pulse count equal to the 4-on and 5-on gaits, i.e. 360 counts. Just one equation is required to specify the radical shape.

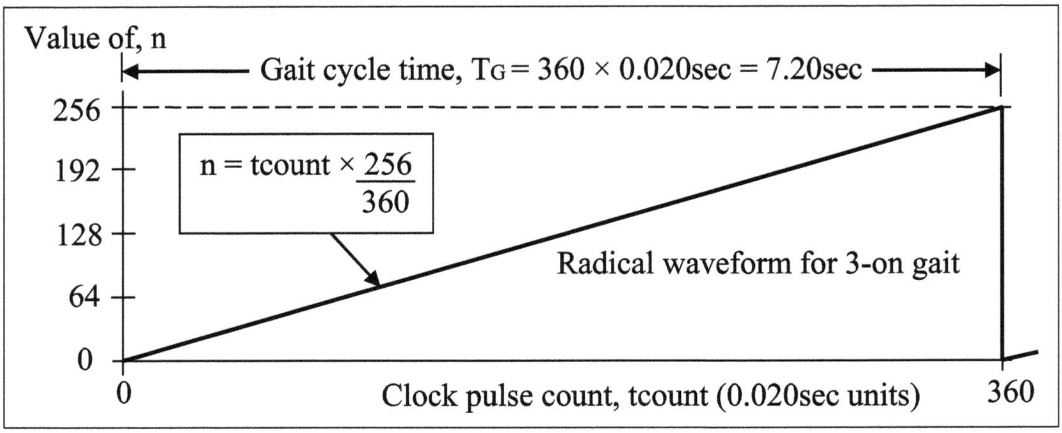

Figure 8.16 The n-value radical waveform for the 3-on, double tripod gait carried out in 360 pulse counts. Note just one equation to specify the radical shape.

The following diagram, figure 8.17 shows how the robot is programmed to walk in double tripod gait.

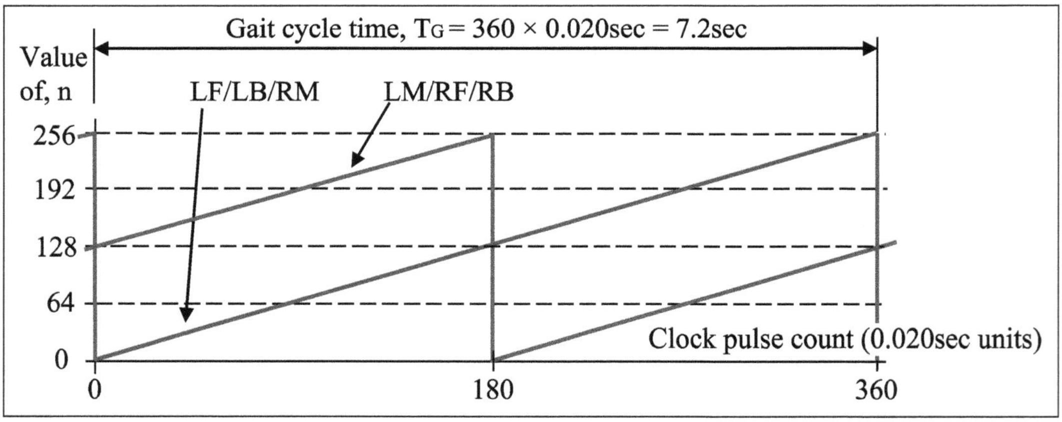

Figure 8.17 n-value waveforms for the two tripods to obtain the 3-on, double tripod gait.

We now conclude our discussion on each of the basic hexapod walking gaits, i.e. the 5-on gait, the 4-on gait and the 3-on gait. It is now interesting to think about how a hexapod can be programmed to transit between gaits. This is left as an interesting project for students especially when there are numerous variations of the 5-on and 4-on gaits. Could it be that the radical waveforms can be morphed during gait transition... let's take a look.

8.8 Morphing of the radical waveforms during gait transition

The following diagram, figure 8.18, may be of help in formulating gait-transitioning algorithms. It can be seen that the 3-on gait radical waveform is a straight line then develops bends at n-values of 64 and 192 as the gait transitions to 4-on and 5-on gaits.

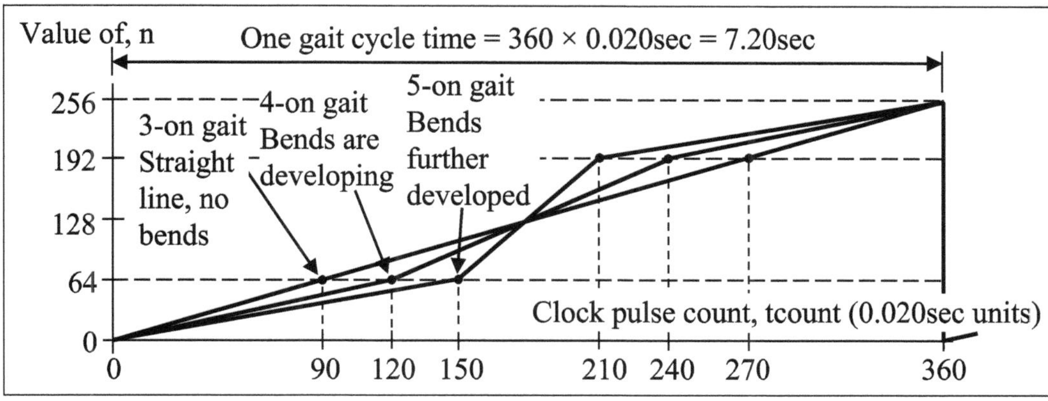

Figure 8.18 The n-value radical waveforms for 5-on, 4-on and 3-on gaits showing how they morph their shapes.

Don't forget though that not only must the radicals morph their shapes but also each of the six leg tips must move to different starting positions as well as the radical phases for each leg must change also, (see figures 8.5, 8.13 and 8.17).

Now we move on to discussing locomotion speed.

8.9 Locomotion speed in each gait

The maximum locomotion speed is dependent on two factors; the first factor concerns the type of gait and the second factor concerns, (i) the performance limit of the electric motor servomechanisms that drive the leg mechanism and (ii) the mass and geometry of the leg mechanism. As regards the first factor, i.e. the type of gait, we will show here that the 3-on gait produces the fastest locomotion speed, the 4-on gait produces the second fastest speed and the 5-on gait produces the slowest locomotion speed. As regards the second factor, let this maximum leg tip speed along the leg tip digital locus be equal to, V_{MAX}.

Now assume that we are using digital locus dimensions of amp and zag. Hence the leg tip distance through the air is, $d_{air} = 2 \times (\text{amp}+\text{zag})$ and the leg tip distance in contact with the ground is, $d_{gnd} = 2 \times \text{amp}$.

Now look at the 5-on gait n-value waveforms in figure 8.4 where it can be seen that 5 sub-cycle times are used to displace the leg tip through the ground distance, d_{gnd} and 1 sub-cycle time is used to displace the leg tip through the air distance, d_{air}.

Hence,

Leg tip ground speed, $v_{ground} = d_{gnd}/(5 \times \text{sub-cycle time}) = 2 \times \text{amp}/(5 \times \text{sub-cycle time})$

....and leg tip air speed, $v_{air} = d_{air}/(1 \times \text{sub-cycle time}) = 2 \times (\text{amp}+\text{zag}) / \text{sub-cycle time}$

From these equations we note now an important point which is, v_{air} will always be faster than v_{gnd}. For example, let amp=40mm and zag=40mm.

Hence, $v_{gnd} = 2 \times 40/(5 \times \text{sub-cycle time}) = 16\text{mm/sub-cycle time}$
and $v_{air} = 2 \times (40+40)/\text{sub-cycle time} = 160\text{mm/sub-cycle time}$

Hence the leg tip speed is 10 times faster in the air than on ground. Recall, now, that the leg tip maximum speed is V_{MAX}, which is thus to be set equal to the maximum leg tip speed experienced which is v_{air}. However, the robot locomotion speed is equal to v_{gnd}.

Thus the locomotion speed is, $V_{MAX}/10 = 0.10V_{MAX}$ for the 5-on gait.
Similar calculations for the 4-on gait and 5-on gaits give:
locomotion speed is, $V_{MAX}/4 = 0.25V_{MAX}$ for the 4-on gait
and locomotion speed is, $V_{MAX}/2 = 0.50V_{MAX}$ for the 3-on double tripod gait

Thus it can be seen that each gait, 5-on, 4-on, 3-on, respectively, permits the robot to travel successively faster. Note that each gait can travel faster if the value of zag is decreased. For example, when travelling on smooth flat surfaces, zag can be decreased to just 10mm. Repeating the above calculations for each gait with amp=40mm and zag=10mm we obtain the following locomotion speeds:

5-on locomotion speed = $0.16V_{MAX}$
4-on locomotion speed = $0.40V_{MAX}$ } 60% speed increase compared with zag=40mm
5-on locomotion speed = $0.80V_{MAX}$

Now let us move on to consider hexapod static stability.

8.10 The support polygon for static stability

For static stability, the support polygon is formed by straight lines joining the adjacent leg tips that are in contact with the ground.

At all times, when viewing the robot in plan view, the centre of gravity of the robot should not move outside the support polygon. This is to stop the robot from toppling over. For example, figure 8.19 shows a plan view of the robot that is walking in double tripod gait with the blue legs in contact with the ground and the brown legs in the air.

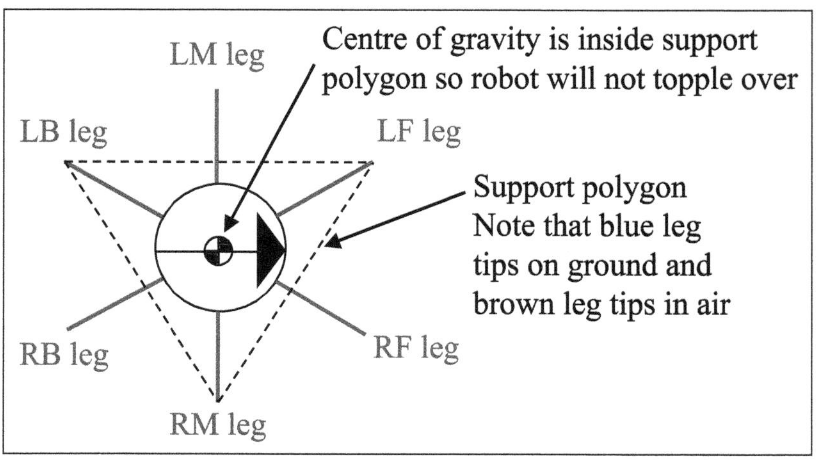

Figure 8.19 The support polygon in double tripod gait.

For the configuration shown above in figure 8.19 can be seen that the robot will not topple over because the centre of gravity lies inside the support polygon. However, as the leg tips shift through their digital locus as the robot walks, there is a danger that the support polygon will expose the centre of gravity beyond the support polygon figure 8.20. Note that the brown legs are removed for clarity.

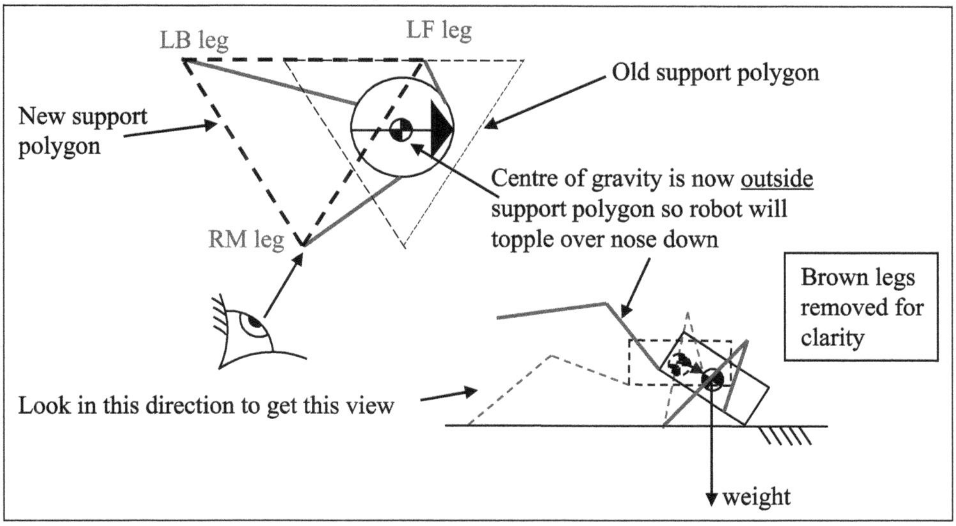

Figure 8.20 If amplitude of leg tips during walking is too much then the c. of g. falls outside the support polygon.

In order to prevent an occurrence of the c. of g. falling outside the support polygon we must place restrictions on the control software of the robot. The simplest way is to restrict the diameter of the cylindrical coordinate frame that is the working volume of the leg tip. The idea is shown in the following diagram, figure 8.21, which is a plan view of the robot.

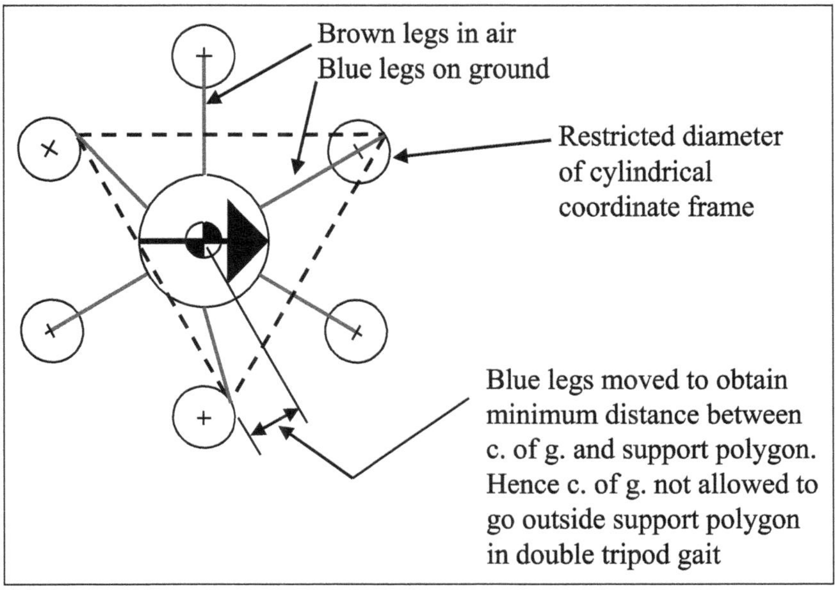

Figure 8.21 restricting the leg tip amplitude to maintain static stability during walking.

The c. of g. can still fall outside the support polygon if the robot is on a slope. The way to keep the c. of g. safely inside the support polygon is to keep the robot body weight low on the ground as shown in Figure 8.22, below. The configuration assumes that the coefficient of friction between the leg tips and the ground is greater than the angle of the slope from horizontal. If not, the robot may slip before toppling.

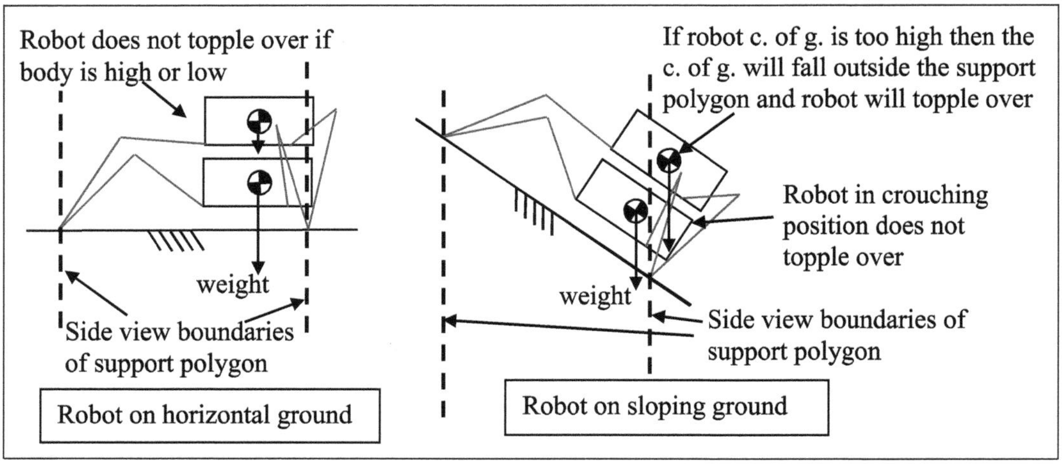

Figure 8.22 Some more conditions for static stability.

However, even with a restriction on the diameter of the cylindrical coordinate frames and with the robot on horizontal ground, if a different double tripod gait is chosen then the c. of g. could fall outside the support polygon. For example if different legs are chosen to form the two tripods in a double tripod gait then the robot can topple over like this in figure 8.23, below.

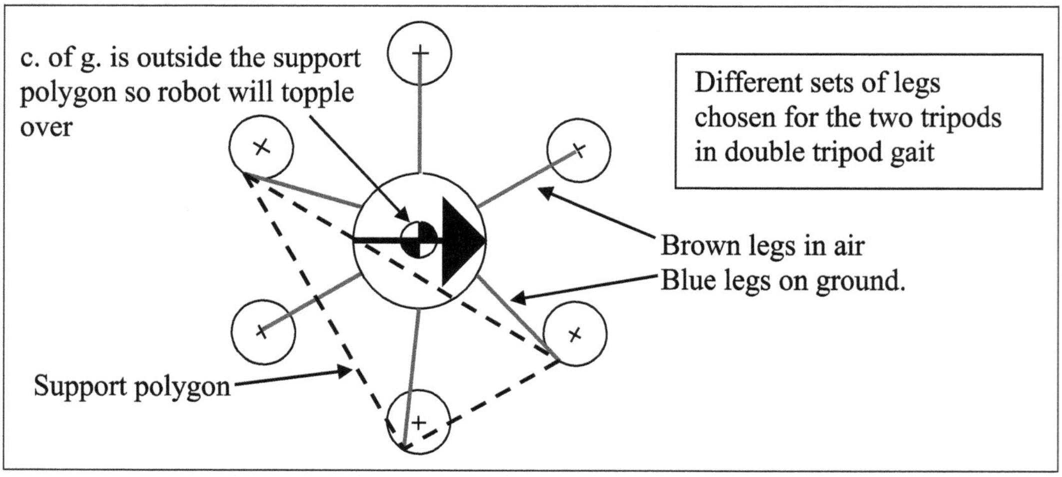

Figure 8.23 Another situation for losing static stability.

8.11 Walking on 4 legs

We need to consider what happens if one or more legs of the robot become unserviceable due to failure or because two legs are required as manipulators. For example, what happens if the robot can walk on a minimum of 4 legs in 3-on gait.

Actually the robot can walk on less than 4 legs but we are only considering statically stable situations where inertial forces are relatively insignificant.

As an example consider a robot that uses its front two legs, LF and RF for manipulators and the remaining legs for locomotion in 3-on gait as in figure 8.24.

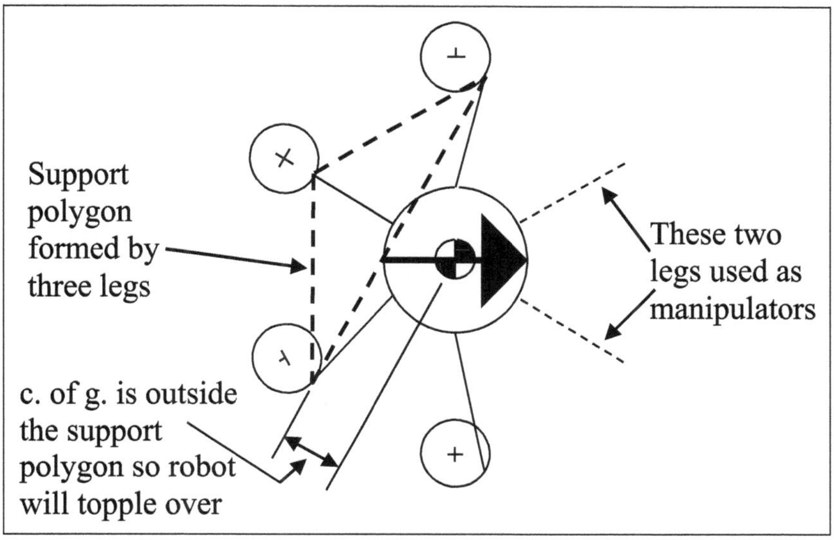

Figure 8.24 Losing static stability when using only four legs.

The only way the robot can stop itself from toppling over is to put four legs on the ground. As soon as it lifts any one of the legs off the ground to start walking then it will start to topple over. Hence the robot cannot walk unless the c. of g. can be brought inside the support polygon. One solution may be to move the cylindrical coordinate frames so that they are equally disposed at 90^0 intervals around the robot body as shown in figure 8.25.

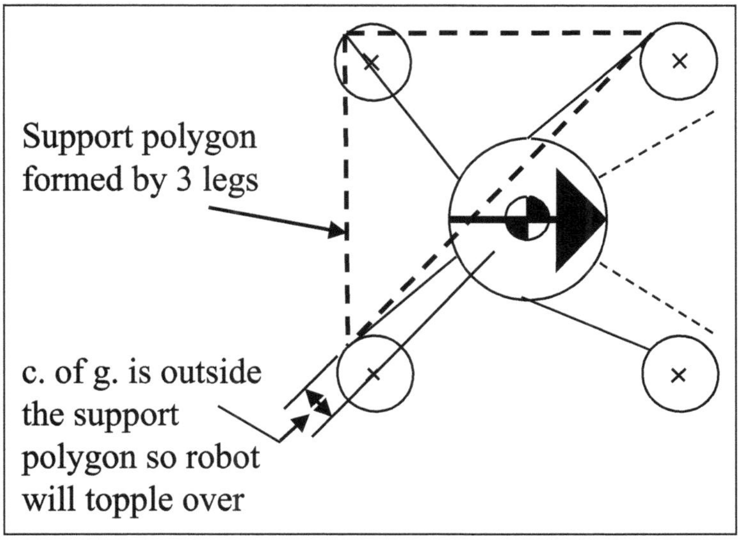

Figure 8.25 The problem of static stability when walking on four legs.

But no, the c. of g. can still fall outside the support polygon as shown above, figure 8.25

A solution is required where the cylindrical coordinate frames, set at 90^0 intervals are not disturbed with respect to the ground. Instead we will shift the c. of g. inside the support polygon. So try this arrangement in figure 8.26.

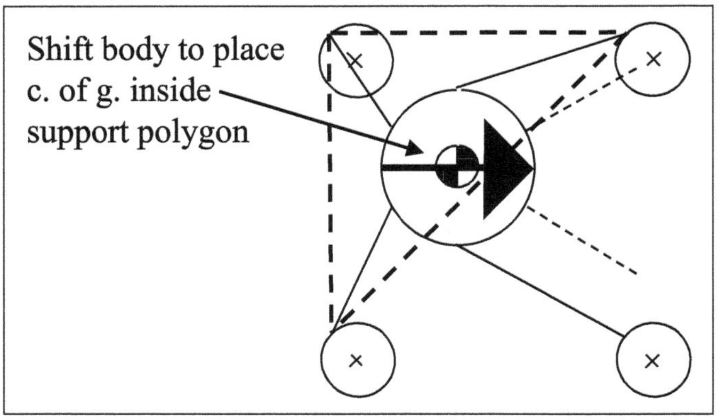

Figure 8.26 Solving the problem of static stability on four legs.

However the support polygon changes its placement as the robot progresses with its 3-on walking gait so the body needs to be constantly shifting away from the leg that is in the air. You can do an experiment by getting down on the floor on your hands and knees and trying to walk slowly by moving one hand or knee at a time (but make sure nobody is looking). Otherwise if pride stops you doing this the idea is shown in figure 8.27 below.

Squat animals like crocodiles walk like this but camels and elephants put their feet on a single line directly beneath their bodies but they are constantly maintaining dynamic stability

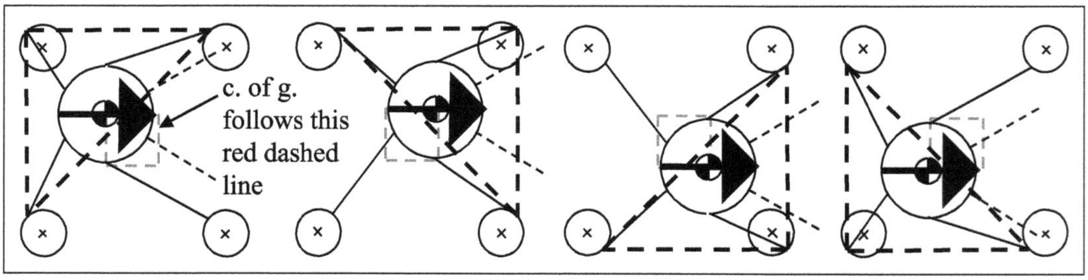

Figure 8.27 3-on gait for 4-legged robot showing c. of g. shifting in synchronism with constantly shifting support polygon so that c. of g. constantly stays inside support polygon. The trajectory of the c. of g. follows the *red dashed line which is fixed with respect to the cylindrical coordinate leg tip frames.*

It is also informative to show how the cylindrical coordinate frames shift with respect to the body. Figure 8.28 shows this as follows.

Figure 8.28 shows the cylindrical coordinate frames of each leg set in their own x-y shiftable Cartesian coordinate axes. The dashed line cylindrical coordinate frames indicate the shiftable limits of the frames. Inverse kinematic equations for each 3dof leg must be worked out as a function of the x-y axes to program this behaviour pattern into the robot.

The following figures 8.29, 8.30 and 8.31 show how the legs are programmed for one gait cycle. The first diagram, figure 8.29, shows the horizontal motion of the leg tips. The second diagram, figure 8.30, shows these motions translated to values of, n, on the digital locus and the final two diagrams in figure 8.31 show how the, x, and, y, axes are displaced during the gait cycle in order to keep the c. of g. inside the support polygon.

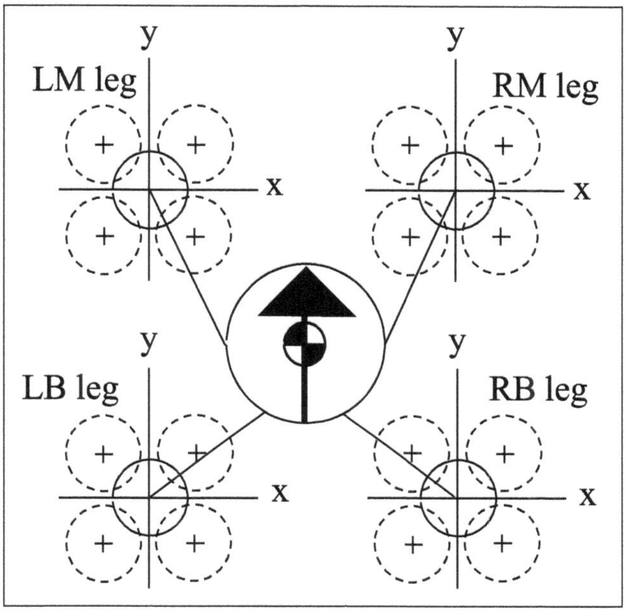

Figure 8.28. 4-legged robot showing the shiftable cylindrical coordinate frames of each leg.

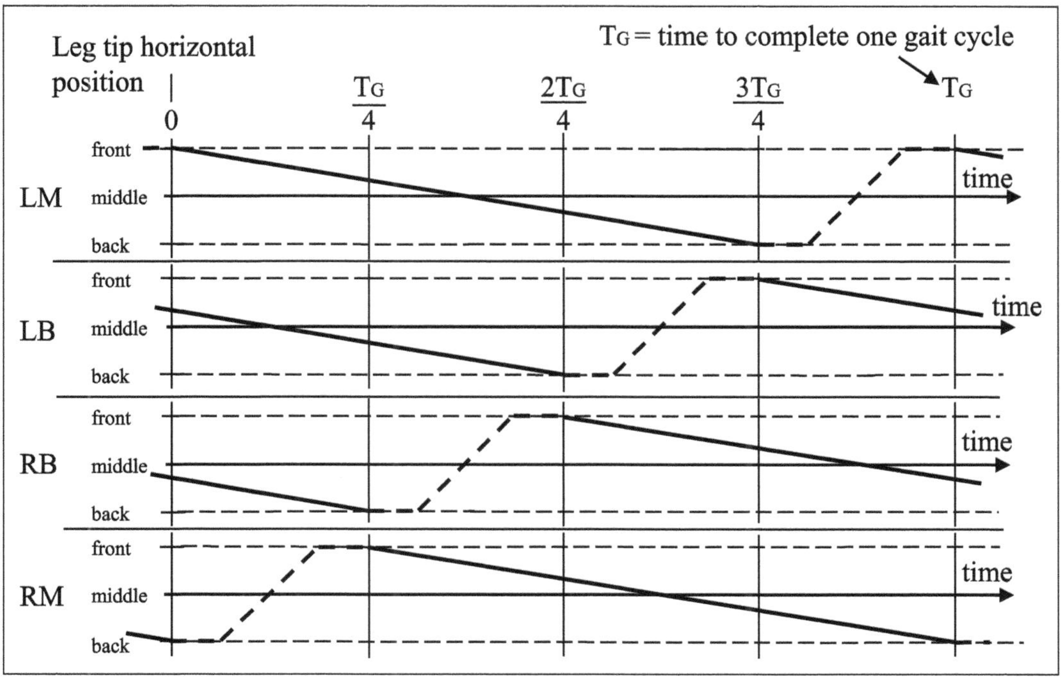

Figure 8.29 Leg tip motion when using four legs and 3-on gait for locomotion.

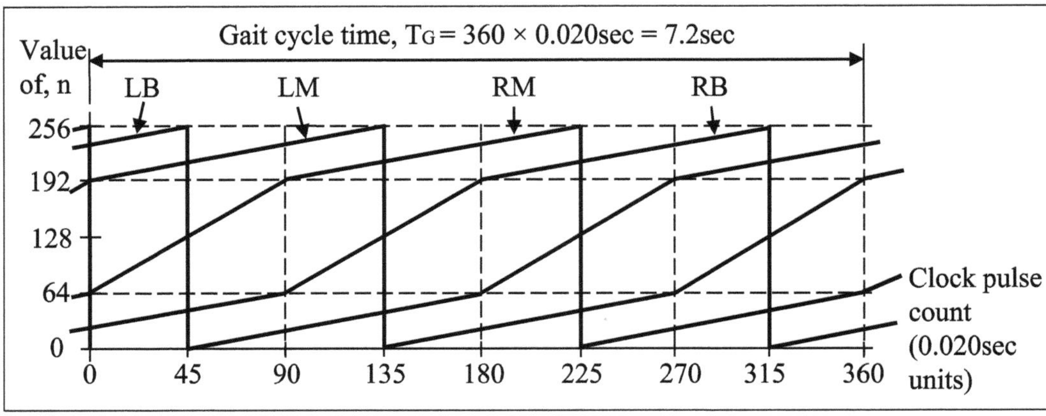

Figure 8.30 n-value waveforms for each of the four legs to obtain 3-on gait.

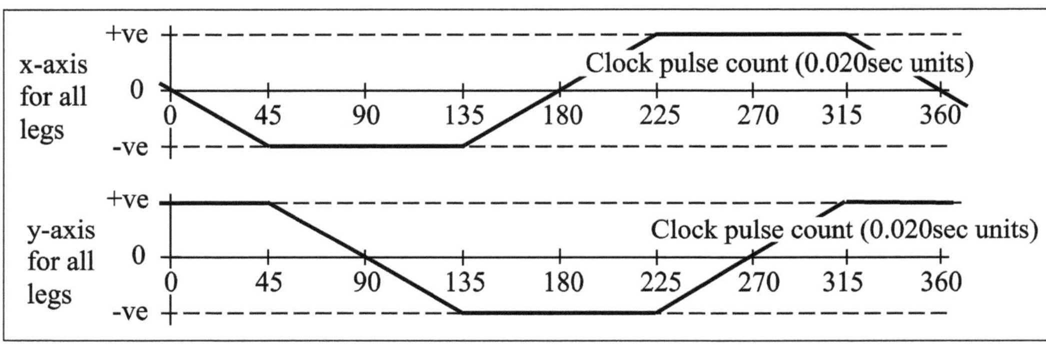

Figure 8.31 The shifting of the cylindrical coordinate frames during the 3-on gait with four legs.

8.12 Walking robot turning about an instantaneous centre of rotation, IC of R

It is convenient now to explain the principle of programming the walking robot to carry out manoeuvres and to follow set paths. A very useful tool to do this is to arrange for the robot to turn about an instantaneous axis of rotation, IAofR, which is an axis referred to the ground on which the robot is walking. However, for the convenience of explanation, and because it applies to most robot manoeuvres, the robot will be assumed to walk on a horizontal ground plane and turn about a vertical IAofR. In this ubiquitous special case it is convenient to view the robot in plan view, in which case the IAofR becomes a point and is now referred to as the instantaneous centre of rotation, the ICofR. The technique has already been described in chapter 1 and has been applied to a tricycle robot in chapter 3. For example, consider a six-legged robot turning about an instantaneous centre of rotation as shown in figure 8.32 below.

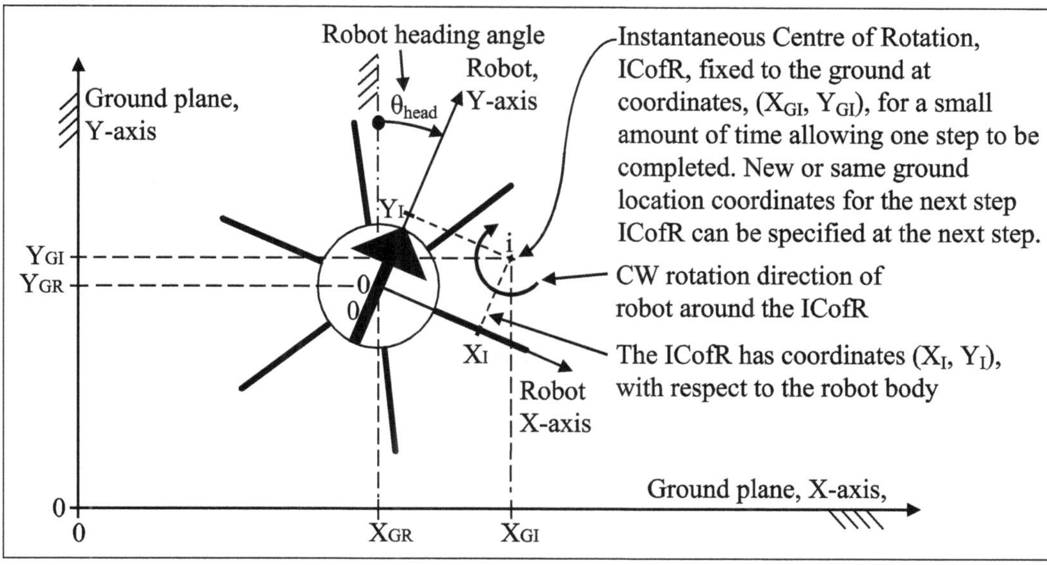

Figure 8.32 A walking robot turning about an Instantaneous Centre of Rotation, IC of R.

The term "instantaneous" is used because its position can change with time or for each walking step. However for the example used here its location remains constant. Suppose the ICofR is specified at ground plane coordinates, X_{GI},Y_{GI}, figure 8.32. Now consider the existence of a local positioning system that can measure the position of the robot relative to the ground plane as coordinates, X_{GR},Y_{GR} and θ_{head}. The local positioning system can be a computer vision system mounted on the robot or it can be an alternative system mounted on the ground and the information transmitted to the robot. The 5 values, X_{GI},Y_{GI}, X_{GR},Y_{GR} and θ_{head} is sufficient information to calculate the position of the IC of R relative to the robot body axes as, X_I, Y_I, coordinates.

The ICofR is used as a method to get the robot to follow a given pathway. However, you can think of the ICofR as a target or a "centre of attention". An example would be an inspection robot that encircles a damaged component at a constant radius with a camera aimed towards the component. The concept can also be applied to intelligent strategies in robot collaborative and competitive game playing.

The next step is to work out how to get the robot to turn about an IC of R that is specified as robot body X, Y, coordinates. Consider the robot shown in the figure 8.33. The robot is shown walking on horizontal ground and turning CW about an instantaneous axis of rotation, which, when viewed from above, becomes the instantaneous centre of rotation. The robot is shown walking in double tripod gait with the LM, RF and RB leg tips shown at, n = 0, in the digital locus and the leg tips LF, LB and RM shown at, n = 128, in the digital locus.

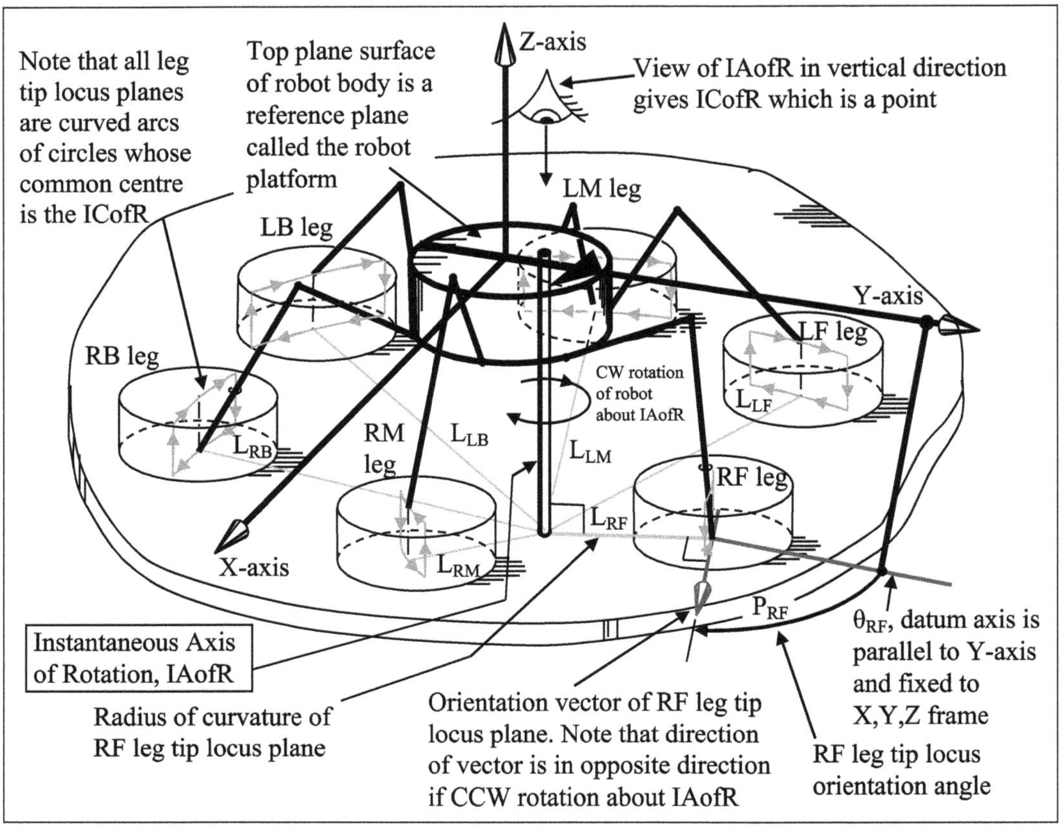

Figure 8.33 Illustration showing the leg tip loci shapes that will achieve robot body rotation about an instantaneous axis of rotation, IA of R.

The diagram contains a lot of information, so it is necessary to take care in reading all the comments. The X, Y and Z axes are rigidly connected to the robot body so if the body moves then the axes will move with the body. Note that, (i) the × and Y axes lie in the robot platform which is a reference plane and is coincident with the top surface of the robot body, (ii) the datum of the × and Y Cartesian axes is located at the robot body plane centre, and (iii) the positive Y-axis is coincident with forward direction of the robot body and the positive X-axis lies on the right hand side of the body.

In order for the robot body to rotate about the ICofR, the leg tips, when in contact with the ground, must describe arcs of circles that all share a common centre which is the ICofR. Thus the plane of each leg tip locus is curved to match the curvature of each of these arcs. The radius of curvature of each arc is equal to the radial distance between the ICofR and each leg tip cylindrical coordinate frame Z-axis.

The following diagram, figure 8.34, takes a closer look at the RF leg. The other 5 legs have a similar arrangement. Note how the, θ_{RF}, datum axis of the leg tip cylindrical coordinate frame, is aligned in parallel with the robot body Y-axis and the positive direction of the Z_{RF} axis is vertically downwards. The downwards direction is chosen as positive because displacing the leg tip in the positive direction will raise the robot body. All the remaining 5 cylindrical coordinate frames are similar.

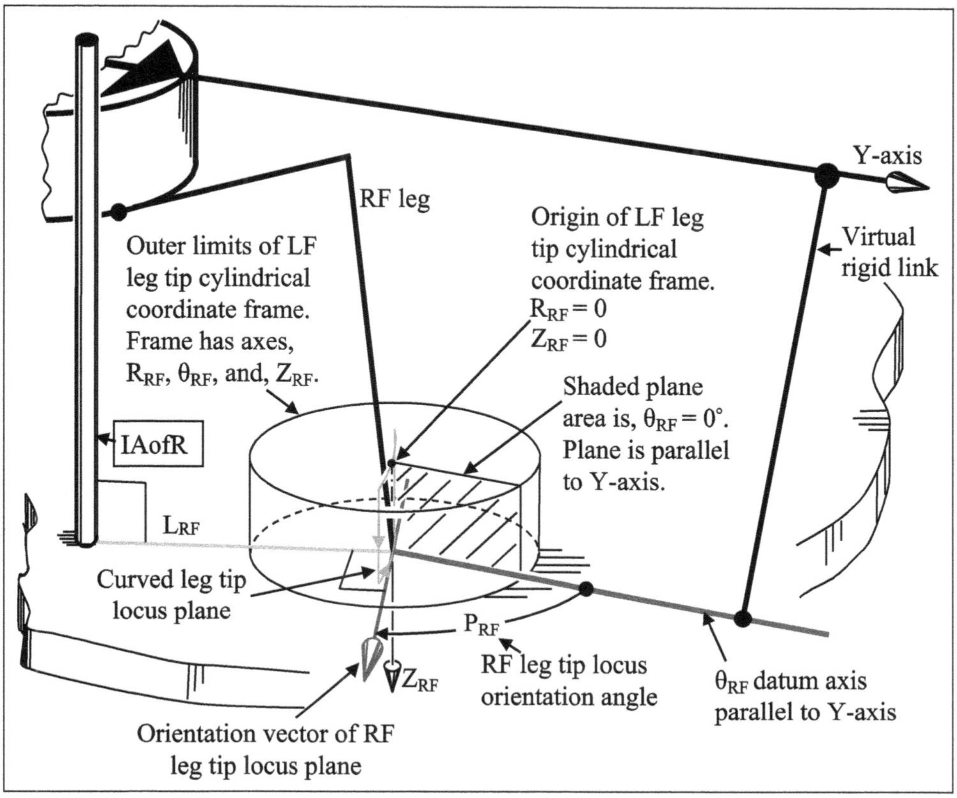

Figure 8.34 Closer look at the LF leg tip locus as the robot rotates about an instantaneous axis of rotation, IAofR.

The green line, labelled, L_{RF}, in figures 8.33 and 8.34, represents the radial distance of RF leg tip from the IAofR. More precisely, it is the perpendicular distance from the IAofR to the Z_{RF} axis of the RF leg tip cylindrical coordinate frame. The radial distances, L_{RM}, L_{RB}, L_{LB}, L_{LM}, and L_{LF}, of the remaining leg tips are shown in figure 8.33. Figures 8.33 and 8.34 also show a blue line which represents the orientation vector of the RF leg tip locus plane. It represents the generalised direction of the leg tip in order to produce the demanded CW rotation about the IAofR The vector lies in the ground plane and is oriented perpendicular to the green line, L_{RF}. If the robot were given wheels then the green line would be the axle of a wheel and the blue line the plane of the wheel. However, here the analogy ends because the leg tip remains in contact with the ground for a duration of time and during that time the robot displaces itself through an angular rotation about the IAofR. Hence the leg tip must describe an arc of a circle whose centre is the ICofR and the angle subtended by the arc is the angular rotation of the robot during that time.

The following two diagrams, figures 8.35 and 8.36, show the general arrangement in plan view and end elevation where the IAofR becomes the ICofR. The brown lines represent the, θ, datum axes belonging to the cylindrical coordinate frames that exist at each leg tip. The origins of the cylindrical coordinate frames are located in plan view at robot body coordinates, (X_{RF}, Y_{RF}), (X_{LF}, Y_{LF}),........., (X_{RM}, Y_{RM}). Each leg tip coordinate frame is fixed rigidly to the X, Y axes and once again if the robot body moves then the X, Y, and cylindrical coordinate frames move together with the body.

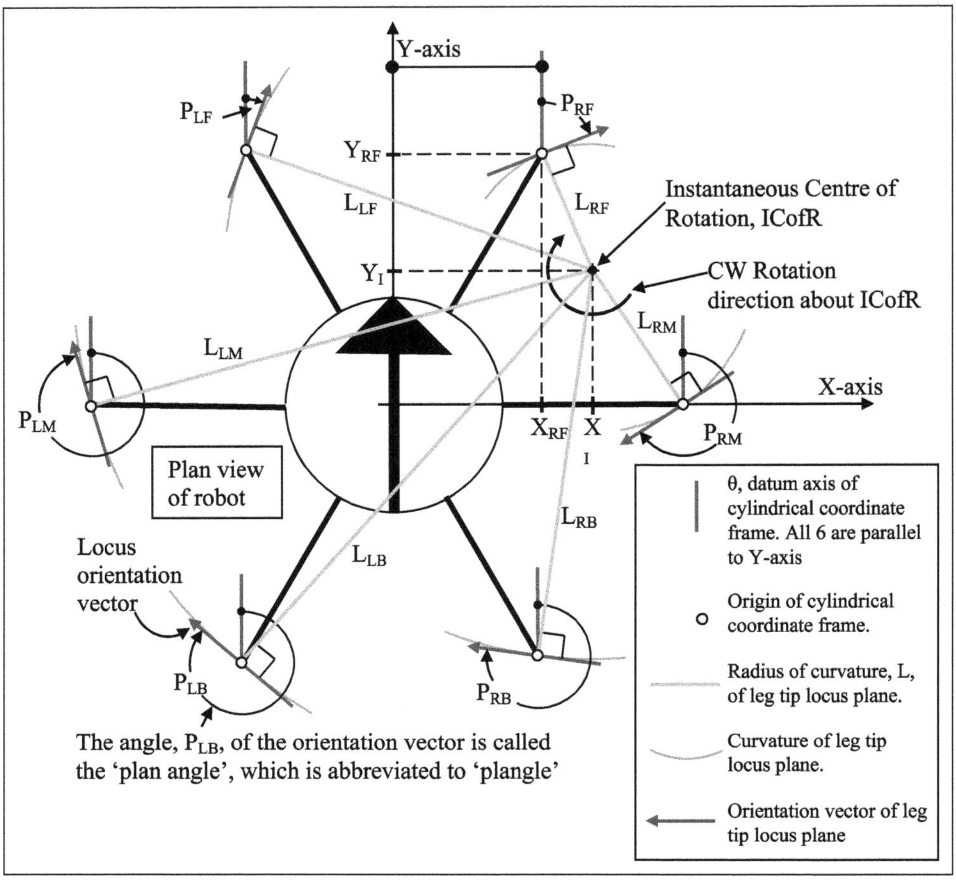

Figure 8.35 Plan view of the leg tip locus planes during rotation about an instantaneous centre of rotation, ICofR.

The following diagram, figure 8.36, shows that all the origins of the leg tip cylindrical coordinate frames are located at an equal displacement, Z_O, below the body platform.

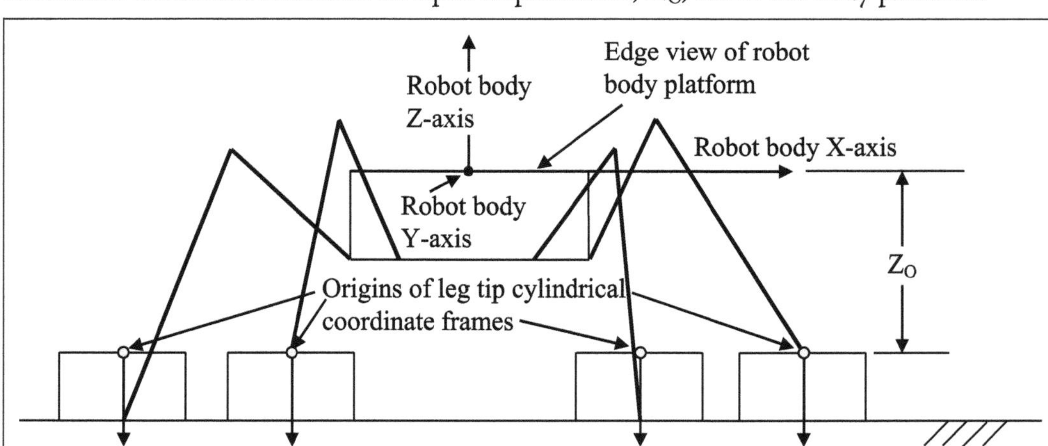

Figure 8.36 End elevation of robot body parallel to robot body Y-axis showing the robot body platform at with respect to the leg cylindrical coordinate frames.

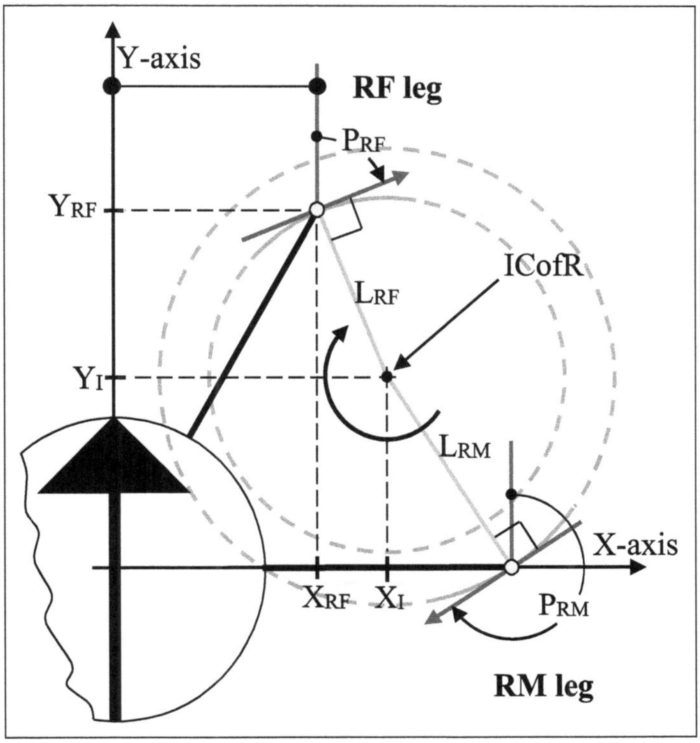

Figure 8.37 Diagram showing CW rotation of robot about the ICofR. Here the locus planes curve away from the orientation vector on the rhs of the vector.

CW rotation about ICofR

The adjacent diagram, figure 8.37, shows just the RF and RM legs. The robot is turning CW about the ICofR, hence the blue orientation vectors are directed CW around the ICofR. Each leg tip locus describes an arc of a circle whose centre is the ICofR. The continuous orange lines represent a plan view of the leg tip loci and the broken orange lines are a continuation of the these loci into circles with the IC of R as the centre The blue orientation vectors are almost pointing in opposite directions due to the chance location of the ICofR lying almost on a line between the origins of the two cylindrical coordinate frames. Despite the almost opposite directions of the orientation vectors, the curvatures of both leg tip locus planes lie on the right hand side of the blue orientation vectors like this, (both blue vectors set vertical for fair comparison)

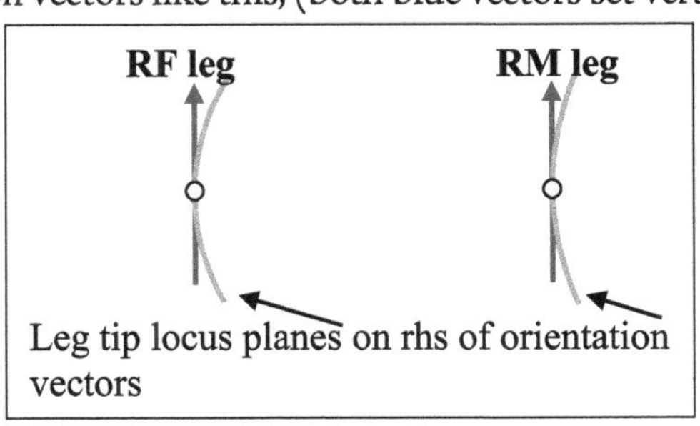

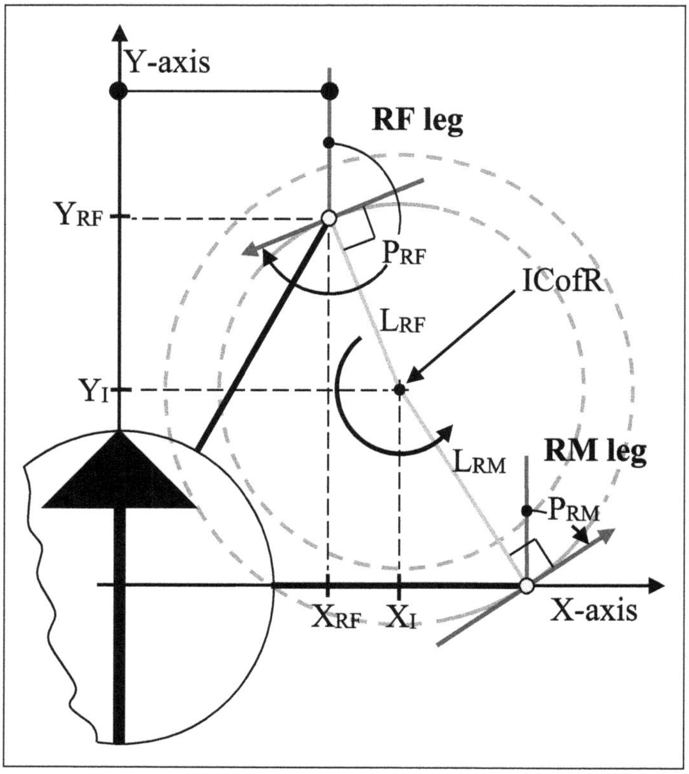

Figure 8.38 CCW rotation of robot about the ICofR.

CCW rotation about IC of R
This diagram, figure 8.38, is similar to figure 8.37 above with the only difference that the robot is turning CCW about the same ICofR. The resulting effect is to, (i) reverse the direction of both blue orientation vectors and, (ii) to shift the leg tip locus planes to the left hand side of the blue orientation vectors.

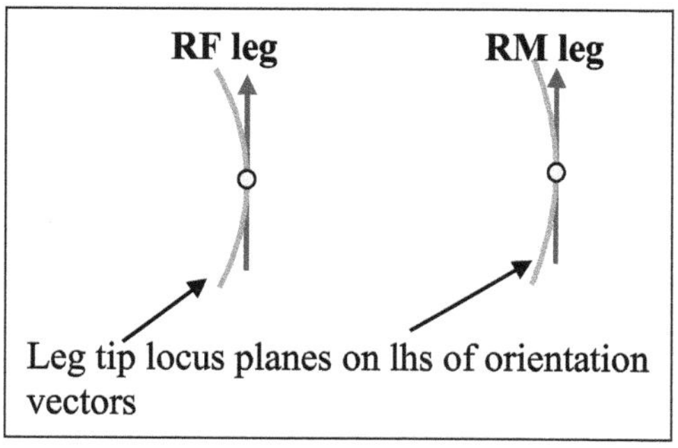

The following diagram, figure 8.39, shows a plan view of the complete robot as all six legs traverse through a single step that rotates the body through an angle, θ_{rot}, about the ICofR. Note that the body is shown fixed to the paper such that the leg motion with respect to the body becomes apparent. Such motion is important since it directly reflects how the legs are programmed.

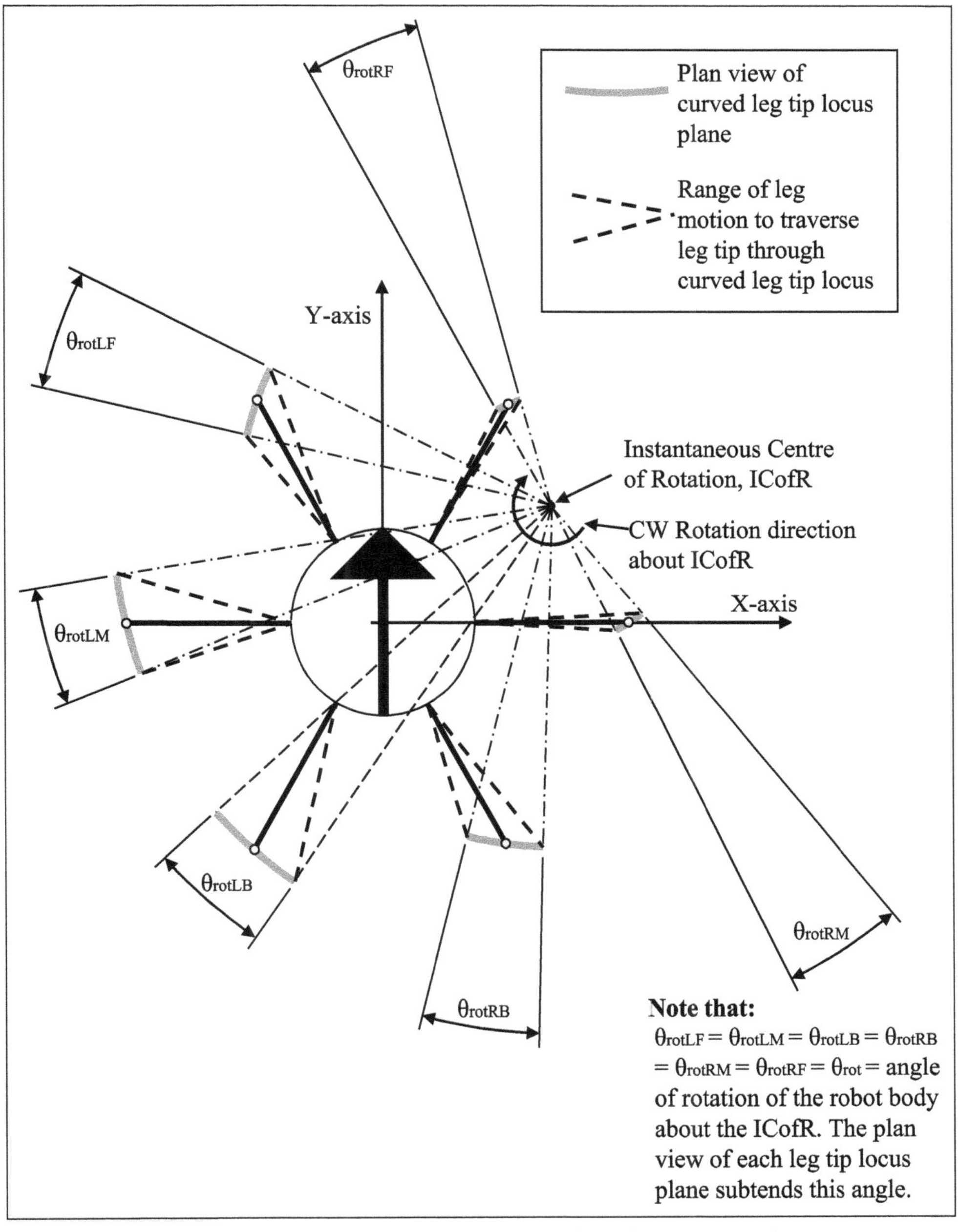

Figure 8.39 Plan view of all six leg tip loci as the robot rotates about an instantaneous centre of rotation, IC of R.

The following two diagrams, figures 8.40 and 8.41, summarise what is to be computed by the on-board microcomputers using the LB leg as an example. Figure 8.40 shows a slanted view and the leg tip and its curved locus plane. Figure 8.41 is a plan view of the same arrangement. Two views are shown to make things more clear. What we have to do is to compute the positions of each of the 3 servomechanisms that actuate the leg tip so that the leg tip tracks around the curved leg tip locus plane. These chapters are concerned with a flat plane leg tip locus with variable amplitude and zag values that define its shape.

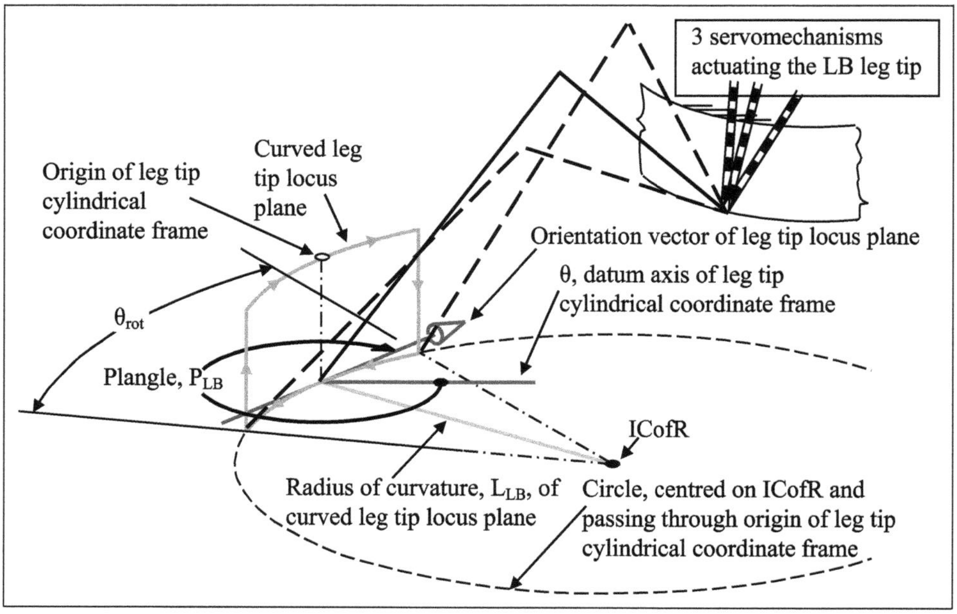

Figure 8.40 LB leg used as an example to show variables concerned with the curved leg tip locus.

However, we now work out how to give curvature to the leg tip locus plane as is required when the robot rotates around an ICofR. Furthermore, the curvature is a variable quantity that is dependent on the distance, L_{LB} between the ICofR and the origin of the leg tip cylindrical coordinate frame.

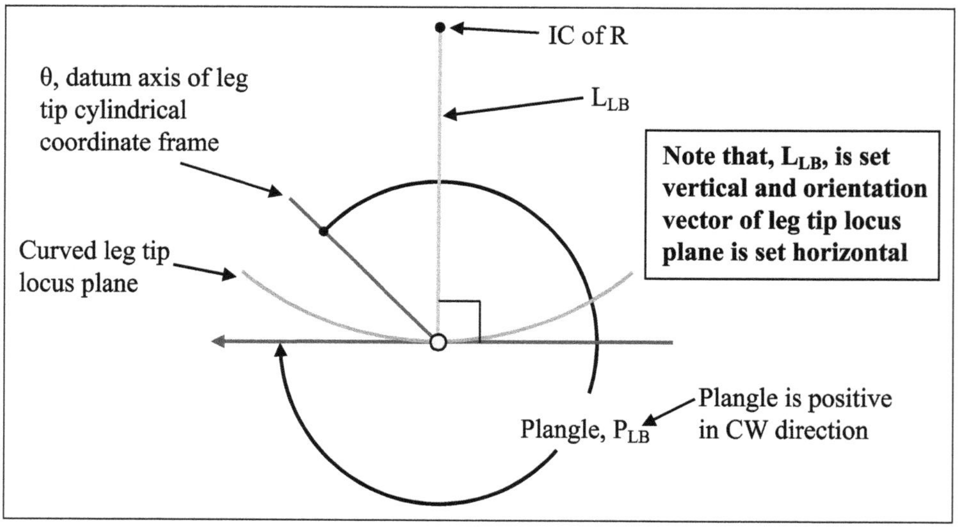

Figure 8.41 Same as figure 7.40 but shown in plan view.

Note that the ICofR is the centre of circle from which the curved leg tip locus plane is composed. The following diagram, figure 8.42, shows a plan view of one typical leg with various leg tip locus planes possessing a range of radii of curvature. The amplitudes of the loci are shown all the same and as a consequence the robot body rotation, θ_{rot}, increases as the radius of curvature decreases. This need not be the case since the amplitude of a locus is dependent on other factors which will be shown later. Notice that the leg tip locus planes can exist on either side of the orientation vector.

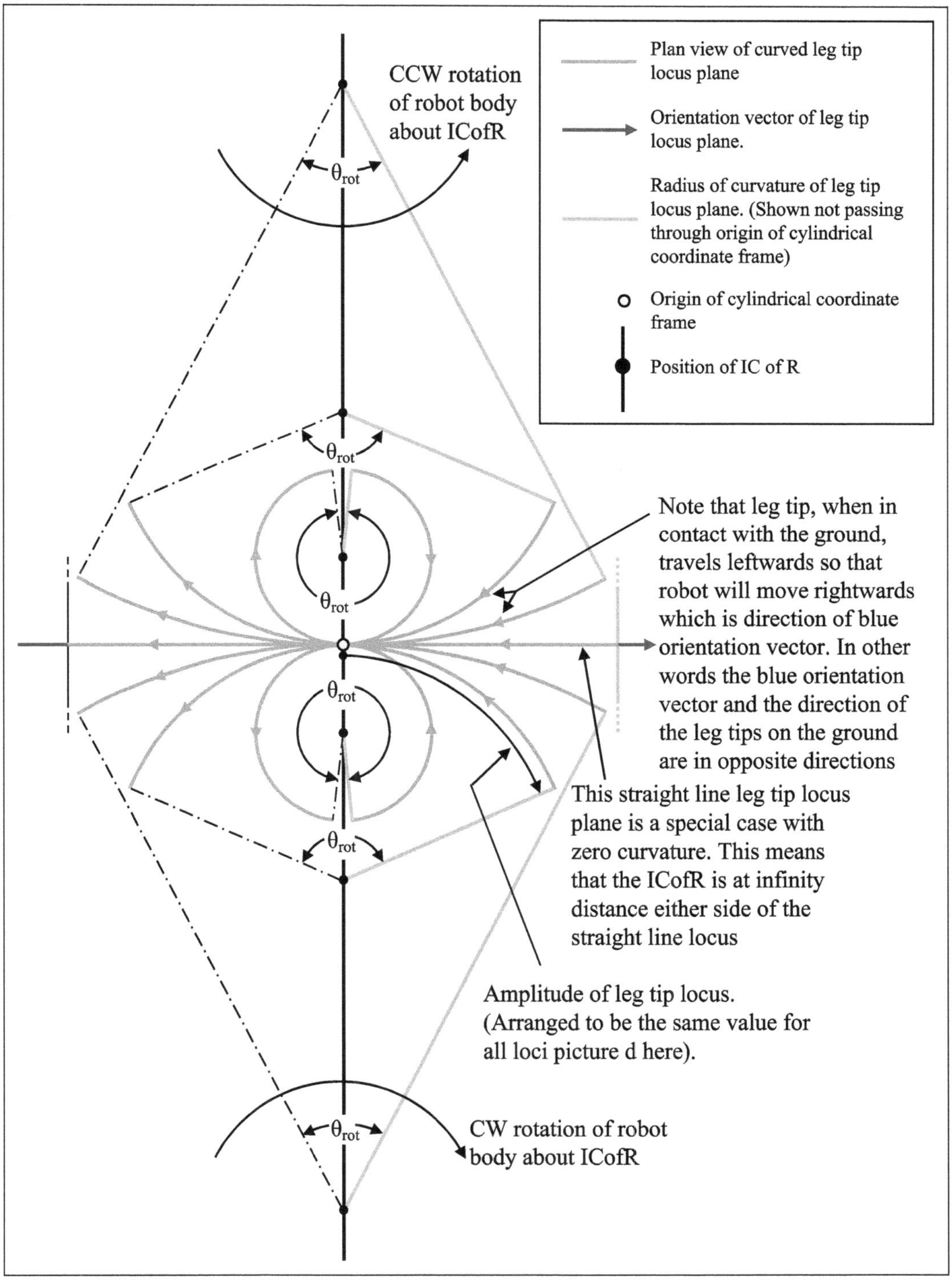

Figure 8.41. Plan view of selection of leg tip loci at various radii of curvature.

The existence of the IC of R either side of the orientation vector means that robot body rotation can be CW or CCW about the IC of R. To make the concept of the curved leg tip locus clearer and to illustrate the locus programming complexity, a 3-d view of the previous diagram is shown below in figure 8.42.

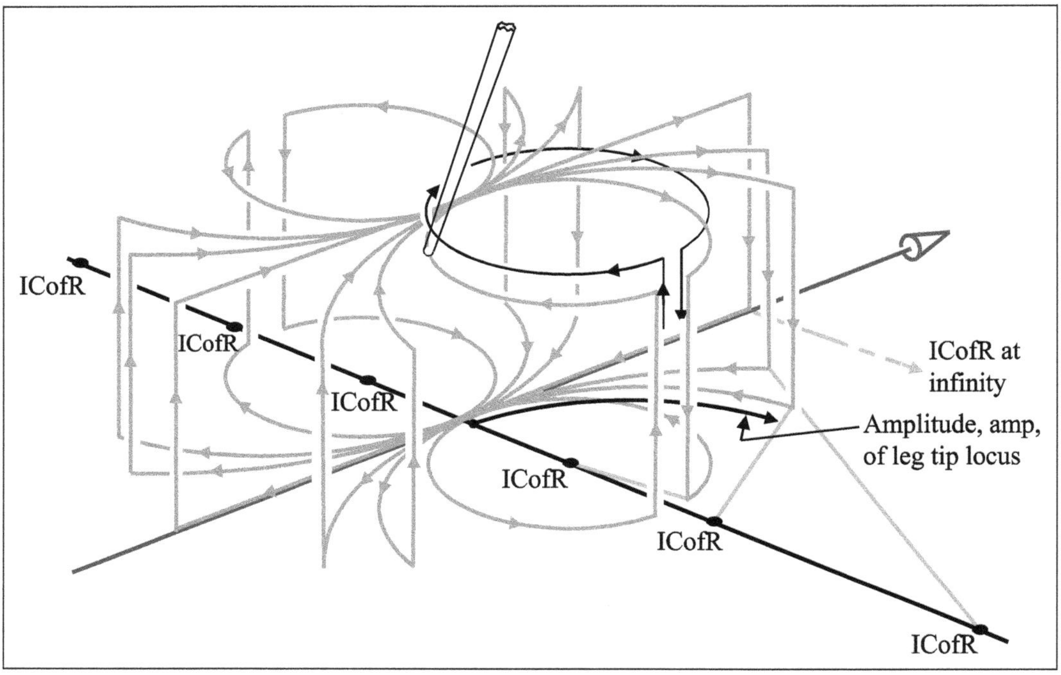

Figure 8.42 3-d view of curved leg tip loci showing the complexity of leg tip paths required by the hexapod robot. Here, the leg tip loci are shown curved around various ICofR. You can see a YouTube video of the motion here: https://youtu.be/bwmp-f7J3Vy0.

8.13 Steps in computing the curvature of the leg tip locus

No we must get down to doing the maths and analyse the curved leg tip locus in more detail. Figure 8.43, below, shows the relevant details. The essence of what we want to do is to force curvature onto the leg tip locus whilst using the existing cylindrical coordinate frame. So curvature of the locus can be carried out by compensating the locus specification values of R and θ to become new values of R_{comp} and θ_{comp} respectively. We use the left back, LB, leg as an example, thus the subscript, $_{LB}$ is used with variable values.

Referring to figure 8.43, consider a point, a, which is situated at an arbitrary distance, R_{LB} along the curved leg tip locus plane. The point, a is to be mapped into coordinates, R, θ, Z, of the cylindrical coordinate frame that compensates for, the curvature of the digital locus. The Z coordinate is unchanged so point, a, is mapped into R, θ coordinate space.

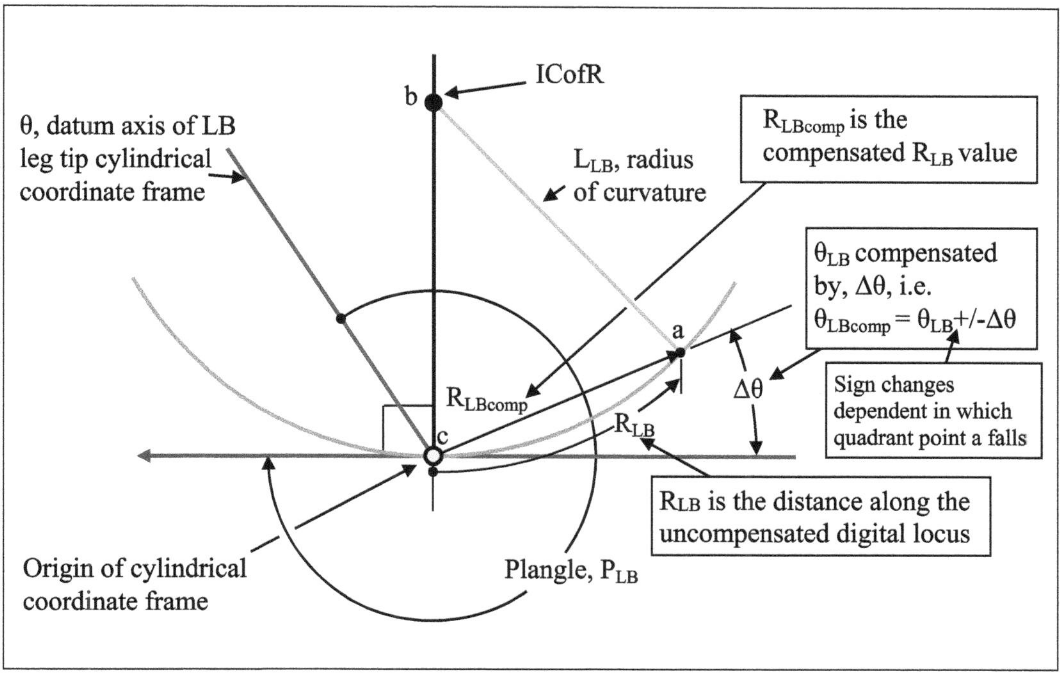

Figure 8.43 Maths analysis of the locus curvature.

8.13.1 Calculation to transform point, a, location into cylindrical, R, θ, coordinates

Referring to figure 8.44, which is a more detailed view of figure 8.43, first find location of point, a, relative to leg tip locus plane orientation vector, (the blue vector). We want to find two functions, fθ, f_R, as follows.

$$\Delta\theta = f\theta(R_{LB}, L_{LB})$$

and, $R_{LBcomp} = f_R(R_{LB}, L_{LB})$

From figure 8.44,
ba = bc = L_{LB} ,so triangle, cba , is an isosceles triangle
angle dca = $90^\circ + \Delta\theta$
angle bca = 180° – angle dca = $90^\circ - \Delta\theta$
angle bca = angle bac (because, bc = ba)
hence, angle cba = $180^\circ - 2 \times (90^\circ - \Delta\theta)$
Thus, angle cba = $2 \times \Delta\theta$
Also, angle cba = R/L_{LB}
hence,

$$\Delta\theta = \frac{R_{LB}}{2 \times L_{LB}}$$

This is, $\Delta\theta = f_\theta(R_{LB}, L_{LB})$

Because, ba = bc, and, eb, is perpendicular to, ca, then triangles bec and bea are identical, so,

$$ea = ec = R_{LB}/2$$

Considering triangle, bea,

$$\sin(\Delta\theta) = (R_{comp}/2)/\ L_{LB} = R_{comp}/(2 \times L_{LB})$$

but, $\Delta\theta = (R_{LB}/2)/\ L_{LB}$

hence, $\sin(R_{LB}/(2 \times L_{LB})) = R_{comp}/(2 \times L_{LB})$

finally, $\boxed{R_{comp} = 2 \times L_{LB} \times \sin\left(\dfrac{R_{LB}}{2\ L_{LB}}\right)}$ or $\boxed{R_{comp} = 2 \times L_{LB} \times \sin(\Delta\theta)}$

This is, $R_{comp} = f_R(R_{LB}, L_{LB})$

Let us now look at how we use these equations to modify the points on the leg tip cylindrical coordinate frame. The use of the equations can be classified into 4 quadrants that will now be shown.

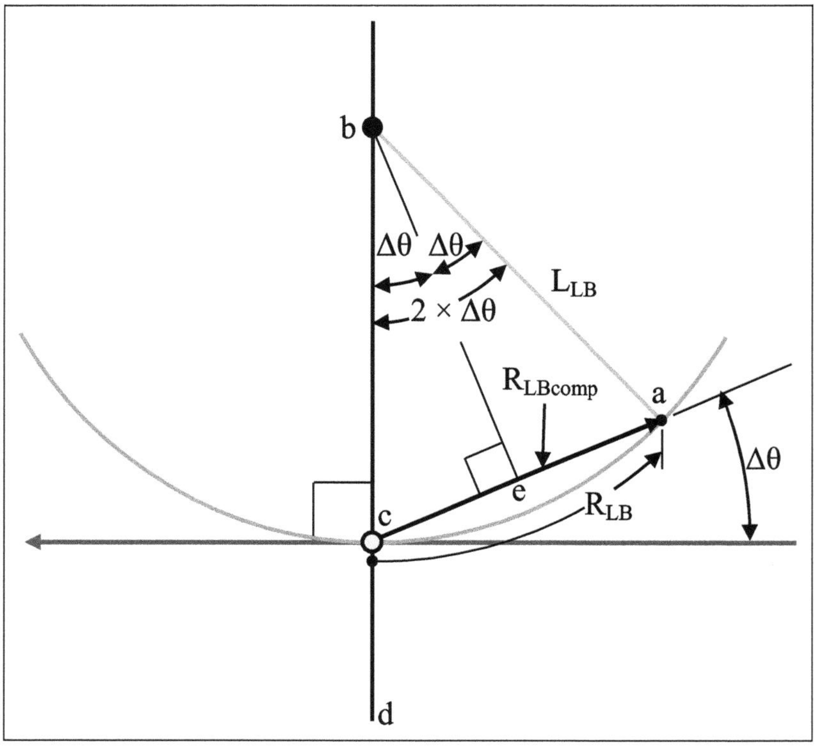

Figure 8.44 Plan view of leg tip locus.

8.13.2 The coordinates of point, a, in 4 quadrants of the leg tip cylindrical coordinate frame

Coordinates of point, a, are called, R_{comp}, θ_{LBcomp}.
The equation for, R_{comp}, is unchanged for all quadrants.
For all quadrants,

$$R_{comp} = 2 \times L_{LB} \times \sin\left(\frac{R_{LB}}{2 \times L_{LB}}\right)$$

However, the equation for θ_{LBcomp} is different for each quadrant of the leg tip cylindrical coordinate frame as shown in figure 8.45 below.
Note in equations below that, $\Delta\theta = \dfrac{R_{LB}}{2 \times LLB}$

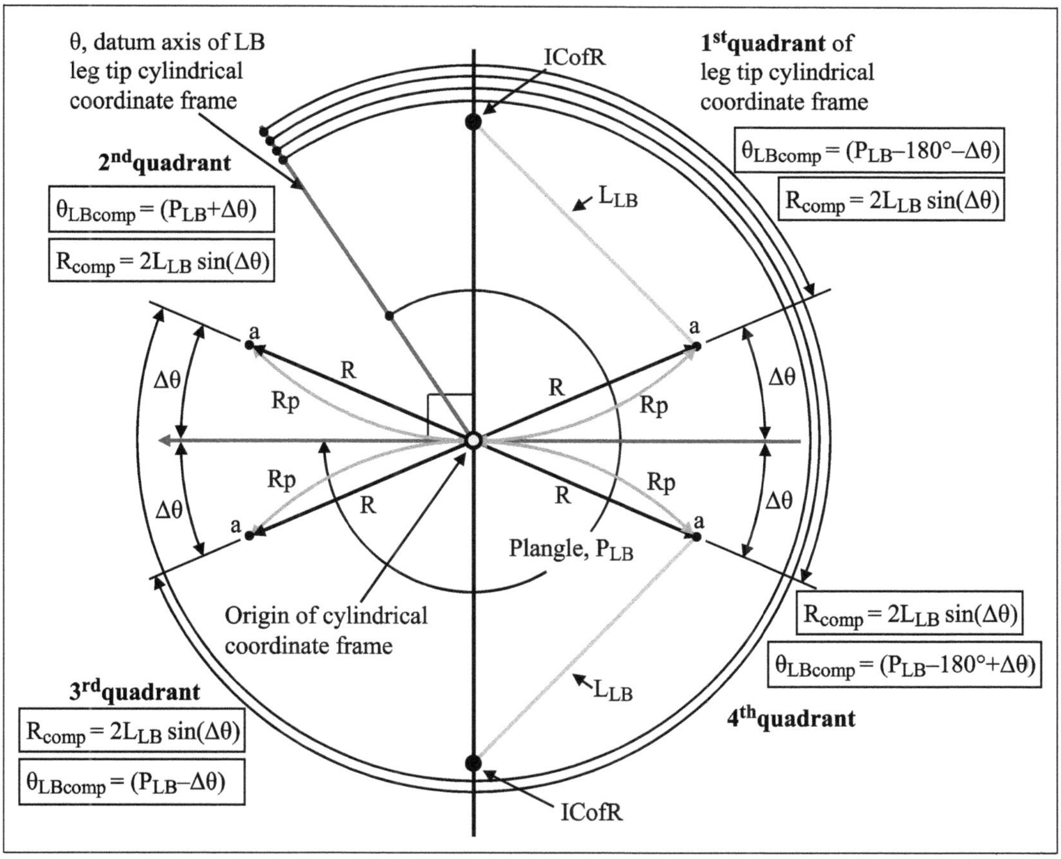

Figure 8.45 Plan view of leg curved tip locus situated either side of the blue locus orientation vector showing the 4 quadrants in which point a can be located.

Now we need to address the problem of computing the values of, R_{comp}, and, θ_{LBcomp}.

8.13.3 First the computation of, θ_{LBcomp}

The value of the compensated plan angle, θ_{LBcomp} in the first quadrant is given by,

$$\theta_{LBcomp} = (P_{LB} - 180^{\circ} - \Delta\theta) \quad \text{..where} \quad \Delta\theta = \frac{R}{2 \times L_{LB}} \text{(rad)} = \frac{573R}{20L_{LB}} \text{(deg)}$$

Note that $\Delta\theta$ has been converted from radians to degrees by multiplying radians by 573/10 (57.3 degrees/radian). We use degree units because they are more intuitive and convenient to use. Hence, the value of, θ_{LBcomp}, in the 4 quadrants is given by the following equations.

1st quadrant (CW rotation about IC of R)

$$\theta_{LBcomp} = P_{LB} - 180^{\circ} - \left(\frac{573R_{LB}}{20L_{LB}}\right)^{\circ}$$

2nd quadrant (CW rotation about IC of R)

$$\theta_{LBcomp} = P_{LB} - 180^{\circ} + \left(\frac{573R_{LB}}{20L_{LB}}\right)^{\circ}$$

3rd quadrant (CCW rotation about IC of R)

$$\theta_{LBcomp} = P_{LB} - \left(\frac{573R_{LB}}{20L_{LB}}\right)^{\circ}$$

4th quadrant (CCW rotation about IC of R)

$$\theta_{LBcomp} = P_{LB} + \left(\frac{573R_{LB}}{20L_{LB}}\right)^{\circ}$$

These 4 equations are straightforward to compute using microcomputers. There is no need to simplify the equations

8.13.4 Now the computation of, R_{comp}

The value of, R_{comp} is given by,

$$R_{comp} = 2 \times L_{LB} \times \sin\left(\frac{R_{LB}}{2 \times L_{LB}}\right)$$

The problem in computing this equation is the sin function so we will resort to using an integer power series, e.g. a Taylor series, for its computation but how many terms will give a suitable accuracy? The maximum value of $\Delta\theta = R_{LB}/2L_{LB}$ is given by the maximum value of R_{LB} divided by two times the minimum value of L_{LB}. These values are 40mm and 2 × 232.5mm respectively. Where do these values come from? Well, the former value is the maximum step amplitude of the walking locus and the latter is given by the lowest value that the ICofR can possibly be away from any one of the leg cylindrical coordinate frame origins and that value is given by the radius of any one of the leg's cylindrical coordinate frame origins from the centre of the robot body. Thus the maximum value of $\Delta\theta$ is given by (40mm/(2 × 232.5mm))radians = 0.0860rads = 4.93° which is a relatively small value. Sufficiently small in fact that sin($\Delta\theta$radian) $=\Delta\theta = (R/2L_{LB})$ to less than 0.2% error which is considered a tolerable error. Thus it is easy to compute R_{comp} accurately because $R_{comp}=2L_{LB} \times R/2L_{LB}= R$. This means that only the first term of the Taylor series of a sine function is required.

In other words:
No compensation of R is required because $R_{comp}=R$ to more than 99.8% accuracy.
We conclude this section by saying that the foregoing equations for, R_{comp}, and, θ_{comp}, for each leg, are computed from the following 3 variables, (note that we use the LB leg as an example).

(i) Orientation of leg tip locus plane, plangle, P_{LB}
(ii) Radius of curvature of leg tip locus plane, L_{LB}
(iii) Digital locus radius from origin of cylindrical coordinate frame, R

The last variable, R, is computed from the amplitude, (=amp), of the leg tip locus plane and the value of, n, in the digital leg tip locus. The value of, amp, is proportional to the value of, L, since the angle subtended by each leg tip locus is equal to θ_{rot}, which is the robot body rotation for each walking step, refer to figure 8.39 above. However, the values of, P_{LB}, and, L_{LB}, are computed from the relative displacement of the ICofR from the origin of the leg tip cylindrical coordinate frame of each leg. We now discuss the computation of these quantities. The value of P_{XX}, where the subscript XX refers to any one of the six legs, LF, LM, LB, RB, RM or RF, is called the plan angle of the digital locus or just 'Plangle' for short. The value of L_{XX} is called the radius of curvature.

8.14 Computation of, plangle, P, and radius of curvature, L

We use the left front, LF leg as an example. Consider the RF leg in figure 8.46 below, for analysis. Assume that the location of the ICofR is given with robot body coordinates, X_I, Y_I. Each leg can thus determine its distance between the ICofR and the leg cylindrical coordinate origin. This distance is called the 'radius of curvature', L_{XX}.

The radius of curvature, L_{RF}, is given by,

$$L_{RF}=\left((X_I - X_{RF})^2 + (Y_I - Y_{RF})^2\right)^{1/2}$$

...and the plan angle, (= plangle), P_{RF}, is given by,
For CW rotation about the ICofR

$$P_{RF} = \operatorname{atan}\left(\frac{Y_{RF} - Y_I}{X_I - X_{RF}}\right)$$

..and for CCW rotation about the ICofR;

$$P_{RF} = \text{atan}\left(\frac{Y_{RF} - Y_I}{X_I - X_{RF}}\right) + 180^\circ$$

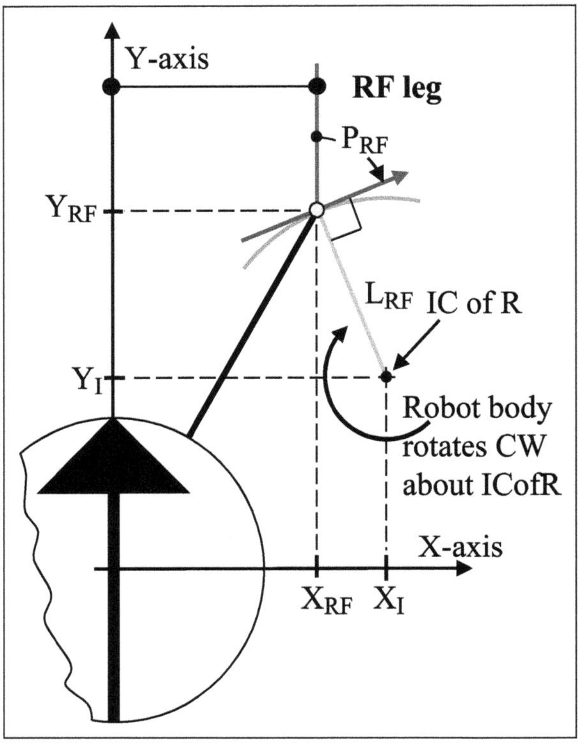

Figure 8.46 Analysis of plangle, P_{LF}, and radius of curvature, L_{LF} for the LF leg.

Similar equations give the radius of curvature, L and plangle, P values for the remaining 5 legs. Rather than explore the computation method for L and P now, we leave this to chapter 11 which is concerned with the programming of a hexapod robot.

9. The complete Basic Stamp PBasic code for one leg of a Roving Robot

Code is generic and can be applied to each and any leg of a Chariot or Hexapod robot. The code is lower level and requires a higher level controller, a Central Pattern Generator, to be described in chapters 10 and 11, to create functionality

'{$STAMP BS2sx}
'This programme is for Futaba S9450 servos.
'The coefficients for the S1, S2 and s3 servo inverse kinematics computation have been taken
'from the tables in figures 6.6, 6.8 and 6.9 respectively
'The whole programme takes less than one page (2kBytes) of Basic Stamp2SX storage
'The whole programme takes less than 15ms to execute

9.1 'The variables

S1	VAR	Word	The three servo angles represented by pulsout integer values that range from 1250 to 2500. This range of numbers represents a servo angle range of 100° approximately
S2	VAR	Word	
S3	VAR	Word	
n	VAR	Byte	'Number that indicates the leg tip location on the digital locus
bp1	VAR	Word	Breakpoint1, breakpoint2. n-values on digital locus for changing from vertical to horizontal directions and vice versa
bp2	VAR	Word	
R	VAR	Byte	The three leg tip cylindrical coordinates, R, Z, θ representing the demanded leg tip location. R= 0mm to 40mm, Z=0mm to 40mm and θ = 0 to 359°. All coordinates in integers.
Z	VAR	Byte	
theta	VAR	Byte	
thetemp	VAR	Word	Two variables used for the classification of which cake slice sector the leg tip resides and the plan angle
twelvesects	VAR	Word	

plangle	VAR	Word	These four integer variables specify the geometry of the leg tip digital locus; plangle is the angle of the locus plane, zag is the 'z-air-to-ground' vertical height of the walking step, amp is the amplitude of the walking step, zbias is the bias shift down of the locus
zag	VAR	Byte	
amp	VAR	Byte	
zbias	VAR	Byte	
thethalf	VAR	Bit	'thethalf=0 is 0°<θ<180°, thethalf=1 is 181°<θ<360°

9.2 'Pin output initialisation

```
DIR5=1   'S1 ┐
DIR6=1   'S2 ├ S1, S2, S3 servo pulsout pins, pin5, pin6 and pin7 set to output
DIR7=1   'S3 ┘
```

9.3 'The constants

'enter bias angle for leg, LF=30, LM=90, LB=150, RB=210, RM=270, RF=330
planglebias CON 30 'Choose 30° plangle bias angle for Left Front leg
' See figure 9.1 below for meaning of planglebias

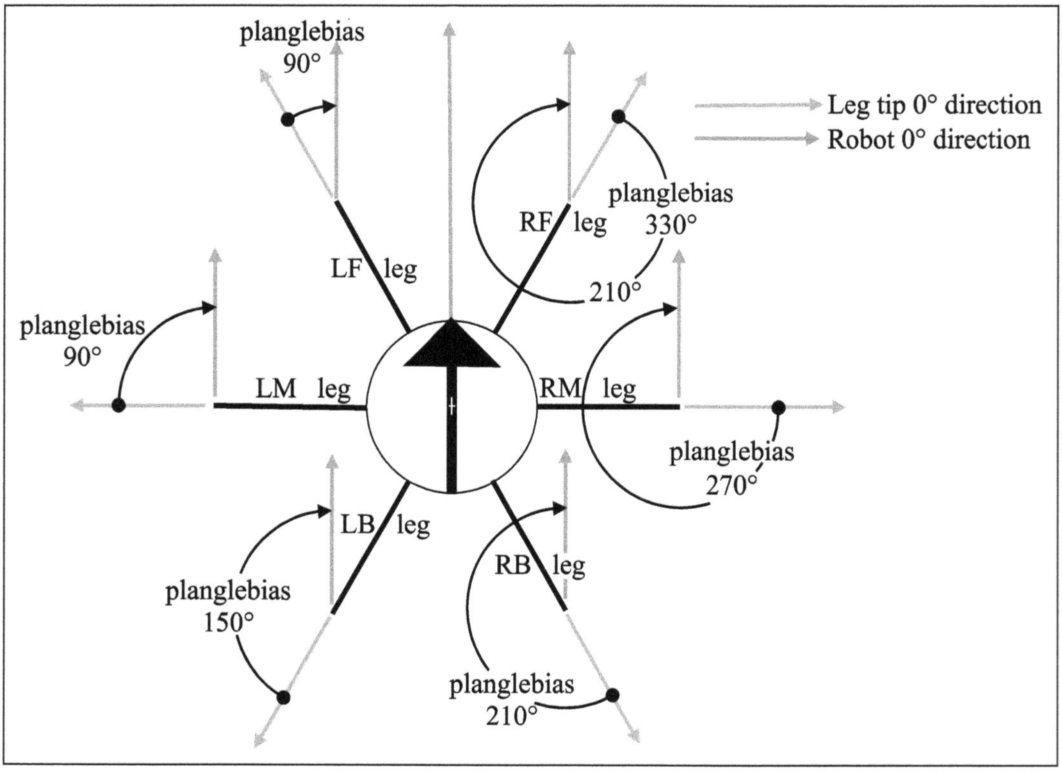

Figure 9.1 Plan view of hexapod showing the planglebias angles for each leg. The planglebias angles align each leg digital locus with the robot forward direction.

9.4 'Servo constants initialisation

'enter values for servo offset (S#off), and servo gain (S#gain) for each of the three servos

```
S1off    CON  285
S2off    CON  -65
S3off    CON  -270
S1gain   CON  101
S2gain   CON  99
S3gain   CON  102
```

} Typical values. Values change according to manufacturing tolerances

9.5 'The main programme starts now

```
again:    SERIN 15,16390,[n,plangle,amp,zag,zbias]
```

'wait for 20ms polled interrupt 'that delivers 5 bytes of data specifying the digital locus geometry. Note that 'the plangle variable has been divide by 2 since 359 degrees is greater than 'one byte

```
'DEBUG DEC n," ",DEC plangle," ",DEC amp," ",DEC zag," ",DEC zbias,CR
```

'optional 'debug statement that checks on the 5 received variable values

```
plangle.HIGHBYTE=0
plangle=plangle*2+planglebias
IF plangle<360 THEN skip10
plangle=plangle – 360
```

} Since plangle is a 2-byte word and the received plangle is one byte with plangle value halved then must multiply plangle by 2 and reset if it overruns 359 degrees

```
skip10:
'Detect six states in the digital locus
```

'See figure 9.2 to recall chapter 4 digital locus specification

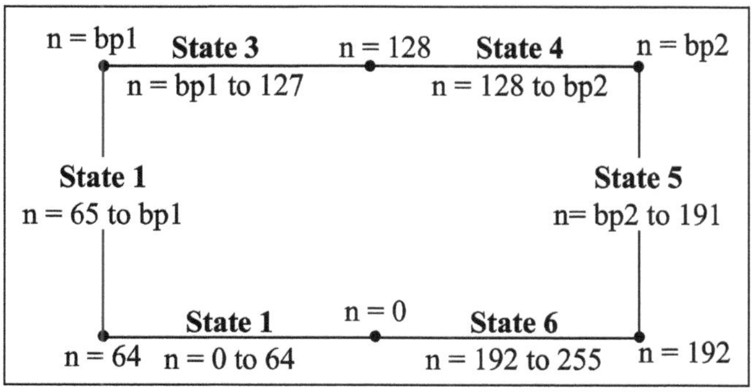

Figure 9.2 The 6 states of the Digital Locus.

```
IF n>191 THEN state6
bp2=amp*64/(zag+amp)+127      'compute bp2 value function of zag and amp
IF n>bp2 THEN state5
IF n>127 THEN state4
bp1=zag*64/(zag+amp)+ 63      'compute bp1 value function of zag and amp
```

```
        IF n>bp1 THEN state3
        IF n> 63 THEN state2
        'default to state 1 immediately following

state1: R=amp*n/64
        Z=zag+zbias                    } Compute R and Z values
        thetemp=plangle+180
        GOTO skip0

state2: R=amp
        Z=zbias+zag-(zag*
        (n-64)/(bp1-64))               } Compute R and Z values
        thetemp=plangle+180
        GOTO skip0

state3: R=amp-(amp*(n-bp1)/
        (127-bp1))
        Z=zbias                        } Compute R and Z values
        thetemp=plangle+180
        GOTO skip0
```

States 1, 2 and 3 plan angles of digital locus, thetemp, are equal to plangle

```
state4: R=amp*(n-128)/(bp2-
        128)
        Z=zbias                        } Compute R and Z values
        thetemp=plangle
        GOTO skip0

state5: R=amp
        Z=zbias+(zag*(n-bp2)/
        (192-bp2))                     } Compute R and Z values
        thetemp=plangle
        GOTO skip0

state6: R=amp-(amp*(n-192)/64)
        Z=zag+zbias                    } Compute R and Z values
        thetemp=plangle
```

States 4, 5 and 6 plan angles of digital locus, thetemp, are 180° added to plangle

```
skip0:      IF thetemp>359 THEN
            adjust
            GOTO skip1
adjust:     thetemp=thetemp–360
```

Reset plan angle if it overruns 359 degrees

```
skip1:      IF thetemp=180 THEN
            skip20
```

'special case

```
            thethalf=thetemp/180
            IF thethalf=0 THEN
            skip21
            thetemp=360–thetemp
```

If leg tip is in the first half of the coordinate cylinder, set thethalf flag to 0 otherwise 1. If thethalf=1 then reset plangle to mirror image value, see chapter 6, figure 6.3

'Determine which cake slice in which the leg tip resides

```
skip21:     twelvesects=thetemp/15
            theta=thetemp-(twelvesects*15)
            BRANCH twelvesects,[T0,T1,T2,T3,T4,T5,T6,T7,T8,T9,T10,T11]

skip20:     theta=15                        'special case, see skip1 above
            GOTO t11
```

'following variables are reusable from now on:-

```
'twelvesects   VAR   word
'thetemp       VAR   word
'bp1           VAR   word
'bp2           VAR   word
```

We need more variables to carry out the inverse kinematics calculations, T0 to T11 that are next to be executed, but have run out of them. So our option is to reuse variables that are no longer required.

9.6 'Twelve Branch vectors to compute the inverse kinematics equations, (see chapter 6)

```
T0:         twelvesects   =   0+(130*theta/15)
            thetemp       =200–( 67*theta/15)
            GOSUB servo1
```

Set inverse kinematics S1coefficients for T0 slice and compute servo S1

```
            twelvesects   =  14+(  5*theta/15)
            thetemp       =103–(  4*theta/15)
            bp1           =753–( 18*theta/15)
            bp2           =   9+(  0*theta/15)
            GOSUB servo23
```

Set inverse kinematics S2 and S3 coefficients for T0 slice and compute servos S2 and S3

```
            twelvesects    =227          }
            thetemp        =566          } Set correction coefficients for
            bp1            =247          } S2 and S3 and then recom-
            bp2            =618          } puted for S2 and S3 servos
            GOSUB corrections            }
            GOTO pout     'now go and compute and output S1, s2 and S3 to servos

T1:         twelvesects    =133+(124*theta/15)
            Thetemp        =133-( 55*theta/15)
            GOSUB servo1

            Twelvesects    = 19+( 14*theta/15)
            Thetemp        = 99-( 12*theta/15)
            bp1            =735-( 52*theta/15)
            bp2            =  9+(  2*theta/15)
            GOSUB servo23

            Twelvesects    =216
            Thetemp        =540
            bp1            =224
            bp2            =561
            GOSUB corrections
            GOTO pout

T2:         twelvesects    =254+(108*theta/15)
            thetemp        = 78-( 34*theta/15)
            GOSUB servo1

            twelvesects    = 33+( 21*theta/15)
            thetemp        = 87-( 15*theta/15)
            bp1            =683-( 83*theta/15)
            bp2            = 11+(  4*theta/15)
            GOSUB servo23

            twelvesects    =197
            thetemp        =493
```

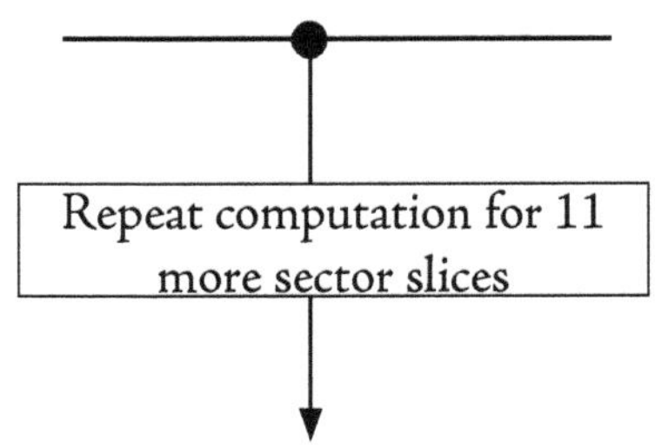

```
        bp1             =184
        bp2             =560
        GOSUB corrections
        GOTO pout

T3:     twelvesects     =362+( 86*theta/15)
        thetemp         = 44+(  3*theta/15)
        GOSUB servo1

        twelvesects     = 54+( 26*theta/15)
        thetemp         = 72-( 17*theta/15)
        bp1             =600-(105*theta/15)
        bp2             = 16+(  8*theta/15)
        GOSUB servo23

        twelvesects     =173
        thetemp         =433        S1, S2 and S3 coefficients
        bp1             =134        listed in tables in chapter 6
        bp2             =334        for all slices T0 to T11
        GOSUB corrections
        GOTO pout

T4:     twelvesects     =448+( 56*theta/15)
        thetemp         = 41+( 33*theta/15)
        GOSUB servo1

        twelvesects     = 80+( 27*theta/15)
        thetemp         = 55-( 15*theta/15)
        bp1             =495-(115*theta/15)
        bp2             = 23+( 11*theta/15)
        GOSUB servo23

        twelvesects     =150
        thetemp         =375
        bp1             = 83
        bp2             =208
```

```
        GOSUB corrections
        GOTO pout
T5:     twelvesects     =504+(22*theta/15)
        Thetemp         = 74+(71*theta/15)
        GOSUB servo1

        twelvesects     =107+( 28*theta/15)
        thetemp         = 40-( 14*theta/15)
        bp1             =380-(117*theta/15)
        bp2             = 34+( 16*theta/15)
        GOSUB servo23

        twelvesects     =132
        thetemp         =331
        bp1             =42
        bp2             =104
        GOSUB corrections
        GOTO pout

T6:     twelvesects     =526-( 17*theta/15)
        Thetemp         =145+(100*theta/15)
        GOSUB servo1
        twelvesects     =135+( 25*theta/15)
        thetemp         = 26-( 11*theta/15)
        bp1             =263-(106*theta/15)
        bp2             = 49+( 19*theta/15)
        GOSUB servo23

        twelvesects     =123
        thetemp         =309
        bp1             =18
        bp2             =46
        GOSUB corrections
        GOTO pout
```

S1, S2 and S3 coefficients listed in tables in chapter 6 for all slices T0 to T11

```
T7:     twelvesects    =509-( 57*theta/15)
        thetemp        =245+(108*theta/15)
        GOSUB servo1

        twelvesects    =160+(22*theta/15)
        thetemp        = 16-( 7*theta/15)
        bp1            =157-(80*theta/15)
        bp2            = 69+(21*theta/15)
        GOSUB servo23

        twelvesects    =124
        thetemp        =310
        bp1            =24
        bp2            =61
        GOSUB corrections
        GOTO pout

T8:     twelvesects    =452-(90*theta/15)
        thetemp        =353+(81*theta/15)
        GOSUB servo1

        twelvesects    =182+(17*theta/15)
        thetemp        =  9-( 3*theta/15)
        bp1            = 77-(42*theta/15)
        bp2            = 90+(21*theta/15)
        GOSUB servo23

        twelvesects    =131
        thetemp        =328
        bp1            =66
        bp2            =166
        GOSUB corrections
        GOTO pout

T9:     twelvesects    =362-(113*theta/15)
        thetemp        =434+( 10*theta/15)
        GOSUB servo1
```

S1, S2 and S3 coefficients listed in tables in chapter 6 for all slices T0 to T11

```
            twelvesects   =199+(11*theta/15)
            thetemp       =  6+( 1*theta/15)
            bp1           = 35-(10*theta/15)
            bp2           =111+(18*theta/15)
            GOSUB servo23

            twelvesects   =139
            thetemp       =348
            bp1           =144
            bp2           =355
            GOSUB corrections
            GOTO pout

T10:        twelvesects   =249-(124*theta/15)
            thetemp       =444-( 85*theta/15)
            GOSUB servo1

            twelvesects   =210+( 7*theta/15)
            thetemp       =  7+( 2*theta/15)
            bp1           = 25+( 7*theta/15)
            bp2           =129+(13*theta/15)
            GOSUB servo23

            twelvesects   =144
            thetemp       =360
            bp1           =230
            bp2           =574
            GOSUB corrections
            GOTO pout

T11:        twelvesects   =125-(125*theta/15)
            thetemp       =359-(159*theta/15)
            GOSUB servo1

            twelvesects   =217+( 2*theta/15)
            thetemp       =  9+( 2*theta/15)
```

```
bp1            =  32+( 8*theta/15)
bp2            =141+( 4*theta/15)
GOSUB servo23

twelvesects    =145
thetemp        =363
bp1            =290
bp2            =726
GOSUB corrections
'default GOTO pout
```

9.7 'Output of the servo values

```
pout:     IF thethalf=1 THEN skip30 <---- Test if digital locus is in 2nd half of
                                          cylindrical coordinate frame

          S1=1875+S1   'Add the S1 component because locus in 1st half of frame
          GOTO skip31
skip30:   S1=1875–S1   'Subtract the S1 component because locus in 1st half of
                        frame
skip31:   IF S2>32768   THEN skip50  'Test if S2 component is negative
          S2=S2/2*S2gain/50+1875      'S2 component is +ve so ok to multiply by
                                       gain
          GOTO skip51
skip50:   S2=1875–(65535–S2+1/2*S2gain/50)<----Make S2 +ve then multiply
                                               by gain then make –ve and
                                               add to 1875


skip51:   IF S3>32768 THEN skip52                 }  Same procedure as for S2
          S3=S3/2*S3gain/50+1875                  }  above, i.e. test if S3 component
          GOTO skip53                             }  is –ve then if yes change to
skip52:   S3=1875–(65535–S3+1/2*S3gain/50)        }  +ve then multiply by gain then
                                                  }  change back to -ve
skip53:   S1=S1+S1off               }
          S2=S2+S2off               }
          S3=S3+S3off               }
          S1=S1 MIN 1000 MAX 2750   }  Send out the final S1, S2, S3 servo
          S2=S2 MIN 1000 MAX 2750   }  values to the servos but first add the
          S3=S3 MIN 1000 MAX 2750   }  offset values then set software limits
          PULSOUT 5,S1              }  for safety
          PULSOUT 6,S2              }
          PULSOUT 7,S3              }
          GOTO again                   'repeat the routine
```

9.8 'The Subroutines to compute inverse kinematic equations for S1, S2,S3

```
servo1:
S1 =(twelvesects*R/  100)+(thetemp*R/25*R/400)-(200*R/25*R/400)*S1gain/100
              RETURN
              servo23:
S2 =0-(753*Z/100)+(twelvesects*R/ 10)+(thetemp*R/25*Z/40)-(15*R)-(R*Z*20/1000)
S3 =0+(151*Z/ 10)+(    bp1*R/100)+(    bp2*R/25*Z/40)-( 2*R)-(R*Z* 6/ 100)
              RETURN
corrections:  S2=S2+(twelvesects*R/100)-(thetemp*R/25*R/400)
              S3=S3-(        bp1*R/100)+(    bp2*R/25*R/400)
              RETURN
```

10 2-wheeled-2-legged rickshaw Robot: replacing some wheels with legs and programming its central pattern generator

10.1 Introduction

The basic theory of motion of vehicles in a ground plane was covered in chapter 1 where a vehicle can be either a legged or wheeled device. Chapter 3 then used this theory to give motion to a Tricycle Robot which is a 3-wheeled vehicle with no legs. Chapters 4 to 9 then build up a theory of legged motion for a single 3 dof leg. The most important operational technique of leg motion is the ability to guide the leg tip through a programmably shaped and directioned locus, which is explained in chapter 4, and then to develop walking gaits as described in chapter 8. This book leads up finally to the programming of a 6-legged omnidirectional walking robot to carry out a behaviour pattern. However this is a big jump due to the complexity of such a robot. In order to bridge this gap we now introduce a 2-wheeled Rickshaw Robot with 2-legged locomotion, figure 10.1. The Rickshaw Robot is not as complex as the 6-legged robot but brings together the theory of the motion of vehicles whilst introducing the programming of two legs. The Rickshaw Robot thus serves as an ideal 'bridge-the-gap' introduction to complex multi-legged walking robots.

First let us show the steering similarity between the Tricycle and Rickshaw Robots.

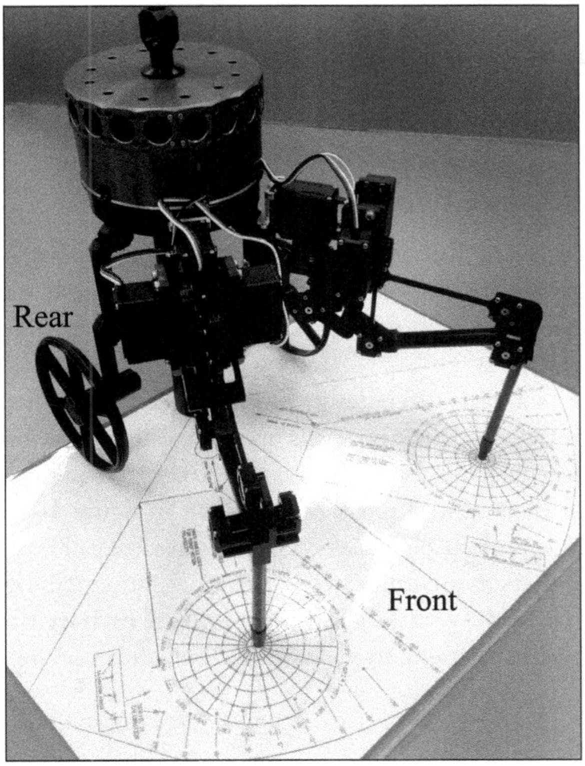

Figure 10.1 The Rickshaw Robot, shown mounted on its calibration table, is equivalent to the Tricycle Robot by replacing steered wheels with steered legs which can also propel the robot.

10.2 Steering kinematics of the Tricycle and Rickshaw Robots

Figure 10.2 shows the similar steering kinematics between the Tricycle and Rickshaw Robots where the former is a wheels-only robot and the latter is a 2-wheeled-2-legged hybrid robot. Note that (i) for both robots, the ICofR lies on an axis projected from the wheel axis and (ii) with the Rickshaw Robot, the leg tips move in an arc centred on the ICofR.

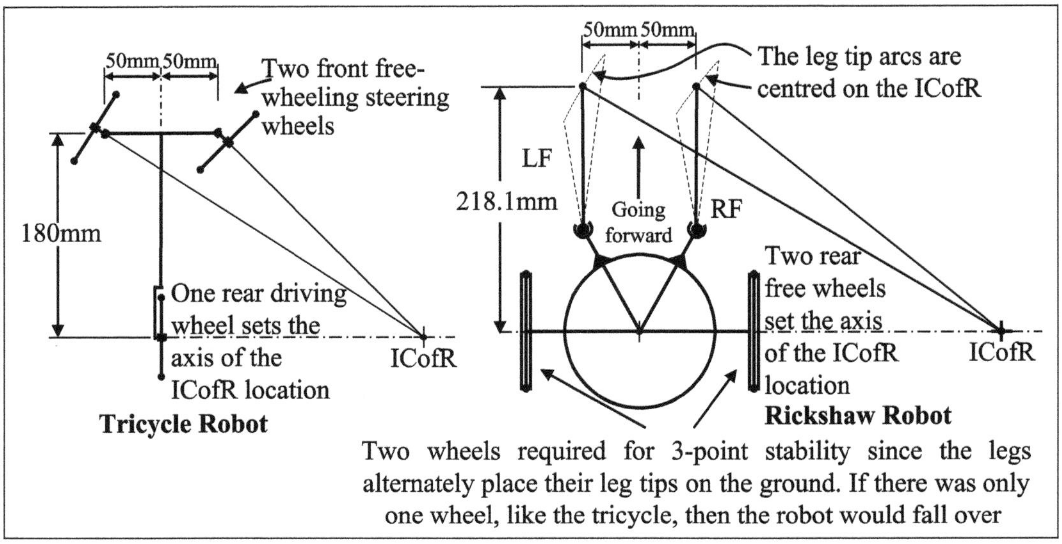

Figure 10.2 The Rickshaw Robot is an analogue of the chapter 3 Tricycle Robot in terms of its principle of steering that replaces steering wheels with steering legs.

Now the important dimensions of the Rickshaw Robot are described that permit the robot to be programmed to follow a figure-of-eight pattern similar to the Tricycle Robot programmed path shown in chapter 3.

10.3 Plan view geometry of the Rickshaw Robot

Figure 10.3 below is a plan view of the Rickshaw Robot mounted on its calibration table showing the basic dimensions for computing its motion kinematics. The robot can be thought of as a 2-wheel carriage given motion by a man *pulling* a Chinese rickshaw or conversely as a man *pushing* a 2-wheeled wheel barrow.

The Rickshaw Robot calibration table is a carry-over design from a Hexapod Robot where the six legs are equispaced at 60° intervals. These intervals are spread too wide for the Rickshaw Robot where it can cause fall-over instability so it is necessary to bring the legs closer together so they are parallel in their central position as shown in figure 10.4. Nonetheless, the calibration table serves its important purpose because it is used to calibrate each 3dof leg as in chapters 7 and 9. Once the legs are calibrated at position A they can be rotated about each of the servo S1 servo axes by 30° to a parallel position where each leg tip will transport its calibration cylindrical coordinate frame to position B. This effect can also be seen in figure 10.5.

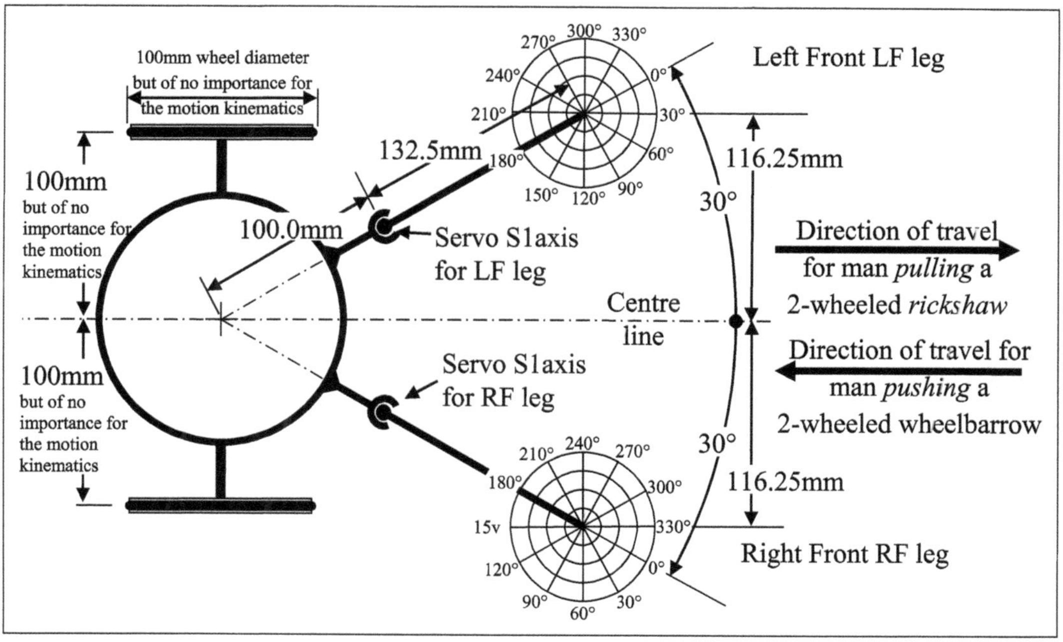

Figure 10.3 Plan view of the Rickshaw Robot mounted on its calibration table showing basic dimensions.

As regards the computer control of the two legs, figure 10.4 shows three on-board Basic Stamp microcomputers which are, (i) the Central Pattern Generator, the CPG, which provides high level walking instructions for negotiating such paths as a figure-of-eight path and (ii) two low-level Basic Stamp Leg microcomputers, one dedicated to the left front (LF) leg and one dedicated to the right front (RF) leg. These LF and RF leg microcomputers have both been programmed with their calibration programmes.

10.4 The Central Pattern Generator, the CPG

The Central Pattern Generator microcomputer sends five 8-bit variables every 20ms (the 'update rate') to each of the LF and RF microcomputers. These variables are, n, amp, plangle, zag and zbias and they (i) are different for each leg and (ii) control the geometry of the digital leg tip locus and the leg tip position for each leg. The CPG, together with the leg microcomputers, thus becomes an extremely powerful hierarchical computer system due to its separated devolved and delegated software and hardware components.

The cylindrical coordinate frames are shifted by 30°from their calibrated virtual frame locations, **A**, on the calibration table, figures 10.3, 10.4 and 10.5. The virtual calibration frames *rotate with the leg tips* towards the centre line by 30° to **B**. The inverse kinematics programmes are stored in the LF and RF microcomputers. The rotation is brought about by adding and subtracting offset values to the S1 servos.

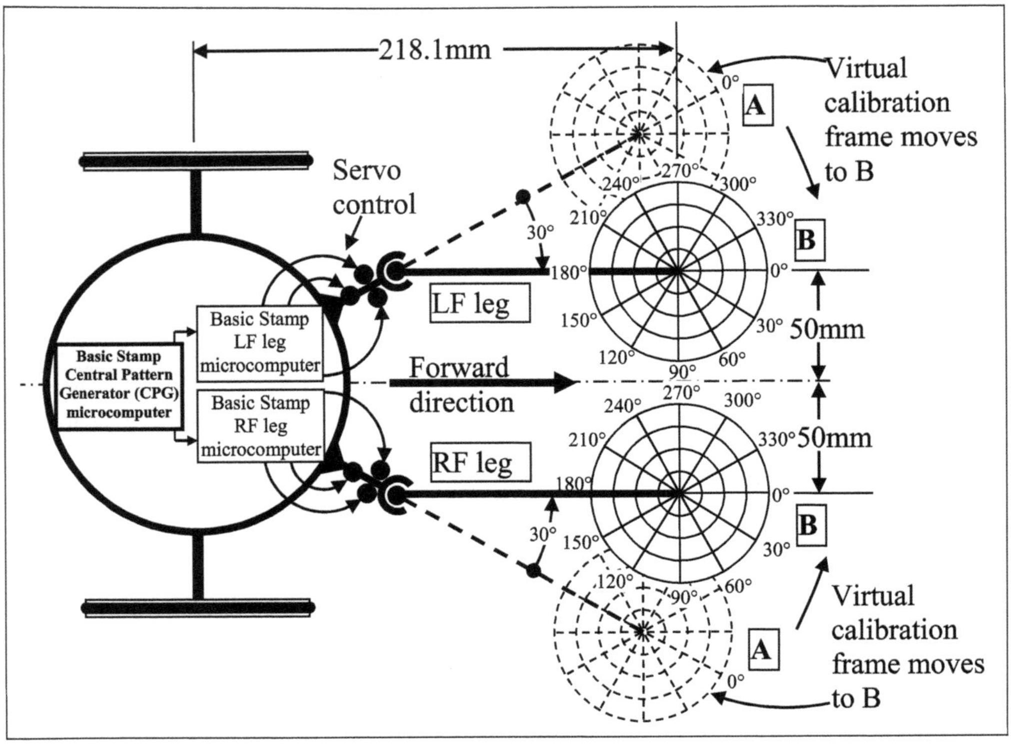

Figure 10.4 The three on-board control microcomputers and the cylindrical coordinate calibration circles A shifted by 30° towards the centre line to B.

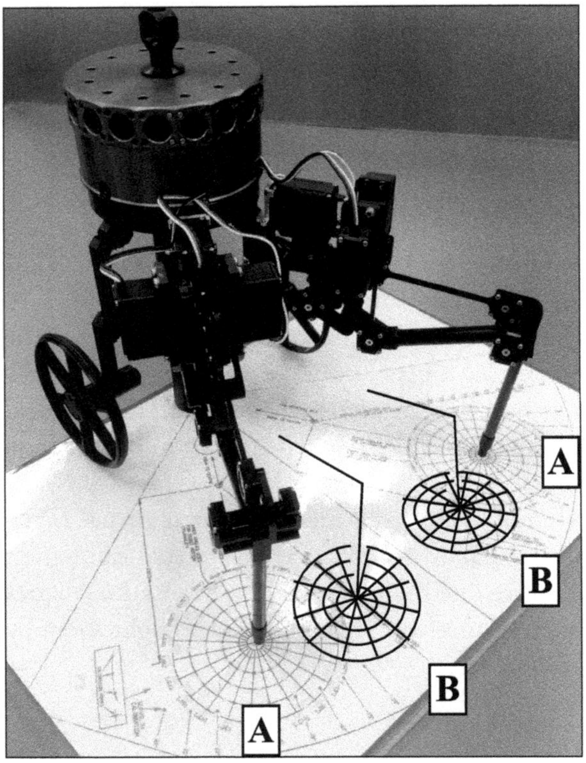

Figure 10.5 Diagram showing more clearly how the virtual calibration frames are shifted towards the robot centre line from A to B.

10.5 Rotating the LF and RF legs by 30° to become parallel

Once calibration has been done at position A, the legs are brought parallel to position B by adding and subtracting offset values to the LF and RF S1 servos, figure 10.6 and 10.7, respectively. These figures extract, and show the modification to, a sample few lines of code from the generic leg programme of chapter 9.

Shift the LF leg CW by 30° by adding 12pulsout units/° × 30° = 12 × 30 pulsout units from servo S1 offset

```
'Servo constants initialisation
'enter values for servo (S#off), and servo gain (S#gain) for each of the three servos
S1off          CON   285+(12*30)
S2off          CON   65
S3off          CON   270
```

Figure 10.6 Shifting the LF leg calibration by 30° CW about its S1 axis.

Shift the RF leg CCW by 30° by subtracting 12pulsout units/° × 30° = 12 × 30 pulsout units from servo S1 offset

```
'Servo constants initialisation
'enter values for servo (S#off), and servo gain (S#gain) for each of the three servos
S1off          CON   285- (12*30)
S2off          CON   65
S3off          CON   270
```

Figure 10.7 Shifting the RF leg calibration by 30° CCW about its S1 axis.

10.6 The figure-of-eight path and how to negotiate it with the rickshaw robot

The figure-of-eight outline for the Rickshaw Robot is the same as for the Tricycle Robot. You can see the outline and the Rickshaw path pattern, together with the programmed algorithm in figure 10.8.

Both the Tricycle Robot and the Rickshaw Robot are non-holonomic vehicles so the Rickshaw cannot 'crab' sideways. In order to turn left or right it has to turn-on-the-spot or do a three-point turn. The Rickshaw is capable of turning on the spot withe ICoR set at a point between the wheels, so that's the method chosen for turning.

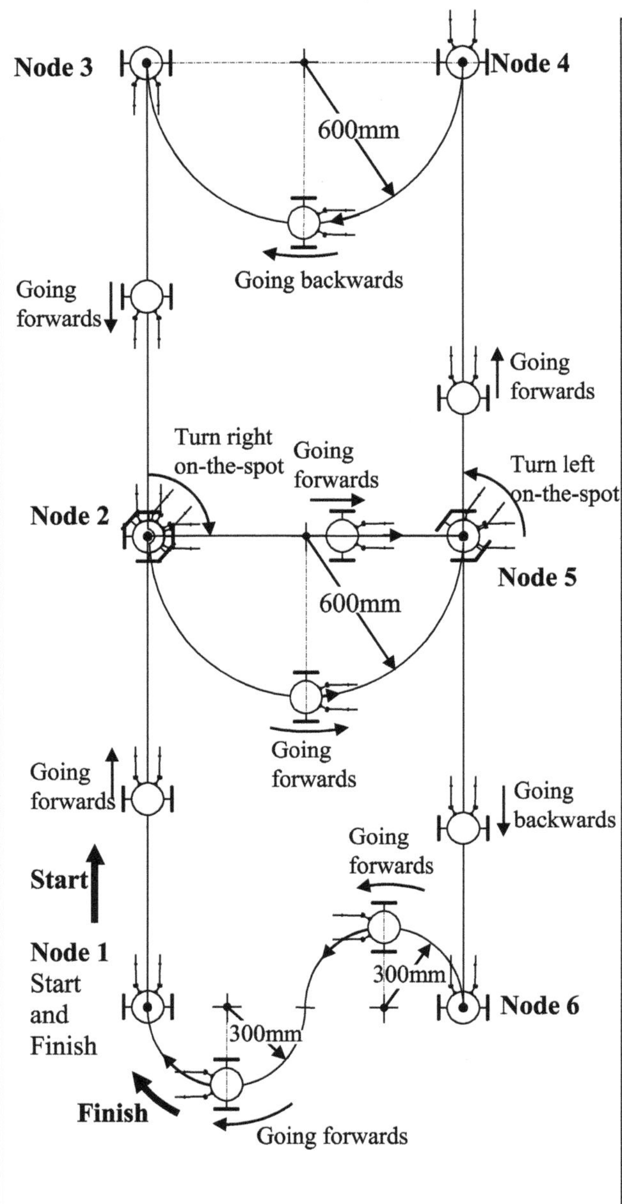

Figure-of-Eight algorithm
Start:Node 1 to Node 2
Go forward for 1800mm
Stop for 5 secs
Turn right on-the-spot 90°.

Node 2 to Node 5
Go forward 1200mm
Stop for 5 secs.
Turn left on-the-spot 90°.

Node 5 to Node 4
Go forwards for 1800mm
Stop for 5 secs.

Node 4 to Node 3
Go backwards for $\pi \times$ 600mm, turning left (turning right in forward direction).
Stop for 5 secs.

Node 3 to Node 2
Go forwards for 1800mm.
Stop for 5 secs.

Node 2 to Node 5
Go forwards for $\pi \times$ 600mm turning left.
Stop for 5 secs.

Node 5 to Node 6
Go backwards for 1800mm.
Stop for 5 secs.

Node 6 to Node 1
Go forwards for $\pi \times$ 300mm turning left.
Go forwards for $\pi \times$ 300mm turning right.
Stop for 5 secs
Programme end.

Node 1
Finish

Figure 10.8 The Figure-of-Eight path and the programmed path pattern for the Rickshaw robot from Node 1, 2, 5, 4, 3, 2, 5, 6, and back to node 1.

10.7 Plan for negotiating the figure-of-eight

The Rickshaw Robot could negotiate the figure-of-eight path solely by straight lines and turns-on-the-spot but that would be a bit boring so we will show how to spice up the path traversal with going forward straight lines, going backwards straight lines, turns-on-the-spot, and turns left and right with different radii of curvature and travelling in different directions. Ok, how are we going to do that? Let's start off with turning right and left about an ICofR that is placed distance R from the centre of the robot, figure 10.9 for right side and figure 10.10 for the left side ICofR.

The LF and RF legs trace out digital loci that are at angles P_{LF} and P_{RF} when viewed in plan. The formulae for these angles are shown in a table in figure 10.11.

Note that the amplitudes of the leg tip loci, amp_{LF} and amp_{RF}, are nominally *set equal* and computed with the limitation of a maximum value of 40mm and as a function of the number of steps and the curve length. These calculations, together with notes, will be shown in the CPG programme code at the end of this chapter.

Be careful to note that the P_{LF} and P_{RF} values are measured CW from the straight ahead 0° position so the turning left plangle values are greater than 180°. Refer to figure 10.10 to appreciate this idea.

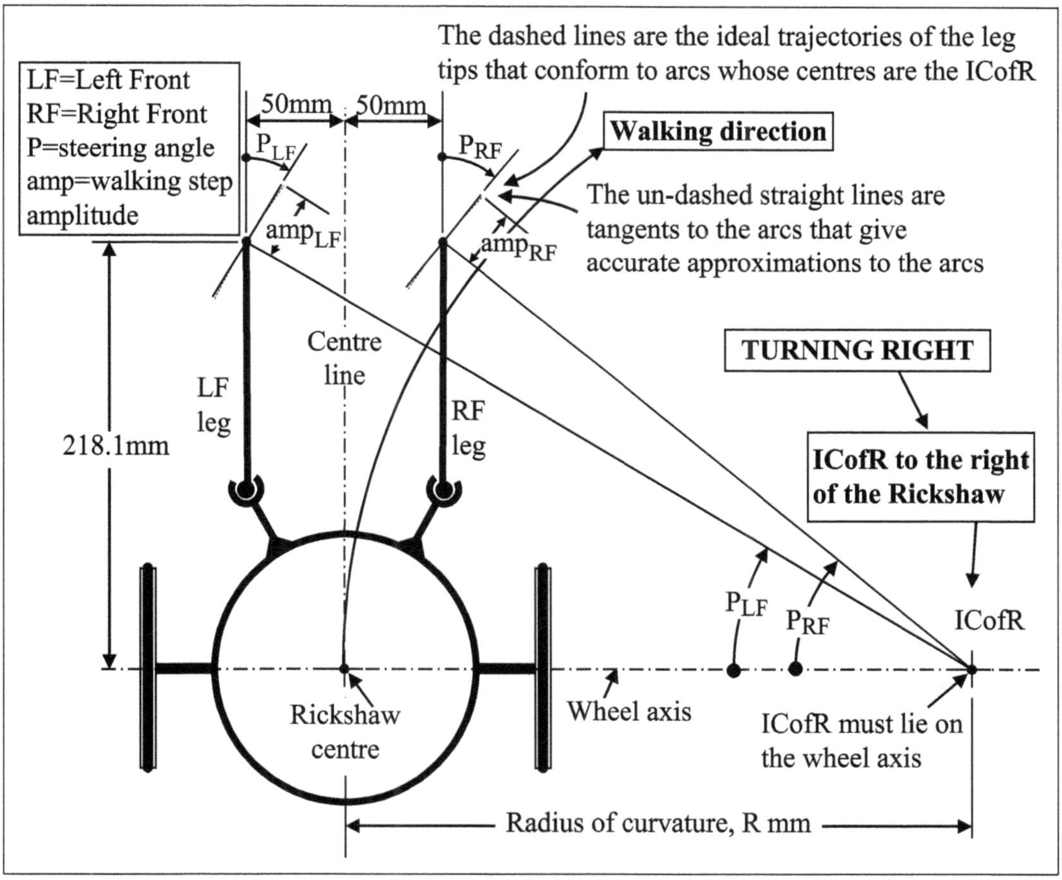

Figure 10.9 Plan view of Rickshaw Robot showing essential dimensions for turning right. We need to calculate the leg steering angles, P_{LF}, P_{RF}, given the location, R of the ICofR on the wheel axis and on the *right side* of the Rickshaw centre.

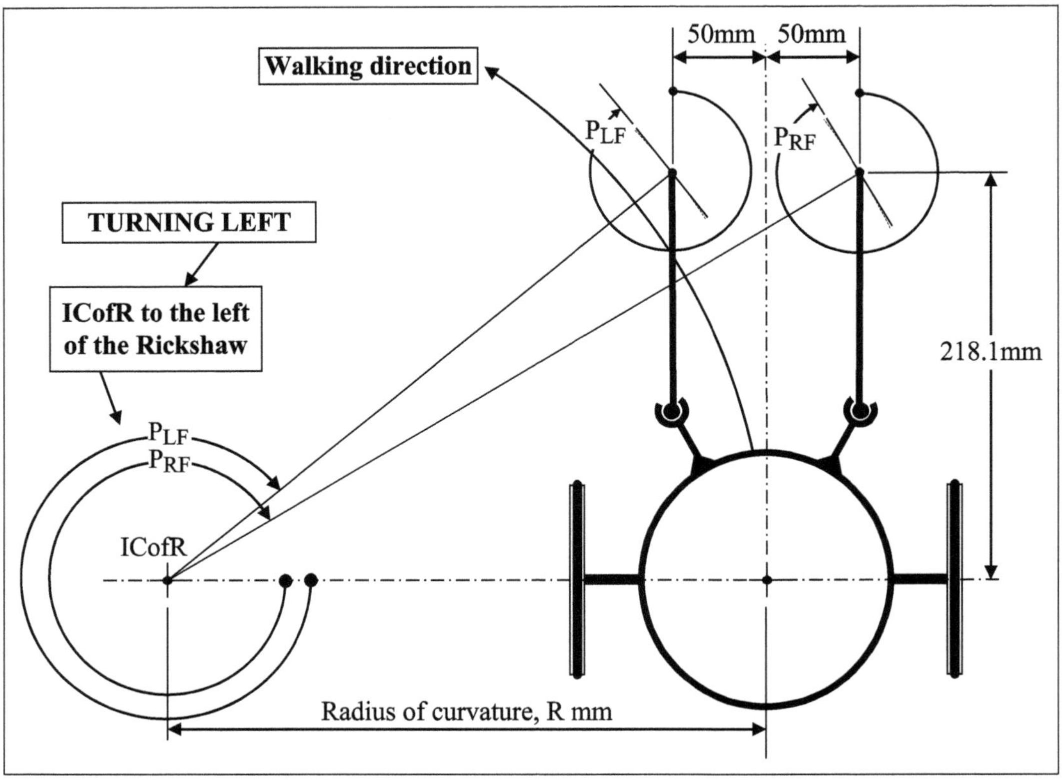

Figure 10.10 Plan view of Rickshaw Robot showing essential dimensions for turning left.

For ICofR on the left side and R≥50mm	For ICofR on the right side and R≥50mm
LF steering angle, plangle, $P_{LF} = 360° - \text{atan}\left(\frac{218.1}{R-50}\right)$	LF steering angle, plangle, $P_{LF} = \text{atan}\left(\frac{218.1}{R+50}\right)$
RF steering angle, plangle, $P_{RF} = 360° - \text{atan}\left(\frac{218.1}{R+50}\right)$	RF steering angle, plangle, $P_{RF} = \text{atan}\left(\frac{218.1}{R-50}\right)$
For ICofR on the left side and R<50mm	**For ICofR on the right side and R<50mm**
LF steering angle, plangle, $P_{LF} = 180° + \text{atan}\left(\frac{218.1}{50-R}\right)$	LF steering angle, plangle, $P_{LF} = \text{atan}\left(\frac{218.1}{R+50}\right)$
RF steering angle, plangle, $P_{RF} = 360° - \text{atan}\left(\frac{218.1}{50+R}\right)$	RF steering angle, plangle, $P_{RF} = 180° - \text{atan}\left(\frac{218.1}{50-R}\right)$

Figure 10.11 Table of formulae for calculating the plangle steering angles, PLF and P_{RF} for the LF and RF legs.

10.8 Computing the algorithm for negotiating the figure-of-eight path

The previous figure 10.8 gives the first impression of the path algorithm that specifies the travel path patterns and the figure 10.11 above gives the formulae for obtaining these patterns. Now putting together these two figures we can tabulate in detail how to programme the algorithm.

This is done in two steps which are, firstly, figures 10.12 to 10.16 inclusive illustrate five plan views of the Rickshaw Robot undergoing the five basic manoeuvres of negotiating the figure-of-eight path and, secondly, figure 10.17 tabulates a compact set of instructions that are to be programmed. Ok, we are almost there; the next step is to programme the Basic Stamp CPG microcomputer with code based on these compact set of instructions.

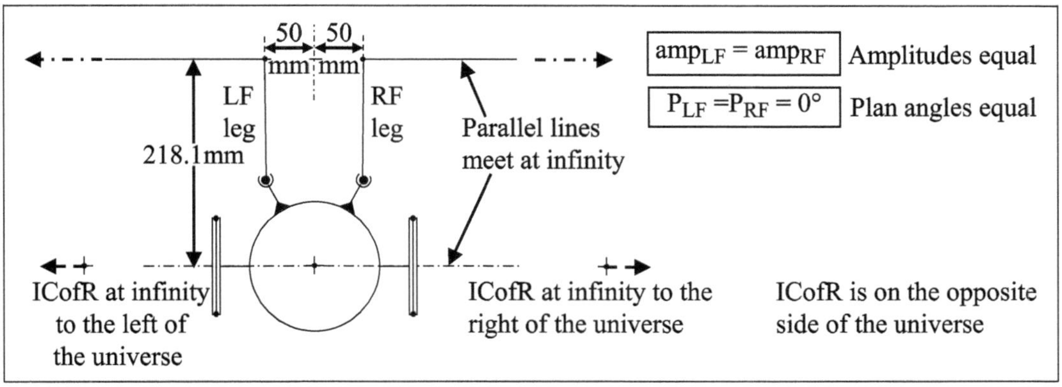

Figure 10.12 **Case 1** Go forwards or Go backwards, i.e. straight ahead or straight reverse.

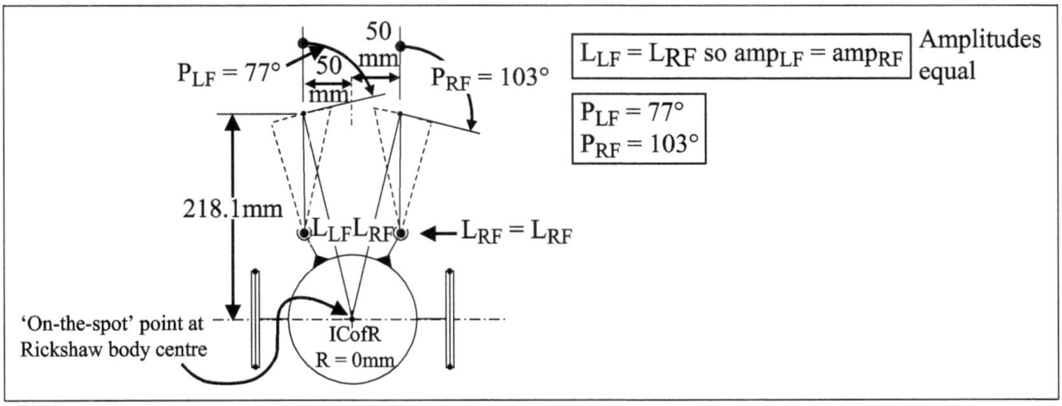

Figure 10.13 **Case 2** Turn right 'on-the-spot' *but can turn left if walking backwards.*

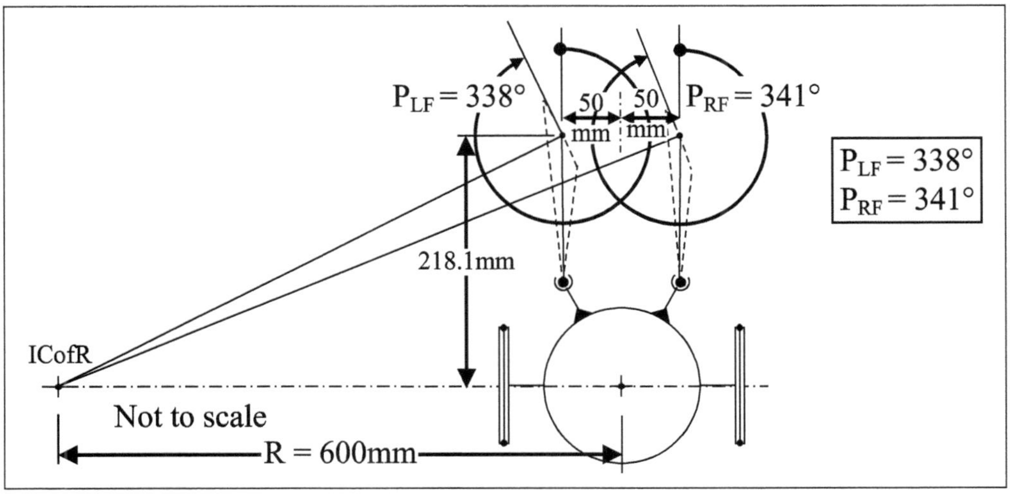

Figure 10.14 **Case3** *Turn left* forwards or backwards with 600mm radius of curvature.

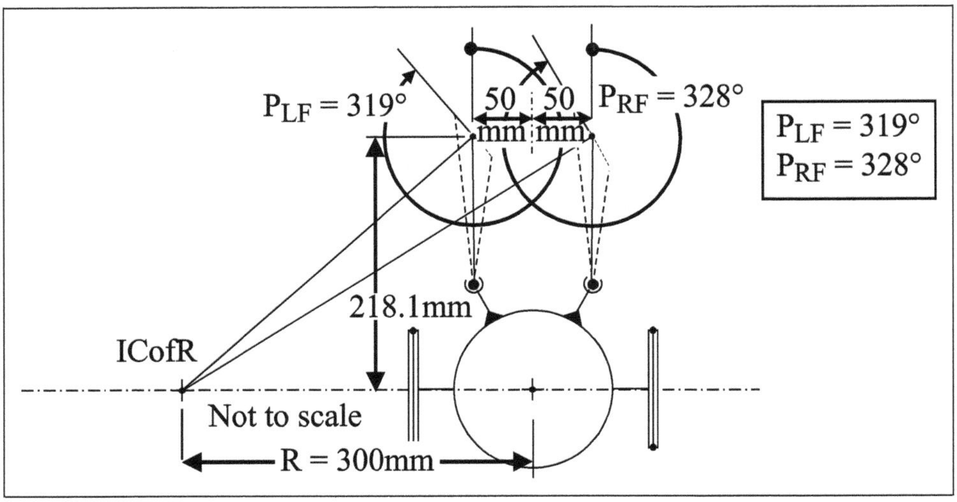

Figure 10.15 **Case4** *Turn left* forwards or backwards with 300mm radius of curvature.

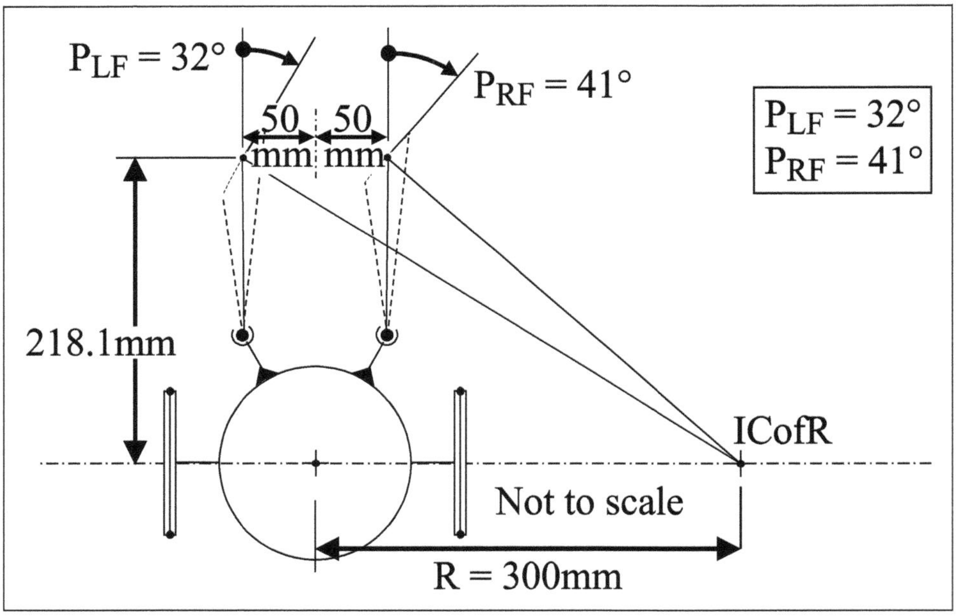

Figure 10.16 **Case5** *Turn right* forwards or backwards with 300mm radius of curvature.

10.9 The basic stamp central pattern generator code

The following Basic Stamp code articulates the Rickshaw Robot by sending walking instructions to the LF and RF legs every 20ms. The code acts as the Central Pattern Generator that facilitates the robot to follow a figure-of eight path and stop at each node for 5 secs. The code is a dead-reckoning programme, i.e. not relying on sensors to measure its path; hence it relies on accurate calibration to stop at each bulls-eye.

Start:Node 1 to Node 2	p_{LF}	p_{RF}	steps	amp_{LF}(mm)	amp_{RF}(mm)	
Go forward for 1800mm	0°	0°	12	37	38	← These give average of 37.5mm step amplitude
Node 2 to Node 5						
Turn right on-the-spot 90°(351.5mm)	77°	103°	4	22	22	
Go forward 1200mm	0°	0°	8	37	38	← These give average of 37.5mm step amplitude
Node 5 to node 4						
Turn left on-the-spot 90°(351.5mm)	77°	103°	4	22	22	
↖ Note that n-variable is to be counted down, 252, 248, 244….8, 4, 0 to turn left with plangle vales set to turn right. In other words legs are walking backwards						
Go forward 1800mm	0°	0°	6	38	37	
Node 4 to node 3						
Go backwards 600π mm turning left	338°	341°	11	40	40	
then	338°	341°	1	31	31	
↖ Note that n-variable is to be counted down, 252, 248, 244….8, 4, 0 to turn left walking backwards						
Node 3 to node 2						
Go forwards 1800mm	0°	0°	12	38	37	
Node 2 to node 5						
Go forwards 600π mm turning left	338°	341°	11	40	40	
then	338°	341°	1	31	31	
Node 5 to node 6						
Go backwards 1800mm	180°	180°	12	37	38	
Node 6 to node 1: Finish						
Go forwards 300π mm turning left	319°	328°	8	29	30	← These give average of 29.5mm step amplitude
Go forwards300π mm turning right	32°	41°	8	30	29	

Figure 10.17 Table of steering parameters giving the detailed algorithm for steering the Rickshaw Robot through the figure-of-eight path.

'{$STAMP BS2px}
'Central Pattern Generator, CPG, for the Rickshaw Robot
'Initialisation of i/o

DIR8=1
DIR13=1 } Set pins 8 and 13 to output. These pins control the LF and RF leg microcomputers

'Define variables

n	VAR	Byte	'important variable that increments the leg tip through locus
phase	VAR	Byte ←	This variable determines phase between leg loci. It will be set to 128 meaning that the n-variable for the RF leg will always be 180° phase advanced from the LF leg. See 'pout1' subroutine at end of programme

'plangles

plf	VAR	Byte	} The plan angle, i.e. the locus direction of each leg tip locus
prf	VAR	Byte	

'amplitudes

alf	VAR	Byte	} The amplitude of each leg tip locus
arf	VAR	Byte	

```
'zbiases
zblf        VAR    Byte  }
zbrf        VAR    Byte  } The zbias of each leg tip locus
'zags
zaglf       VAR    Byte  }
zagrf       VAR    Byte  } The zag of each leg tip locus
'general purpose 8-bit variables
i           VAR    Byte
j           VAR    Byte
k           VAR    Byte
l           VAR    Byte
```

Note on how to calculate distance travelled per step per leg

One step is equal to the n-variable *increasing* from n=0 back to n=0 for forward motion or *decreasing* from n=0 back to n=0 for backward motion. The distance travelled is 4 × amplitude of the step.

```
'Set values of constants
    lf       CON  08
    lm       CON  09   (struck through)
    lb       CON  10   (struck through)
    rb       CON  11   (struck through)
    rm       CON  12   (struck through)
    rf       CON  13
```

The Central Pattern Generator, CPG, has 6 output pins for independently controlling the six legs of the Hexapod Robot. that are the LF, LM, LB, RB, RM and RF leg micro-computers. Here we use the CPG to control only two legs, which are the LF and RF legs.

```
    speed    CON  4 <—this constant is the how much the n-variable is skipped. So
                      for example, if speed=4 then n=0, 4, 8, 12.....248, 252, 0,.....
'Initialisation of variables
    zblf=0       }
    zbrf=0       } No leg tip locus bias values necessary so set both to zero
    zag=40          'set the step height to its maximum value of 40mm
    phase=128       'set the phase to enable Rickshaw 2-legged walking
    n=0             'fix leg tip locus at starting point
    PAUSE 2000      'delay for 2 secs before starting programme

'Start the Figure-of-Eight sequence
'Freeze Rickshaw Robot position for 5 secs at starting position, node 1
    FOR i=1 TO 250  }
    GOSUB pout      } Hold legs fixed in space at starting point for 5 secs.
    NEXT            } (250 × 20ms)=5secs)
```

'Start: Node 1 to Node 2
'go straight for 1800mm

```
alf=37
arf=38
plf=0
prf=0
zaglf=40
zagrf=40
zblf=0
zbrf=0
FOR i=1 TO 12
FOR n= 0 TO 252 STEP speed
GOSUB pout1
NEXT
NEXT
```

Set the step amplitudes and directions according to table in figure 10.X for travel between node 1 and node 2

Set the z-air-to-ground values to maximum. Note there is no change throughout programme

Set the zbias values to zero because walking on flat ground

This short section of programme shows how easy it is to walk a number of defined steps with the Central Pattern Generator. All the complex inverse kinematic equations and the digital locus computation are hidden 'downstream' in the leg microcomputers. Thus we have developed a hierarchical computer system which, because it is easy to use with separated devolved, (delegated), parts, becomes very powerful in terms of controlling a complex robot with many degrees of freedom. Currently the Rickshaw Robot has only two 3dof legs and so is not so complex but we will see shortly that a complex hexapod robot is still easy to programme using the CPG.

```
n=0
FOR i=1 TO 250
GOSUB pout1
NEXT
```

Hold legs fixed for 5 secs

'Node 2 to node 5
'Rotate right 90deg

```
alf=22
arf=22
plf=77/2
prf=103/2
FOR i=1 TO 4
FOR n= 0 TO 252 STEP speed
GOSUB pout1
NEXT
NEXT
```

New set of leg tip locus parameters

The CPG controls the real time (every 20ms) positioning of the leg tip through the locus of each leg tip independently according to 5 variables, n, plangle, amp, zag and zbias as shown in figure 10.18 below.

Important Note
The plangle values have a maximum value of 359 which exceeds the maximum value of a byte, i.e. 255. To solve the problem the plangle values are halved then outputted to the leg microcomputers which then multiply the values by 2. The down side of this technique is that the resolution of the plangle values is halved from 1° to 2°, i.e. plangle values are specified as, 0°, 2°,...356°, 358°, with a consequent slight loss of accuracy

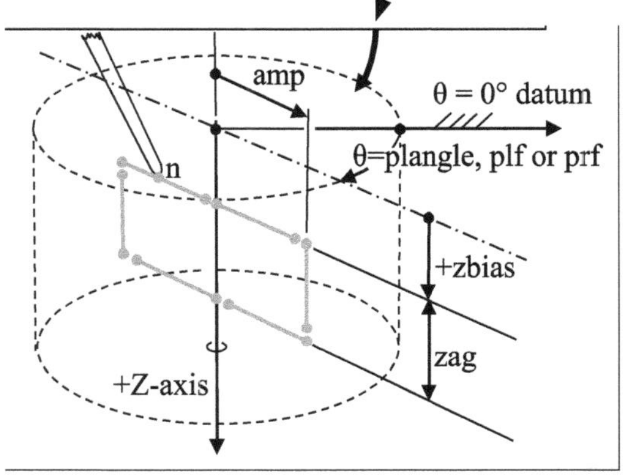

Figure 10.18 The 5 variables that specify the leg tip locus, n, plangle, amp, zag and zbias.

'now go straight forward for 1200mm

```
alf=37
arf=38
plf=0
prf=0
FOR i=1 TO 8
FOR n= 0 TO 252 STEP speed
GOSUB pout1
NEXT
NEXT
```

(alf, arf, plf, prf: New set of leg tip locus parameters)

Travel forward with average value of amplitude set to 37.5mm. Since we cannot have decimal point numbers for variables then setting amplitude for the left leg equal to 38mm and the right leg set to 37mm then we obtain the required amplitude of 37.5mm

```
n=0
FOR i=1 TO 250
GOSUB pout1
NEXT
```

Hold legs fixed for 5 secs

'Node 5 to node 4
'Rotate left 90deg

```
alf=22
arf=22
plf=77/2
prf=103/2

FOR i=1 TO 4
FOR n= 252 TO 0 STEP speed
GOSUB pout1
NEXT
NEXT
```

(alf, arf, plf, prf: New set of leg tip locus parameters)

Note: here the plangle values are set to turn right but.....

...the legs are driven backwards so the robot turns right.

'now go straight forward for 1800mm

```
alf=38
arf=37
plf=0
prf=0
FOR i=1 TO 12
FOR n= 0 TO 252 STEP speed
GOSUB pout1
NEXT
NEXT
```

```
n=0
FOR i=1 TO 250
GOSUB pout1
NEXT
```

Hold legs fixed for 5 secs

'Node 4 to node 3

'Go backwards 180°, i.e. 600π mm distance turning left, i.e. the ICofR is on the left side of the robot

```
alf=40          }
arf=40          } New set of leg tip
plf=338/2       } locus parameters
prf=341/2       }

FOR i=1 TO 11
FOR n= 252 TO 0 STEP speed
GOSUB pout1
NEXT
NEXT

alf=31  } Reset amplitudes
arf=31  }
FOR i=1 TO 1
FOR n= 252 TO 0 STEP speed
GOSUB pout1
NEXT
NEXT
n=0              }
FOR i=1 TO 250   } Hold legs fixed
GOSUB pout1      } for 5 secs
NEXT             }
```

Note 1:
This section consists of walking for 11 steps with amplitude of 40mm followed by 1 step at amplitude of 31mm. The reason is to obtain greater accuracy since we can only work with integer values of amplitudes.

Note 2:
Once again the plangle values have been set for turning left and going forward with 600mm radius of curvature so we have to walk the tricycle in reverse. An alternative is to walk forwards with the plangle values set to;
plf=338°+180°=158°
prf=341°+180°=161°

'Node 3 to node 2

'Go straight for 1800mm

```
alf=38                        }
arf=37                        }
plf=0                         }
prf=0                         }
FOR i=1 TO 12                 } Travel forward with average value of
FOR n= 0 TO 252 STEP speed    } amplitude set to 37.5mm.
GOSUB pout1                   }
NEXT                          }
NEXT                          }
n=0              }
FOR i=1 TO 250   } Hold legs fixed
GOSUB pout1      } for 5 secs
NEXT             }
```

'Node 2 to node 5
'Go forwards 600π mm turning left for 180°

```
alf=40
arf=40
plf=338/2
prf=341/2
FOR i=1 TO 11
FOR n= 0 TO 252 STEP speed
GOSUB pout1
NEXT
NEXT
alf=31
arf=31
FOR i=1 TO 1
FOR n= 0 TO 252 STEP speed
GOSUB pout1
NEXT
NEXT
```

Average amplitude required is:

$$\frac{600\pi}{12 \times 4} = 39.27\text{mm}$$

Now calculate actual value programmed:

$$\left(\frac{11}{12}\times 40\right)+\left(\frac{1}{12}\times 31\right) = 39.25\text{mm}$$

So it is possible to get high resolution step amplitude by playing around with ratios of integers! But be careful because to get 1mm accuracy over a distance of 600π=1885.0mm with 12 steps, we need an amplitude resolution of 1mm/(4 × 12)=0.02mm which we just got, i.e. compare actual 39.25mm with required 39.27mm

```
n=0
FOR i=1 TO 250
GOSUB pout1
NEXT
```

Hold legs fixed for 5 secs

'Node 5 to node 6
'Go straight backwards 1800mm

```
alf=37
arf=38
plf=180/2
prf=180/2
FOR i=1 TO 12
FOR n= 0 TO 252 STEP speed
GOSUB pout1
NEXT
NEXT
```

Note that leg steering plangle values set to go backwards but we nontheless, still *count up* the n-variable

```
n=0
FOR i=1 TO 250
GOSUB pout1
NEXT
```

Hold legs fixed for 5 secs

```
'Node 6 to node 1: Finish
'Go forwards 300π mm turning left
    alf=29
    arf=30
    plf=319/2
    prf=328/2
    FOR i=1 TO 8
    FOR n= 0 TO 252 STEP speed
    GOSUB pout1
    NEXT
    NEXT

'Go forwards300π mm turning right
    alf=30
    arf=29
    plf=32/2
    prf=41/2
    FOR i=1 TO 8
    FOR n= 0 TO 252 STEP speed
    GOSUB pout1
    NEXT
    NEXT
```

We need 300π mm=942.5mm travel distance specified to an accuracy less than 1mm so that's why the distance is specified to the nearest 0.1mm. What we get after programming is:

$$\left\{\left(\frac{30+29}{2}\right)\times 4\times 7\right\}+\{29\times 4\times 1\}=942\text{mm}$$

So error is less than 1mm which is considered accurate

```
    n=0
    FOR i=1 TO 250
    GOSUB pout1
    NEXT
```

Hold legs fixed for 5 secs

```
again: GOTO again
```

End programme with an infinite loop which will stop sending signals to the servos

'This is the subroutine that sends control variables to the leg microcomputers

```
'******************************
'Subroutine to send leg data to LF and RF leg microcomputers
pout1: PAUSE 16       'delay 18ms to give 20ms update rate, (2ms for calculations & output)
SEROUT lf,16390,[n,         plf, alf, zag, zblf]  'send leg data to LF leg BS2
SEROUT rf,16390,[n+phase, prf, arf, zag, zbrf] 'send leg data to RF leg BS2
RETURN
'******************************
```

This number specifies the data rate and 8-bit data protocol

This number is set to 128 which means that the RF leg is 180° in advance of the LF that means the legs are walking like a human being, i.e. the left foot placed on ground followed by the right foot

As a final note: you can see the Rickshaw Robot following this path on YouTube here, (this is the author's attempt: students do a much better job): *https://youtu.be/CIl0kkhGQ9c*

11. Omni-directional hexapod robot: programming its central pattern generator

11.1 Introduction

This chapter is concerned with programming the Central Pattern Generator, the CPG, of an omnidirectional six-legged robot which is called The Hexapod Robot, figure 11.1. The CPG will be programmed to walk the Hexapod Robot through the same figure-of-eight path that was used in preceding chapters for the Tricycle and Rickshaw Robots. Note that chapter 8 should be reviewed because this current chapter delves straight into programming walking gaits that walk the robot CW or CCW around an ICofR.

Figure 11.1 The Hexapod Robot.

The Hexapod Robot is endowed with 3dof legs. As a consequence it is bestowed with omnidirectional motion ability which means the ICofR location can be programmed to be anywhere in an infinitesimally sized walking plane irrespective of the robot current location and orientation, figure 11.2. A practical way of appreciating this concept is that the robot can, (i) translate in any direction and (ii) rotate about any ICofR *and* (iii) can simultaneously combine translation and rotation which means that the ICofR location is changing its position relative to the robot body at every step cycle or even within a step cycle.

As a bonus the Hexapod Robot can be programmed as a limited range 6dof platform and furthermore it can exhibit limited range 6 degree-of-freedom articulation whilst undertaking walking motion, figure 11.3, thus it can be thought of as a mobile Stewart Platform. Such motion is useful as a stabilised body platform whilst travelling across uneven terrain. Thus it is an extremely versatile roving robot. As an example, some ingenious roboticists on YouTube add a milling cutter to the robot belly and programme the legs to do 6-axis lightweight machining operations on low density plastic material.

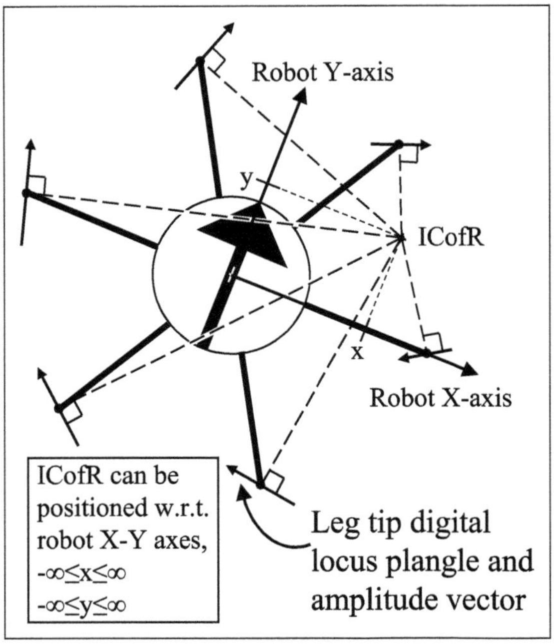

Figure 11.2 The Hexapod Robot is omnidirectional thus the ICofR can be anywhere in its walking plane.

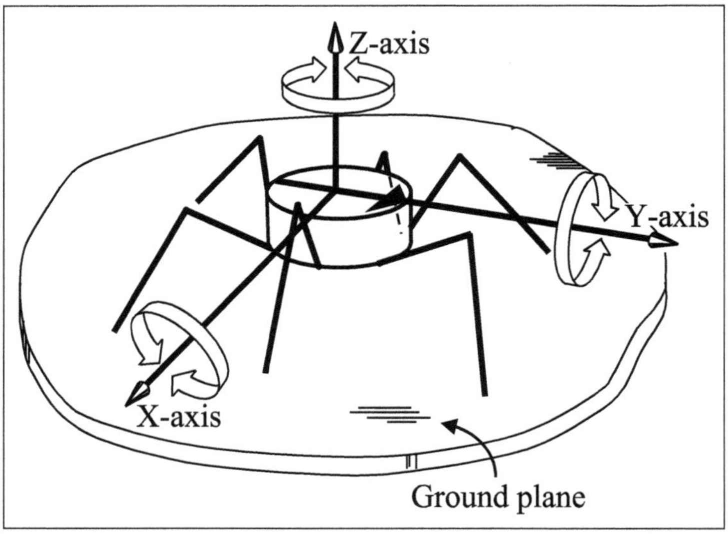

Figure 11.3 The Hexapod Robot is a limited-range 6dof platform that is mobile in a ground plane.

The versatility of the Hexapod Robot can be further illustrated by its emulation of the Tricycle and Rickshaw Robots, (equivalent to traditional front wheel steering 4-wheeled automobiles), by programming the ICofR placement anywhere along a virtual back axle axis, figure 11.4. The virtual axle axis can be shifted to the robot centre, figure 11.5, which emulates a balanced front and rear steering system.

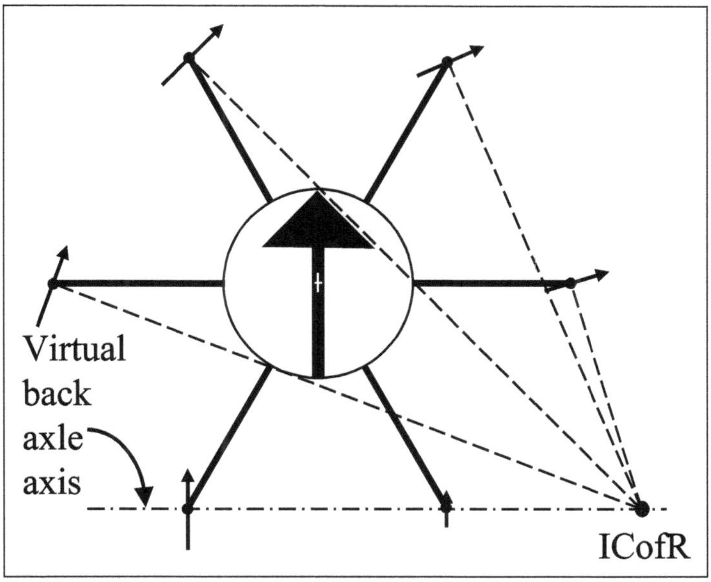

Figure 11.4 The Hexapod Robot can emulate the steering kinematics of the Tricycle and Rickshaw Robots.

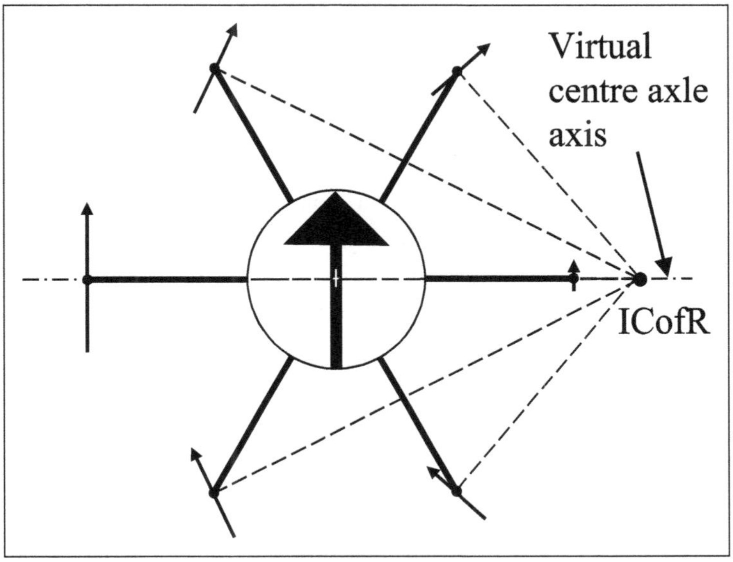

Figure 11.5 ...and here is the Hexapod Robot emulating a balanced front and rear wheel steering system.

11.2 The hierarchical, multi-cored, parallel processing microcomputer control system of the hexapod robot

The control system of an omnidirectional 6-legged robot is complex and is simplified by computing its leg mobility instructions in hierarchical levels which are successive stages of computation, figure 11.6.

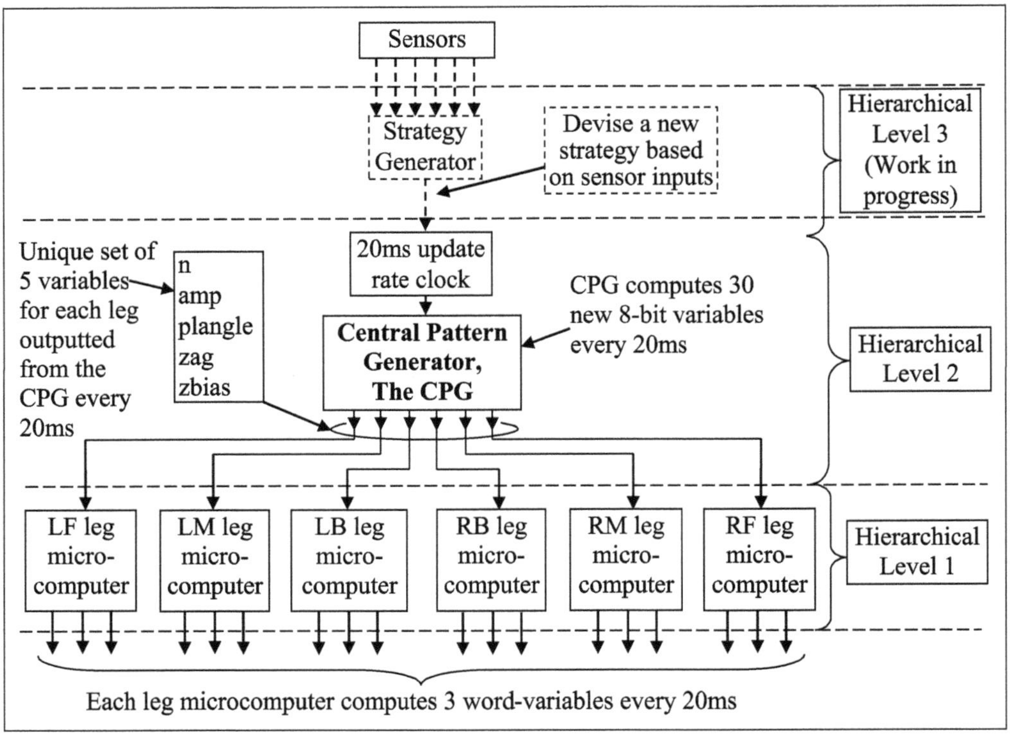

Figure 11.6 The hierarchical, multi-cored, parallel processing microcomputer control system of the Hexapod Robot.

The lowest level in the hierarchical system is the level that directly controls the 18 servos that articulate the 6 legs. It consists of six microcomputers; one microcomputer controlling one leg which consists of 3 servos. The next higher level, level 2, consists of just one microcomputer. It computes behaviour patterns for the robot body that are encoded as five 8-bit variables sent to each leg every 20ms. In chapter 5, figure 5.6 with an inset photograph, shows the geometrical arrangement of the seven Basic Stamp microcomputers that conform to a 2-level hierarchical computer system. A 3rd hierarchical level, a strategy level, is destined as work in progress for students. This level would consist of a single microcomputer reading sensors such as non-contact infrared range sensors mounted, facing outwards, at each leg and then determining a strategy for negotiating, for example, a maze.

11.3 Precision programming the hexapod robot to follow a figure-of-eight path

Figure 11.7 shows the figure-of-eight path together with the algorithm that the Hexapod Robot follows. This is the same path followed as the Tricycle and Rickshaw Robots. The algorithm chosen is the simplest and fastest possible and one that shows the omnidirectional translation (not rotation) capabilities of the robot. The rotation and simultaneous combined translation and rotation capabilities will be shown later.

A programme to implement the algorithm simply means that the plangle values of each leg are set, all equal, to values of 0° for straight ahead motion, 90° for translate right (crab right) motion, 270° for translate left (crab left) motion and 180° for backwards motion. Similarly, the amplitude values for each leg are set, all equal for each section of the path. The figure-of-eight pattern consists of three 1200mm sections and four 1800mm sections. We can achieve 1200mm distance with 15 steps and all leg tip amplitudes set to 20mm, calculated as, $15\text{steps} \times (20 \times 4)\text{mm/step} = 1200\text{mm}$. Similarly, to achieve 1800mm requires 15 steps with all amplitudes set to 30mm, calculated as, $15\text{steps} \times (30 \times 4)\text{mm/step} = 1800\text{mm}$. These combinations of number of steps and amplitudes are far from unique since student programmers can devise other solutions, for example as shown in the Rickshaw Robot chapter. Remember, also, that calibration of number of steps and step amplitude is required for ultimate accuracy, and, accuracy itself is limited by the distance repeatability error so experiments are needed to measure this error and then to correct the step number and amplitude.

11.4 Programme to implement translational figure-of-eight programme

```
'{$STAMP BS2sx}
'Initialisation
   DIRd=%00111111  'set pins 8,9,10,11,12,13 to output. These pins go to the leg
                   computers

   n  VAR   Byte   'the variable that drives the leg tips through their digital loci

   i  VAR   byte   'counter used to set the number of steps between nodes

'plangles
   plangle    VAR      Byte  }  The plan angle and amplitude digital locus vec-
                             }  tors for each leg. No need for individual values
'amplitudes                  }  for each leg because we are only doing transla-
   amp        VAR      Byte  }  tion motion

'zbiases
   zb         CON      0     'all zbias's will be set to zero so no need for vari-
                             ables for each leg
'zags
   zag        CON      40    'all zags will be set to max value
   phase      CON      128   'we will use double tripod gait so phase is set to
                             128, (180°)

   lf   CON   08  }
   lm   CON   09  }
   lb   CON   10  }  The six output pins that send 5 variables to
   rb   CON   11  }  each leg microcomputer
   rm   CON   12  }
   rf   CON   13  }
```

```
speed          CON   4  <------  On a scale of 0 to 8, where 0 is zero speed and 8 is max-
                                 imum speed, this speed is 'brisk'

'Start of Figure-of-Eight                        \
'Node 1 to node 2                                |
'Go forwards for 1800mm                          |
'Stop for 5 secs                                 |
      GOSUB goforwards1800mm                     |
                                                 |
'Node 2 to node 5                                |
'Crab sideways right 1200mm                      |
'Stop for 5 secs                                 |
      GOSUB gosideright1200mm                    |
                                                 |
'Node 5 to node 4                                |
'Go forwards for 1800mm                          |
'Stop for 5 secs                                 |
      GOSUB goforwards1800mm                     |
                                                 |
'Node 4 to node 3                                |
'Crab sideways left 1200mm                       |
'Stop for 5 secs                                 |
      GOSUB gosideleft1200mm                      > Main Programme
                                                 |
Node 3 to node 2                                 |
'Go backwards 1800mm                             |
'Stop for 5 secs                                 |
      GOSUB gobackwards1800mm                    |
                                                 |
'Node 2 to node 5                                |
'Crab sideways right 1200mm                      |
'Stop for 5 secs                                 |
      GOSUB gosideright1200mm                    |
                                                 |
'Node 5 to Node 6                                |
'Go backwards for 1800mm.                        |
'Stop for 5 secs                                 |
      GOSUB gobackwards1800mm                    |
                                                 |
'Node 6 to Node 1                                |
'Crab sideways left 1200mm                       |
'Stop for 5 secs                                 |
      GOSUB gosideleft1200mm                     /

again:         GOTO again          'End of Figure-of-Eight programme
```

```
pout1:  PAUSE 14        'create 20ms update rate
        SEROUT lf, 16390,[n+phase,plf, alf, zag,zb]
        SEROUT lm,16390,[n        ,plm,alm,zag,zb]
        SEROUT lb, 16390,[n+phase,plb, alb, zag,zb]
        SEROUT rb, 16390,[n        ,prb, arb, zag,zb]
        SEROUT rm,16390,[n+phase,prm,arm,zag,zb]
        SEROUT rf, 16390,[n        ,prf, arf, zag,zb]
        RETURN
```

Subroutine to output 5 variables to each leg microcomputer every 20ms

```
goforwards1800mm:   plangle=0/2
                    amp=30
                    FOR i=1 TO 15
                    FOR n=0 TO 252 STEP speed
                    GOSUB pout1
                    NEXT
                    NEXT
                    GOSUB stopforfivesecs
                    RETURN
```

Subroutine that goes forward 1800mm then stops for 5 secs

```
gosideright1200mm:  plangle=90/2
                    amp=20
                    FOR i=1 TO 15
                    FOR n=0 TO 252 STEP speed
                    GOSUB pout1
                    NEXT
                    NEXT
                    GOSUB stopforfivesecs
                    RETURN
```

Subroutine that crabs right 1200mm then stops for 5 secs

```
gosideleft1200mm:   plangle=270/2
                    amp=20
                    FOR i=1 TO 15
                    FOR n= 0 TO 252 STEP speed
                    GOSUB pout1
                    NEXT
                    GOSUB stopforfivesecs
                    RETURN
```

Subroutine that crabs left 1200mm then stops for 5 secs

```
gobackwards1800mm: plangle=180/2
                    amp=30
                    FOR i=1 TO 15
                    FOR n=0 TO 252 STEP speed
                    GOSUB pout1
                    NEXT
                    NEXT
                    GOSUB stopforfivesecs
                    RETURN
```

Subroutine that goes backwards 1800mm then stops for 5 secs

```
stopforfivesecs: n=0
        FOR i=1 TO 250
        GOSUB pout1
        NEXT
        RETURN
```
} Subroutine to hold legs fixed for 5 secs

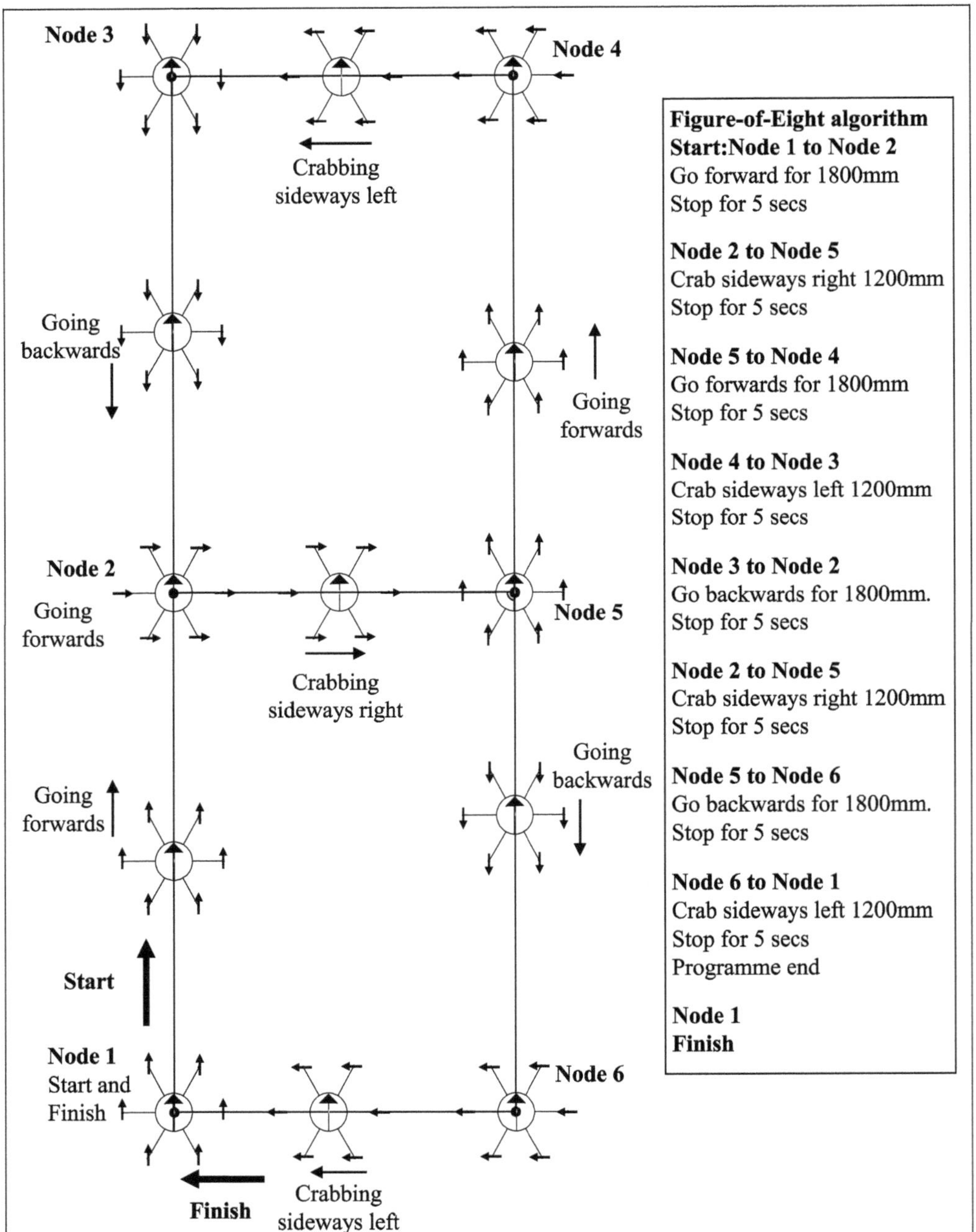

Figure 11.7 Path planning algorithm for negotiating the figure of eight pathway. The path is characterised by being purely translational without any rotation.

11.5 Programme to follow a curved figure-of-eight path

In order to demonstrate the programming of the Hexapod robot to walk CW and CCW around an ICofR, a curved figure-of-eight path is devised and figure 11.8 shows its geometry. Figure 11.9 analyses the geometry so as to give sufficient information to derive a path following algorithm.

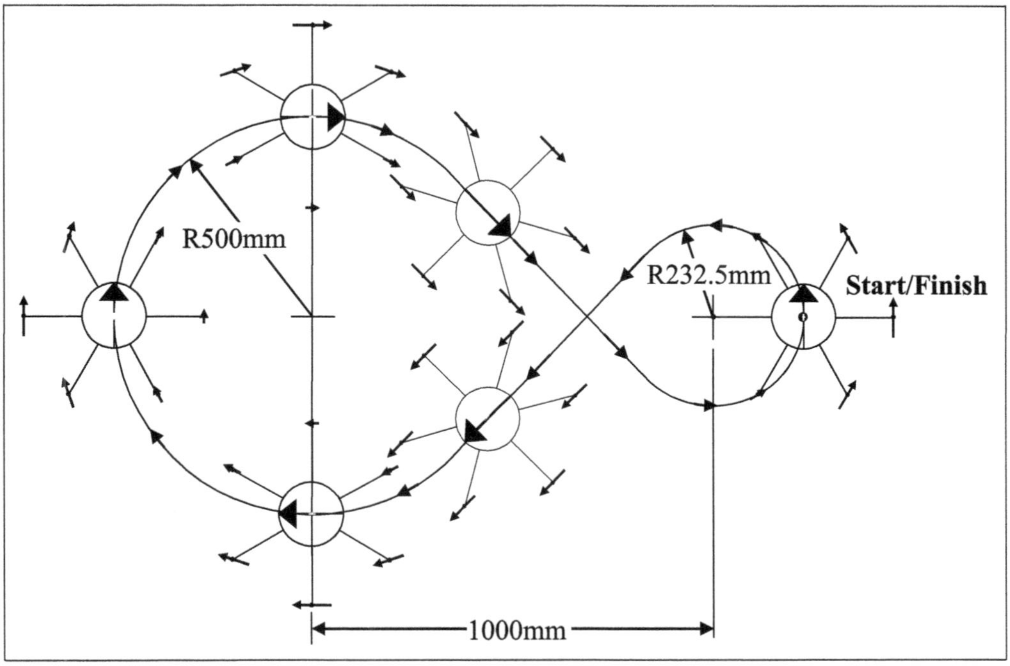

Figure 11.8 Geometry of the curved figure-of-eight path.

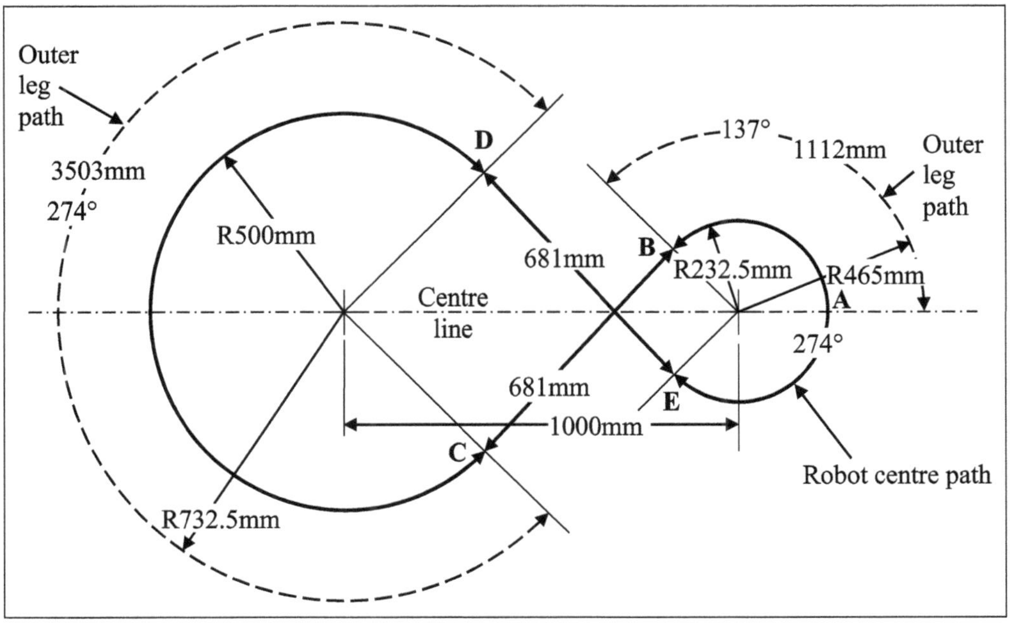

Figure 11.9 Geometrical analysis of the curved figure-of-eight path giving sufficient information to devise a path following algorithm.

11.6 Calculations of, plangle, and, amp, for each leg to negotiate the curved figure-of-eight. (Refer to chapter 8, figure 8.35 for the theory)

Figure 11.10 and figure 11.11 show the geometrical construction that is used to derive the leg tip locus plangle and amplitude values for walking around the curved figure-of-eight path from point **A** to point **B** and from point **C** to point **D** respectively. Calculation of the locus amplitude and number of steps between each point will be shown in the programme code.

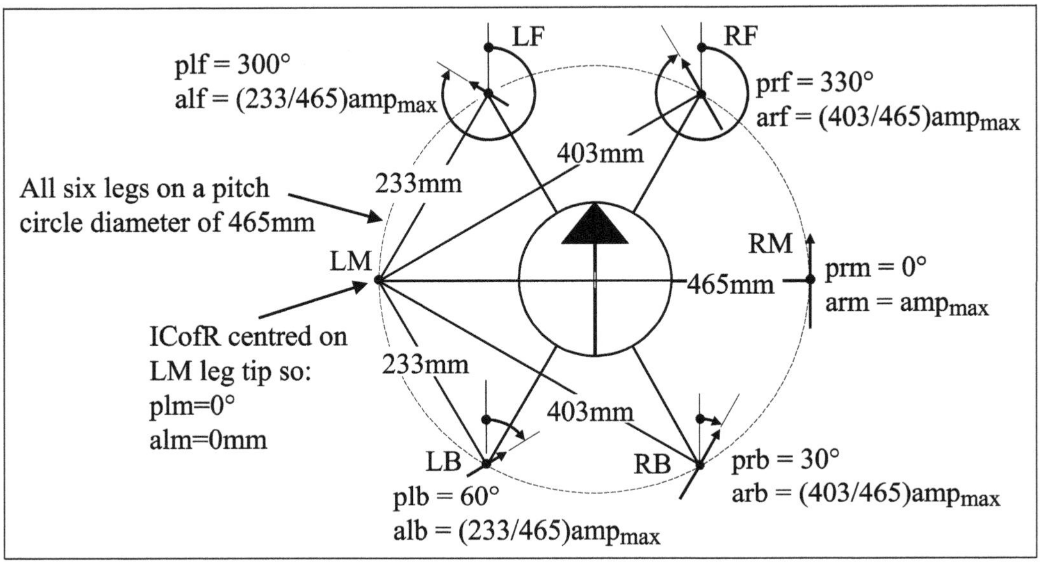

Figure 11.10 Plangle and amp values for each leg to turn CCW from point **A** to point **B** along the figure of eight path.

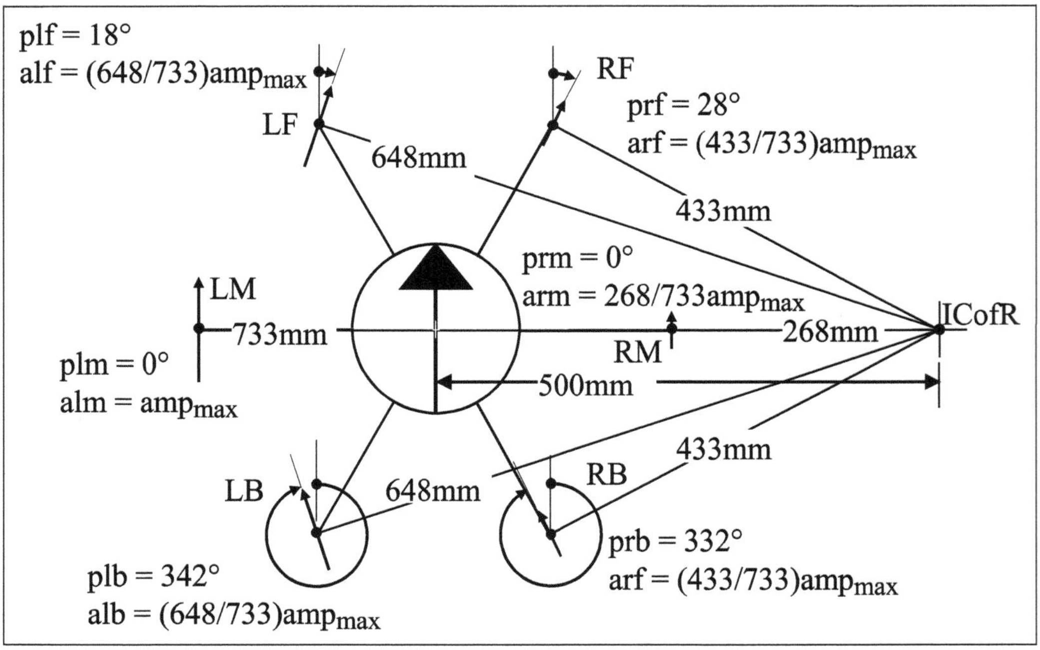

Figure 11.11 Plangle and amp values for each leg to turn CW from point **C** to point **D** along the figure of eight path.

11.7 Programme code to implement the curved figure-of-eight walking pattern

```
'Initialisation
DIRd=%00111111  'set pins 8,9,10,11,12,13 to output. These pins go to the leg
                 computers
n  VAR  Byte    'the variable that drives the leg tips through their digital loci
i  VAR  byte    'counter used to set the number of steps between nodes

'plangles
plf   VAR   Byte  }
plm   VAR   Byte  }
plb   VAR   Byte  }  For this programme we need individual
prb   VAR   Byte  }  control of the plan angle digital locus vec-
prm   VAR   Byte  }  tors for each leg.
prf   VAR   Byte  }

'amplitudes
alf   VAR   Byte  }
alm   VAR   Byte  }
alb   VAR   Byte  }  Once again we need individual control of
arb   VAR   Byte  }  the amplitude digital locus vectors for each
arm   VAR   Byte  }  leg.
arf   VAR   Byte  }

'zbiases
zb     CON   0     'all zbias's will be set to zero so no need for vari-
                    ables for each leg
'zags
zag    CON   40    'all zags will be set to max value
phase  CON   128   'we will use double tripod gait so phase is set to
                    128, (180°)

lf  CON  08  }
lm  CON  09  }
lb  CON  10  }  The six output pins that send 5 variables to
rb  CON  11  }  each leg microcomputer
rm  CON  12  }
rf  CON  13  }

speed  CON  4  <------  On a scale of 0 to 8, where 0 is zero speed and 8 is maxi-
                        mum speed, this speed is 'brisk'
```

```
'Start of main programme
'Point A to Point B
        GOSUB AtoB
'Point B to C
        GOSUB BtoC
'Point C to D
        GOSUB CtoD
'Point D to point E
        GOSUB BtoC              'D to E same as B to C
'Point E to point A      ' end of programme
        GOSUB AtoB              'E to A same as A to B
```

The Main programme calls 3 subroutines to accomplish 5 walking patterns because 2 sections are common

```
again: GOTO again      ' stop programme in infinite loop

'Subroutines
AtoB:
'1112mm distance, 5 steps of outer RM leg at 40mm amp plus 2 steps at 39mm
  amp
'amps of the other legs in proportion as in figure 11.10 above
'Distance will be (5 × 4 × 40)+(2 × 4 × 39)=1112mm so acceptable error <1mm

  plf=300/2
  plm=0/2
  plb=60/2
  prb=30/2
  prm=0/2
  prf=330/2
```

Values copied from figure 11.10 above

```
'amplitudes
  alf=40*233/465
  alm=40*0/465
  alb=40*233/465
  arb=40*403/465
  arm=40*465/465
  arf=40*403/465
```

Once again we need individual control of the amplitude digital locus vectors for each leg. Note the proportioned amplitudes according to figure 11.10 above

```
    FOR i=1 TO 5
    FOR n=0 to 252 STEP speed
    GOSUB pout1
    NEXT
    NEXT
```

'amplitudes

```
alf=39*233/465
alm=39*0/465
alb=39*233/465
arb=39*403/465
arm=39*465/465
arf=39*403/465
```

Reset the amplitudes for 2 more steps

```
FOR i=1 TO 2
FOR n=0 to 252 STEP speed
GOSUB pout1
NEXT
NEXT
RETURN
```

BtoC:
'681mm distance, 5 steps in a straight line forward with 34mm amp for each leg
'That will give 1mm error which is considered acceptable, 5 × 4 × 34mm=680mm

```
plf=0/2
plm=0/2
plb=0/2
prb=0/2
prm=0/2
prf=0/2
```

Straight ahead so all plangles set to °0

'amplitudes

```
alf=34
alm=34
alb=34
arb=34
arm=34
arf=34
```

Straight ahead so all amplitudes are the same

```
FOR i=1 TO 5
FOR n=0 to 252 STEP speed
GOSUB pout1
NEXT
NEXT
RETURN
```

CtoD:

'3503mm distance, 21 steps of outer LM leg at 40mm amp plus 1 step at 36mm amp

'amps of the other legs in proportion as in figure 11.10 above

'Distance will be (21 × 4 × 40)+(1 × 4 × 36)=3504mm which is acceptable error of 1mm

```
plf=18/2
plm=0/2
plb=342/2
prb=332/2
prm=0/2
prf=28/2
```

} Values copied from figure 11.11 above

'amplitudes

```
alf=40*648/733
alm=40*733/733
alb=40*648/733
arb=40*433/733
arm=40*268/733
arf=40*433/733
```

} Once again we need individual control of the amplitude digital locus vectors for each leg. Note the proportioned amplitudes according to figure 11.11 above

```
    FOR i=1 TO 21
    FOR n=0 to 252 STEP speed
    GOSUB pout1
    NEXT
    NEXT
```

'amplitudes

```
alf=36*648/733
alm=36*733/733
alb=36*648/733
arb=36*433/733
arm=36*268/733
arf=36*433/733
```

} Reset the amplitudes for 1 more step

```
    FOR i=1 TO 1
    FOR n=0 to 252 STEP speed
    GOSUB pout1
    NEXT
    NEXT
    RETURN
```

```
pout1: PAUSE 14        'create 20ms update rate
       SEROUT lf, 16390,[n+phase,plf, alf, zag,zb]
       SEROUT lm,16390,[n        ,plm,alm,zag,zb]
       SEROUT lb, 16390,[n+phase,plb, alb, zag,zb]
       SEROUT rb, 16390,[n       ,prb, arb, zag,zb]
       SEROUT rm,16390,[n+phase,prm,arm,zag,zb]
       SEROUT rf, 16390,[n       ,prf, arf, zag,zb]
       RETURN
```

Subroutine to output 5 variables to each leg microcomputer every 20ms

So that's the end of the programme. You can see an early prototype version of the Hexapod Robot walking this curved figure-of-eight on YouTube here;
https://www.youtube.com/watch?v=Oz3ZcjM9Gqg

11.8 Closing comments

It can be seen that programming the CPG microcomputer to control behaviour patterns of the Hexapod Robot is straightforward. In fact the calculations for the plangle and amplitude values are rather tedious. What would be great would be to create a programme where just the ICofR X-Y robot coordinates are specified along with distance and speed and then all the plangle and amplitude values are computed. A future project could do just that where the 3rd level Strategy microcomputer specified these values and the 2nd level CPG microcomputer did the computation. The Hexapod Robot thus provides a rich source of interesting machine intelligence projects.

Brief Biographies of the Authors

Francis (Frank) M.J. Nickols

Ph.D. Manufacturing Mechatronics, University of Wales, College of Cardiff
M.Sc. Electronic Industrial Systems, Cranfield University
B.A. Engineering Science, Cambridge University

Frank spent his initial 16 career-formative years in U.K. industry working initially as an avionics technician on the Buccaneer fighter/bomber aeroplane. This work concerned solving problems on its air radar, navigation instruments and its bomb delivery computer system. He then completed his tertiary education and progressed to a product designer doing research, design and prototype-to-production manufacture of engineering products for the world market. These products included automotive fuel injection sensors, CNC machine tool components, laser metrology components, manufacturing assembly systems and Electromagnetic Flowmeters. Since that time Frank felt the calling of academia and has spent 25 years doing teaching and research at universities in the U.K., Hong Kong, Singapore, Brunei and China. His primary research areas are Mechatronics Systems Design, Desk Top Educational Robots and Dynamic Agile Animal Robotics. He is a strong advocate of experiential learning for engineering students, focusing on fundamental theoretical principles and precision economical design. He is constantly pursuing improvements to the student teaching and learning experience by designing and building new types of robots and intelligent machine systems. This pursuit has resulted in the winning of numerous design prizes, e.g. his team won the Shell Asia Eco Marathon 2014 Design 1st prize for an all-wood chassis-body construction diesel engine powered automobile. Currently Frank is an engineering consultant in the U.K. developing new mechatronics and robotics practice-based experiential educational products together with novel product design hardware solutions in industry. His main research pursuit is a biomimicking muscle for dynamic agile animal robots for which he has been awarded a US patent.

Yueh-Jaw (YJ) Lin

Ph.D. Mechanical Engineering, University of Illinois, Chicago, (UIC)
MSME University of Illinois, Chicago, (UIC)
BSME Tsinghua University in Taiwan

YJ worked as a design engineer in Taipei before going to the U.S. for his postgraduate education. He then worked on a hospital walking robot project and this was followed by an academic career doing teaching and research for 20 years at the University of Akron. In 2007, he served as acting director of the Innovative Learning Research Division, NASA Safety Center. Later, from 2008 to 2013, YJ joined the University of Texas in Tyler as Professor and Department Chair of Mechanical Engineering. In 2014, he became Professor and Dean of Faculty of Science and Engineering at the University of Nottingham Ningbo China. Here, Professor Lin was also appointed as Associate Provost from 2016 to 2018 in charge of internationalization. In 2018, he was recruited by Valparaiso University to serve as Professor and Director of the Engineering Dual Degree Program with Asian institutions, mainly to manage and teach for the ongoing 2+2 engineering program development with Jiao Tung University.

In his academic career over two decades, Dr. Lin has built up rich research and teaching experience in mechanical design, structural mechanics and control engineering that has grown into engineering application areas such as aircraft health monitoring, de-icing, and more recently, to the design of MEMS-devices for energy harvesting. Professor Lin has been on the editorial advisory boards of three international journals in automation, sensors, and robotics. He has also served as PI for a number of NASA sponsored key projects in the development of aircraft mechatronic de-icing systems that was commercialized by the aerospace industry. He has authored or co-authored more than one hundred papers published in refereed journals and conference proceedings.

In the summer of 2019, YJ takes up the tenured post as Professor of Mechanical Engineering at Northern Illinois University, College of Engineering and Engineering Technology (CEET), where he is leading the newly established Mechatronics Engineering Program as its Director. Here he is setting a new vision for the Mechatronics Engineering program to advance education, enrollment management, expansion of research opportunities, and industrial partnerships.

Index

E

F

G

H

I

L

M

N

O

P

R

S

T

V

W

Z

www.ingramcontent.com/pod-product-compliance
Ingram Content Group UK Ltd.
Pitfield, Milton Keynes, MK11 3LW, UK
UKHW050946290926
15470UKWH00005B/1663

9 780791 861912